THE FIRST AMERICANS:

ORIGINS, AFFINITIES, AND ADAPTATIONS

THE FIRST
AMERICANS:
ORIGINS, AFFINITIES,
AND ADAPTATIONS

Edited by

William S. Laughlin

and

Albert B. Harper

Gustav Fischer New York · Stuttgart 1979

William S. Laughlin
Albert B. Harper

Laboratory of Biological Anthropology
The University of Connecticut
Box U-154
Storrs, CT 06268

Library of Congress Cataloging in Publication Data

Main entry under title:

The first Americans.

Proceedings of a conference held at the European Conference Center of the Wenner-Gren Foundation for Anthropological Research, Burg-Wartenstein, Austria, Aug. 21–30, 1976.
    Bibliography: p.
    Includes index.
    1. Indians—Origin—Congresses.　2. Indians—Physical characteristics—Congresses.　3. Human population genetics—Congresses.　I. Laughlin, William S. II. Harper, Albert B., 1950–
E61.F57   970'.004'97   79-19722
ISBN 0-89574-103-2

Printed in the United States of America.

ISBN 0-89574-103-2 Gustav Fischer New York, Inc.
ISBN 3-437-30295-7 Gustav Fischer Verlag, Stuttgart

Preface

A conference leading to the production of this book was held at Burg-Wartenstein, Austria, the European Conference Center of the Wenner-Gren Foundation for Anthropological Research in 1976. The sagacious contributions of Mrs. Lita Osmundsen, Director of Research, of the Wenner-Gren Foundation and of the very able staff, Dr. Karl Frey, Judith Webb, Arlene Sheiken, Miriam Casanova, Monica Parrott, and Janet Pumphrey, have been deeply appreciated.

Some of the roots of this volume may be traced back to the Fourth Viking Fund Summer Seminar in Physical Anthropology held at the New York center in 1949, and to the resulting volume, *Papers on the Physical Anthropology of the American Indian*, published in 1951.

Throughout this period Mrs. Osmundsen has been alive to and supportive of new research strategies and of the international implications of research.

William S. Laughlin

Albert B. Harper

Contributors

VALERI P. ALEXSEEV
Institute of Ethnography
Academy of Sciences
117036
Moscow, B-36, USSR

BARUCH S. BLUMBERG
The Institute for Cancer Research
The Fox Chase Cancer Center and
University of Pennsylvania
Philadelphia, PA 19111

BRUNO FRØHLICH
Department of Biobehavioral
Sciences
The University of Connecticut
Storrs, CT 06268

JAMES B. GRIFFIN
Museum of Anthropology
The University of Michigan
Ann Arbor, MI 48104

KAZURO HANIHARA
Department of Anthropology
Faculty of Science
The University of Tokyo
Bunkyo-ku, Tokyo, Japan

ALBERT B. HARPER
Department of Biobehavioral
Sciences
The University of Connecticut
Storrs, CT 06268

DAVID M. HOPKINS
US Geological Survey
Menlo Park, CA 94025

FRANCIS E. JOHNSTON
Department of Anthropology
The University of Pennsylvania
Philadelphia, PA 19174

JØRGEN B. JØRGENSEN
Laboratory of Physical Anthropology
Department of Human Anatomy
The University of Copenhagen
Universitetsparken
2000 Copenhagen, Denmark

ROBERT L. KIRK
Department of Human Biology
John Curtin School of Medical
Research
Canberra, Australia

MICHELLE LAMPL
Department of Anthropology
The University of Pennsylvania
Philadelphia, PA 19174

WILLIAM S. LAUGHLIN
Department of Biobehavioral
Sciences
The University of Connecticut
Storrs, CT 06268

JOËLLE ROBERT-LAMBLIN
Centre de Recherches
Anthropologiques
Musée de l'Homme
75116 Paris, France

FRANCISCO ROTHHAMMER
Department of Cellular Biology and
Genetics
The University of Chile
Santiago, Chile

LAWRENCE M. SCHELL
Department of Anthropology
The University of Pennsylvania
Philadelphia, PA 19174

WILLIAM J. SCHULL
Center for Demographic and
Population Genetics
The University of Texas
Houston, TX 77025

JAMES N. SPUHLER
Department of Anthropology
The University of New Mexico
Albuquerque, NM 87131

T. DALE STEWART
Smithsonian Institution
National Museum of Natural History
Washington, DC 20560

EMÖKE J.E. SZATHMARY
Department of Anthropology
McMaster University
Hamilton, Ontario
Canada L8S 4L9

SUSAN I. WOLF
The Institute for Cancer Research
The Fox Chase Cancer Center and
University of Pennsylvania
Philadelphia, PA 19111

Contents

Introduction

The First Americans:
Origins, Affinities, and Adaptations

William S. Laughlin and Susan I. Wolf

The essential rationale for this inquiry is provided by the rich body of fact and theory, of science and history, embedded in the last great migration of our single species. The origins and evolution of the First Americans can only be appreciated as an event of intercontinental or interhemispheric proportions with international and interdisciplinary dimensions. The entry migrations took place early enough to provide irrefutable proof that our species could deploy into a new suite of environments and remain a single species. Arctic cold, Andean altitude, and desert heat imposed a full range of climatic and ecologic challenges, each with nutritional and disease correlates, which were successfully mastered, with no external intervention. A full repertoire of human behavior was demonstrated long before the arrival of the first Norse visitors. The Middle American Indians developed civilizations with agriculture, monumental architecture, astronomy, mathematics, a concept of zero, and writing. This achievement of civilization is all the more significant when it is appreciated that the genetic background of these Indians is overwhelmingly similar to those of North and South America. The First Americans provide good evidence that any sample of the human species can adapt to any and all environments and is capable of any intellectual achievement.

The single entry point into North America automatically filtered the candidates for migration and served as a small bottleneck for a very large size bottle. There were likely few migrants because Siberia was sparsely occupied and moving farther north meant more cold and generally fewer resources. The earliest immigrants were obviously skilled hunters. Those who emanated from the Soviet Far East had to wend their way between carnivorous members of an unusually aggressive fauna, Siberian tigers, wolves, bears, boars, and wolverines. Mammoth hunters farther north probably had no fewer problems with food, fuel, and clothing. Movements across the Bering Land Bridge interior were likely intermittent and without the foreknowledge that an entire continent was available to them. The movement along the southern coast of the land bridge may have been more congenial and permitted some feedback to those populations that remained behind.

We do not know the critical population parameters of the first entrants. If the general picture seen in Eskimo populations is applied, the immigrants most likely were characterized by a relatively high fertility and high mortality with a very slow rate of population increase.

To make matters more difficult, the entire connecting area has been thoroughly submerged. Although the rise of sea level was fairly rapid, beginning some 14,000 years ago, it is not likely that anyone drowned as an immediate consequence. However, villages, campsites, skeletons, artifacts, and faunal remains are underwater. The tangible evidence of skeletons and teeth over the entire North and South American continents firmly proves that all early Siberian immigrants were Mongoloids. The skeletons are clear evidence that the American Indians were the First Americans, and the blood group evidence is equally unambiguous. This crucial event in human history, the crossing of the land bridge, took place recently enough to preserve the morphologic configurations, with limited alteration, characteristic of the original populations of the north Asian regions from which they first came.

This book is a momentary cross-section of research strategies and their consequences generated by original researchers concerned with the First Americans and with the processes of adaptation, the analysis of variability, affinities, origins, and health, and the history of their successful occupation of the Western Hemisphere. The topics discussed have such diversity that only the most embracive categories can be used to order them in serial arrangement. The contributions have been placed into three domains: origins, affinities, and adaptations.

Origins

All entrants into the New World passed through a cold filter. This experience may truly be said to have been trying to the individuals involved and death on the populations. The cost of living was very high because the Arctic is unfriendly. The beneficent effects, whether in challenge to invent or in epidemiologic amelioration, are poorly known.

The difficulties of life in the Arctic are sometimes dismissed by the prevailing belief that Eskimos are somehow immune to cold, hunger, and darkness by virtue of a mystical physiologic adaptation (for which evidence indicates only a slight advantage), an attitude toward suffering that relieves them of true suffering, and the ability to see as well in the dark as other peoples see in full daylight. Actually, Eskimos freeze exposed members of their body, they drown as do other people in cold water (although they may survive longer), and they become very hungry. Occasionally, all too frequently from the viewpoint of the Eskimos, whole bands and groups became extinct because they did not have enough to eat when they needed it. Their individual longevity is relatively low, the life tables indicate a lamentably short life expectancy, and they suffer from early and severe loss of intrinsic bone material.

The ingenuity and fortitude of the Eskimos is justifiably legendary. However, neither they nor any other hunting people in the Arctic are exempt from the basic physics of cold, light, and the scourging need for food. We are on sound ground when we infer that the effects of cold water immersion were precisely the same 14,000 years ago as they are today. All the people who lived anywhere at anytime on the Bering Sea, whether hunting on ice or from boats on water, were at high risk whenever they passed through the surface barrier and tarried too long in the water. There are many kinds of cultural interventions that worked to prevent immersion, to shorten the periods of immersion, and to mitigate the effects of hypothermia. In this respect only, the effects of cold could be buffered, whereas the decreased oxygen effects of high altitude were inexorable.

The Beringian region, its ground cover, temperatures, and surface characteristics, is slowly becoming better known. The combination of the chronology of deglaciation with the biotic communities that reflect climatic conditions and that are economically useful to hunters provides the essential picture for understanding the possibilities and the constraints upon immigrant bands of hunters. Beringia during the last glaciation (20,000–14,000 years ago) and Beringia during deglaciation (14,000–10,000 years ago) are dramatically different in many ways, and it is both the continuities and the discontinuities that D.M. Hopkins considers in his foundation chapter. The disappearance of the Arctic steppe biome appears to have eliminated a large number of large gregarious herbivores, leaving only the reindeer. The distribution of the Pacific walrus, various seals, porpoises, and small whales is not known. Anadromous fish may have served as ecologic magnets where rivers entered the Bering Sea. The Eskimos of north Greenland provide useful models in appreciating the ability to survive in colder regimes than now exist in the Bering Strait region. The current view that the Beringian entry route was in fact a real cold filter is well substantiated and epidemiologic aspects of such a filter require searching consideration.

J.B. Griffin supplies a substantial body of facts in his archaeologic synopsis of the origin and dispersion of American Indians. It articulates well with the preceding chapter on Beringia and with many other points later in the sequence. What he refers to as the implication of population interaction is handsomely illustrated in the genetic analysis by J.N. Spuhler. It is particularly timely to have a reasonable list of stipulations for the acceptance of an "Early Man" site in the New World. The combination of geologic context, adequate sample of material culture items, faunal remains, pollen data relevant to climate and useful vegetable products, human skeletons, and an adequate series of radiocarbon dates may not always be met, but several of the elements must be discovered in order to assess such a site rather than to accept or reject it on more metaphysical grounds. The concern here lies with analyzable information rather than with antiquity for its own sake.

It is not possible to discuss the peopling of the New World from Siberia without a firsthand consideration of the people of Siberia. Siberia is the vineyard where the First Americans were grown; it was not simply a cask from which they were decanted. Therefore, the continuing changes in the populations of Siberia must be traced back where possible to secure a picture of the ancestors of the American migrants. V.P. Alexseev usefully calls attention to the known changes in distribution since the seventeenth century, and then to basic problems in the affinities of the major groups in dynamic terms of differentiation, admixture, and migration. With major attention to the problems of standardization of measurements and observations, there is increasing confirmation of the existence of relic groups. The hypothesis of I.M. Zolotareva that surviving examples of ancient morphologic complexes can be discerned is genuinely important as well as attractive. Although the American Indians have been evolving separately for some 14,000 years or more, the discovery of paleoanthropologic data and single traits that can be referred to one Siberian area rather than to Siberia as a whole gives great promise for problems of affinities raised by Lampl and Blumberg and Szathmary. Increase in brachycephalization is importantly recognized as a trend that can occur separately in populations. The perceptible intensification of Mongoloid characters along a west to east gradient is a valuable feature to aid both in the assessment of European admixture, especially in the Uralic groups, and in the migration of peoples from the Asiatic coast of the Pacific Ocean, from south to north into Alaska.

Laughlin, Jørgensen, and Frøhlich demonstrate the impressive continuity in the large Aleut–Eskimo population or stock. The genetic integrity of this stock is well attested and the amount of divergence between the two principal divisions, Aleut and Eskimo, appears to reflect their early separation with major adaptations to winter ice conditions for Eskimos, and year-round open-water hunting and greater access to intertidal foods for Aleuts. An enduring commitment to the marine coast has placed these people uniquely in a distribution which is long, linear, and with no one living to the north of them. Greenland and the Aleutians have served as refugia at the opposite ends of this long human chain. These two population anchors contain a faithful and large sample of the original linear chain of isolates that was confined to the eastern Bering Sea coasts prior to the northern and eastern expansion of the northern segment of the early Bering Land Bridge population. The dichotomy between Aleuts and Eskimos on the one hand, and Indians on the other, appears to be as old as it is extensive. In one of his lectures to the 1938 Smithsonian Field Expedition to the Aleutian and Commander Islands, Aleš Hrdlička remarked that the relation of Eskimos to Indians was best understood as the relation of the thumb to the other fingers of the hand, and that for the Aleuts a second thumb was needed.

Affinities

The succeeding five chapters, gathered together under the rubric of "affinities," illustrate another suite of research strategies as well as different kinds of data. The geographic range is enormous, encompassing the entire Pacific arc from America to Australia.

An implicit assumption upon which several of the papers are based is the idea that one can determine the origins of New World populations by comparing the genetic similarities of contemporary New World groups to contemporary Siberian (or other Mongoloid) groups. Populations having more recent common ancestry should be measurably more alike than those having more remote common ancestry. An example of the use of this approach is the elucidation of the relationships between the two major divisions of New World natives—American Indians on the one hand and Aleuts and Eskimos on the other—and the relationship of these two divisions to Siberian populations. Did all New World groups share an undifferentiated "Siberian" ancestry or did they arise from previously separated Siberian groups which might still be recognizably distinct today?

The chapter by Lampl and Blumberg presents data on some particularly informative polymorphic genetic markers having alleles found only among certain groups of American Indians and not in Eskimos or having alleles with extremely different frequencies in Indians and Eskimos. Four albumin variants have been identified that appear to be unique markers for certain American Indian groups. Albumin Naskapi is found only among some North American groups and Albumin Mexico is found only among some populations in the Valley of Mexico and related groups. Albumins Makiritare and Yanomamo-2 are found only in restricted areas of South America. These variants have not been reported in other populations even though very large numbers of sera have been examined. The highly restricted distributions of these polymorphic albumin variants make them ideal research tools for studying affinities between Siberian and New World natives. Since polymorphic variants of albumin have been found only in American Indians, their discovery in Siberian populations would provide very strong evidence of a historical genetic relationship. Importantly, none of the polymorphic albumin variants has been found in Eskimos or in Aleuts. If in future testing of Siberian groups albumin variants are found with a sharp division as in the New World, they might be construed as evidence for premigration differentiation between groups that gave rise to New World populations.

Dental data are invaluable in studies of human populations for their inherent wealth of information, their applicability to several kinds of studies ranging from growth to population affinities, and for their amenability to rigorous statistical treatments. In addition, dental data importantly supply information on a relatively independent portion of the human genome, and these data are directly comparable between skeletal populations and living populations. Definition of the Mongoloid dental complex has been instrumentally useful in analyzing and characterizing a major division of mankind as a whole. K. Hanihara's paper examines the results of various distance computations on several populations of long-standing anthropologic interest. It was only a few years ago that Ainu were linked with Australian Aborigines, and an evanescent group of Amurian archaic whites was postulated as ancestors to explain the similarities noted, such traits as head form, hirsuteness, a large fleshy ear, and a high frequency of blood type N. One of the real achievements in the analysis of affinities has been recognition of the Mongoloid ancestry of the Ainu, firmly founded on the analysis of many traits as well as the definitive dentition. The correspondences between R. Kirk's chapter and the findings of K. Hanihara are genuinely complementary.

Still another implication of the data and their analyses by Hanihara is the relevance to the evolutionary thesis that tools have replaced teeth, in the sense that small front teeth suggest greater dependence upon the use of tools and less in the manipulatory functions of the front teeth. The Mongoloid cluster (Japanese–Ainu–Pima) is characterized by larger front teeth and smaller molar teeth in contrast to the Australian Aborigine–Caucasian–American Negro cluster, which is characterized by smaller front teeth and larger molar teeth. This can well be interpreted to mean that these contrasting size proportions have no essential relevance to tool use.

Numerical methods used to infer biologic history and affinity from the study of gene frequencies are well developed and have been applied to population units of successfully larger inclusiveness ranging from tribes to regions, continents, the major races of man, and to the entire world. The chapter by J. Spuhler makes an important contribution to research strategies by also addressing a problem implicit in nearly all studies of the genetic affinities of populations. The problem resides in the criteria used to define the population units being compared. This question of definition or delimitation has an important effect on the generation of the expected genetic relationships against which hypotheses are tested. An additional point of value is that the simultaneous comparison of many different groups who are in part in contact with some of the other groups, with no key groupings omitted, utilizes a complete matrix. The Arctic, as anticipated, shows the highest concordance of groups' affiliation to their culture area.

Two observations of critical importance for many kinds of studies employing different research strategies are given modest space. First, only a small part of the total Indian, Eskimo, and Aleut biologic diversity was sampled by the 53 groups, and second, the results represent biologic relationships at the time the samples were taken. With appropriate rephrasing, these two strictures of course apply to all studies and strict observance of them is necessary to the objective interpretation of them.

Linguistic classifications are commonly used to define population units for distance analysis. For example, we might say that on the basis of a particular hypothesis, Eskimo speakers would be expected to be genetically closer to Algonkian speakers than to Athapaskan speakers. The implicit assumption is that language and biology are very highly correlated. If it should be that the populations within a linguistic group are not closely related genetically or have highly variable genetic relationships, a calculation of genetic distance between that group and any others may lead to dubious conclusions.

To explore the degree to which biology, language, and culture are correlated, Spuhler analyzed the genetic distances among 53 North American groups that were categorized by both language phylum and culture area. He calculated the probability of each group's being most closely related genetically to its own language phylum and culture area. The results showed that genetic distances classified tribes into their own language phyla in 64.7 percent of cases and into their culture area in 58.5 percent.

These results indicate that biologic, linguistic, and cultural inheritance are significantly but not highly correlated among the groups tested. This finding is an important contribution to genetic distance investigations because it shows the effect of the definition of the units of analysis.

The problem of admixture with Europeans, and its potential effects on distance derived from blood group gene frequencies, is given prominent attention in E.J.E. Szathmary's chapter. Her point that admixture in excess of 13 percent may be the amount that distorts relationships between closely related groups is analytically useful.

The configuring effects of admixture and of genetic drift are expectedly greater in regions with small population size. The depression of the frequencies of the genes for blood group B in the central and eastern Arctic, including the Polar Eskimo of Thule, is highly suggestive of genetic drift associated with small isolates with large geographic distances between them, in contrast to the larger populations of the Aleutians, Alaska, and Greenland. Admixture with Indians lacking the gene for B cannot be distinguished from reduction of the frequency by drift. Population history, especially the use of skeletal and dental evidence, may prove to be especially strategic in such areas.

The common descent of Eskimos and Indians from populations in northeastern Siberia is certainly well attested and their basic similarities appear to be evidence of their recency. The findings by Szathmary are consonant with a maximum time depth estimate of 14,000 to 20,000 years.

The Australian and American continents are the terminal ends of a great Pacific arc. The populations inhabiting them originally came from the Asiatic mainland but they moved in opposite directions. R.L. Kirk uses a research strategy based on the distribution of genetic markers and genetic distances with an eye on the linguistic and archaeologic data. Importantly, he finds no specific Mongoloid genetic markers in the Australian Aborigines, in the Melanesians, or in Polynesians and Micronesians. Thus, the penetration of Pacific Mongoloid marker genes southward as far as Indonesia did not continue across into Australia, New Guinea, Island Melanesia, or apparently into Micronesia or Polynesia. The American Indians are most clearly aligned with populations in the northwest Pacific. The Ainu of Japan fit comfortably into the Mongoloid–American Indian portion of humanity.

On a map scale of smaller relief, Kirk also discusses the evidence of more local alignments within Australia and Melanesia. Archaeologic discoveries in Australia are bearing out an earlier time depth estimate of Kirk based on transferrin marker genes. The complex groupings in this area are of great significance for both that end of the Pacific arc and for the other end as well. An actual time depth in excess of 30,000 years appears to be well established. It may eventually prove to be twice as great as the occupation of America.

Adaptation

Adaptation is a key concept, as attractive as it is elusive. It is as ubiquitous as life itself, and it unambiguously ceases with the termination of life. The custom of asking people on their 100th birthday to what they attribute their longevity has resulted in a vast number of responses that often recommend habits either forbidden or discouraged in young people. Discussions of measures of genetic fitness, like discussions of measures of physical fitness, tend to ascend into intellectually stratospheric altitudes already occupied by discussions of devotees of the strategy of dry-fly versus wet-fly fishing. Fortunately, some parameters of human adaptation are amenable to precise measurement and to rigorous testing. All members of the human species must breath air containing a reasonable amount of oxygen if they are going to grow, and they must grow if they are to proceed to reproductive ages, and finally they must and will die. Although some people succeed in living remarkably long in spite of outstanding pathologies, what is good for the pathogen is seldom good for the host. The five chapters of this section differ markedly in research strategy and in the kinds of data utilized. They are complementary to each other with respect to the theme of adaptation and they adumbrate many points that appeared in the preceding chapters.

A novel research strategy suggested by Schull and Rothhammer is the use of the process of deadaptation (the relaxation of selection) to study adaptation (the process of selection). Genetically controlled traits important in the adaptation of a population to an environmental stress should show decreased variation as selection pressure removes less favorable alleles from the gene pool. When selection pressure is relaxed, new mutations would no longer be eliminated and could accumulate, thereby increasing the variation in the trait.

Schull and Rothhammer suggest the use of this strategy to examine adaptation of Andean populations to low oxygen tension at high altitudes. In this case, the presumed adaptation took place several thousands of years ago and therefore is inaccessible to study. However, with the contemporary migration of many highland natives to lower elevations, the reverse process of deadaptation can be observed. The demonstration of increasing variation (deadaptation) at progressively lower elevations in heritable traits considered important to high-altitude survival would strongly imply a previous adaptive process that was at least partially genetic.

The traits used in such a study should be both heritable and related in some reasonable way to the environmental variable considered to be the agent of selection. In the case of hypoxic stress, the enzymes of respiration and oxygen metabolism are parts of a critical pathway through which low oxygen tension is mediated, thus making them an ideal system for investigation. Further, enzyme structure is usually under simple genetic control and can therefore be used to substantiate the genetic nature of the response to selection.

Skeletons provide hard evidence of disease in the lifetime of the individual and those skeletons that belong to periods prior to the arrival of the Europeans provide our best evidence of the nature of pathologies prior to the alteration introduced by the Europeans. T.D. Stewart calls attention to significant differences in the expression of pathologies, their frequencies, and their patterning between the Aleut–Eskimo stock and the Indians. The studies by M.F. Eriksen on osteoporosis cited by Stewart are supported by the direct photon scanning measurement of bone mineral in both living Eskimos and skeletal series. The early onset of bone mineral loss and the continued rapid loss is especially evident in female Eskimos. It is important to note that this problem of bone loss, which has serious clinical implications, predated the nutritional and behavioral changes attendant upon European intrusion. A number of the pathologies are age related and will become even more clear with the development of osteon–photon methods of determining age at death in skeletons. After noting the near incredible effect of smallpox in dramatically reducing the numbers of the First Americans, Stewart poses, and answers in the negative, a fundamental question. Did the First Americans, have any acute diseases for which the Second Americans had no immunity? The effect of smallpox in the population history of the Aleuts is contained in A.B. Harper's chapter.

The growth period between birth and adulthood is a period of high risk for the individual and of high information for the researcher. Attention is focused on trends and processes, on sequence and patterning. Growth and development are affected by environmental pressures, and indices founded on this sensitive period provide predictive measures of adult proportions. In evolutionary perspective, it is significant that a high degree of variation within groups and between groups, well documented for height and weight, has been introduced into America by small bottleneck samples, drawn from small samples of sparsely populated northeastern Asia.

F.E. Johnston and L.M. Schell provide interesting evidence that Native Americans are in greater jeopardy of obesity in the process of urbanization than are European-derived populations. This has serious clinical implications because poorer health is often associated with obesity. There are substantial differences between sexes both in amounts of fat and in the patterning of fat distribution.

It is unfortunate that at this date good normative standards for the many First Americans do not yet exist. The effects of nutrition and various diseases on growth and development must still be studied by hypotheses more applicable to the laboratory animals on which they have been developed than to the humans to which they are applied.

The study of marriage patterns and socioeconomic structure presented by J. Robert-Lamblin enjoys the advantage of study by the same observer in both the Aleutians and in Greenland. One of the basic contrasts between the eastern Aleutians and east Greenland lies in the originally large population size in the Aleutians and the low population size in east Greenland. Today, and for some time in the past, this situation has been diametrically reversed. The Aleuts have declined and the Eskimos have increased sevenfold in only 75 years. Choice of mates from outside populations is increasing in both areas.

The Danes were among the first to develop an effective medical and research program at the village level and their success is well registered in the population increase.

In the final chapter, A.B. Harper brings the formidable power of life tables to bear on a consideration of adaptation and on two major groups of high contrast in life expectancy. Compared with many other populations, not only with Eskimos, the Aleuts have moved toward high life expectancy at least three times in their population's history. In marked contrast, the relocated Pribilof Aleuts, who are genetically the same as their Aleutian progenitors, suffered a low life expectancy. This is also an excellent example of the way in which skeletal data are important to population history in establishing the existence of a condition or direction prior to the disrupting effects of European intervention. Another point that emerges from this chapter is that all peoples are entitled to calculation of life tables based upon their own life experience, just as they should have child growth tables prepared for them specifically and thereby be relieved of the inaccuracies of extrapolation from European tables designed for Europeans. In reviewing the rationale of the various assumptions used in life expectancy analysis, the relevance of population growth rate to population expansion, migration, and extinction of megafauna also emerges. It seems increasingly unlikely that there was a mega-population sufficiently large to effect the extinction of the megafauna in the New World. It is clearly significant that Aleuts established a genuinely high life expectancy prior to non-Aleut intervention. It may be expected that other populations in the New World enjoyed a similar accomplishment.

Summary

The existing data base on the origins, affinities, and adaptation of the First Americans is good. Much of it can be profitably reanalyzed and other parts can be used in their existing forms. New kinds of research design and strategies are needed, however, to provide new kinds of verifiable data and to improve the existing categories of data.

It is indeed fortunate that the Western Hemisphere was the last major continental area to be occupied, and that it was entered recently enough to fall within the effective range of radiocarbon dating. Another stroke of luck is that the Western Hemisphere was solidly joined to the Eastern Hemisphere by only one connection, the Bering Land Bridge. Had there been similar land bridges in recent times connecting the Western Hemisphere with Australia, Africa, and Europe, the multiplicity of research problems resulting from multiple entry ports would easily exceed our multidisciplinary analytical capabilities. Another stroke of good fortune is the fact that the antecedent Mongoloid populations of eastern Asia have also survived and increased.

In the generic sense at least, a scientific family reunion is possible. A.P. Okladnikov, following upon his researches in the Aleutian Islands, made the sapient observation, "The First Americans were Siberians." The scientific and historical elucidation of this solid fact should occupy the attention of researchers for many years to come. The results of these continuing researches should answer basic questions in the processes of human adaptation and should confer tangible benefits on all the people who are studied. It should provide them with information that enables them to improve their health, including their longevity, and therefore to have an option of the perpetuation of their populations.

Part 1

ORIGINS

Chapter 1

Landscape and Climate of Beringia during Late Pleistocene and Holocene Time

David M. Hopkins

Introduction

The landscape and climate of arctic regions was dominated, 20,000 to 14,000 years ago, by the presence of enormous continental ice caps on either side of the Atlantic Ocean. Ice stored on land resulted in a reduction in sea level that displaced the shoreline hundreds of kilometers northward from the present northeast Siberian coast and exposed the floors of the Chukchi and Bering Seas as extensive low-lying plains connecting Siberia and Alaska. The climate was dry, and a now extinct biome, arctic steppe, supplanted present-day tundra and taiga in Beringia. A rich and varied land mammal fauna provided an economic base for nomadic Late Paleolithic hunters, the ancestors of modern American Indians.

The remnant continental shelf of the Bering Sea was then, as the southern shelf is now, an area of high productivity. Bird cliffs on headlands at the site of the present Pribilof and St. Matthew Islands and anadromous fish entering the mouths of major streams were a food resource potentially available to hunters reaching the south coast of the Bering Land Bridge. Scouring by sea ice probably inhibited the development of an intertidal fauna and discouraged the establishment of large shore-based seal and sea lion rookeries; because of this, a resource base capable of supporting permanent coastal settlements probably was not yet available.

Beringia was isolated, although perhaps not completely, by the merged or closely adjoining Laurentide and Cordilleran ice sheets; although an interior ice-free corridor may have existed during much of the interval 20,000 to 14,000

The First Americans: Origins, Affinities, and Adaptations.

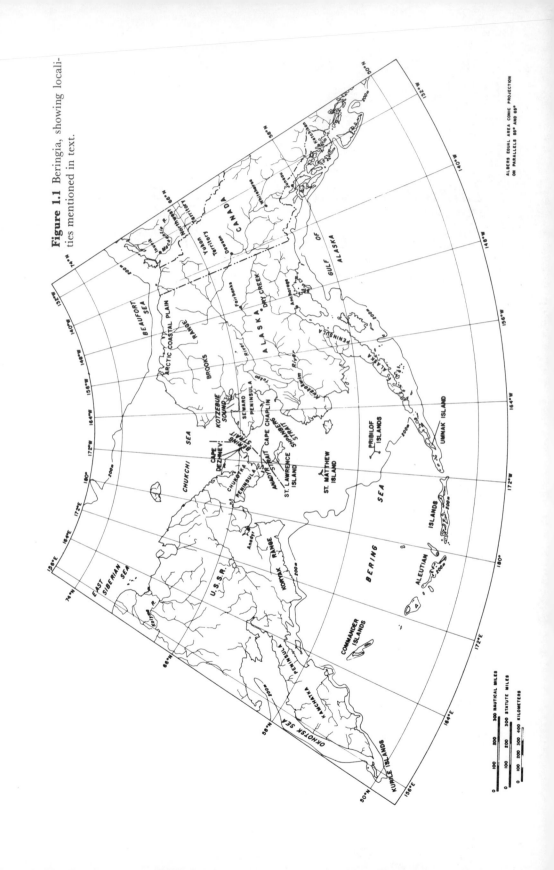

Figure 1.1 Beringia, showing localities mentioned in text.

years ago, its great length and narrow width would have restricted potential gene flow between populations in Beringia and any that might have been present in the Americas south of the ice sheets. Glacierized mountains extending to the continental shelf along the Canadian and Alaskan coasts of the Pacific Ocean and along the Koryak and Kamchatka coasts of the Bering Sea would have inhibited dispersals of coast-oriented people, if any existed during the 20,000 to 14,000 years time interval.

An abrupt climatic warming about 14,000 years ago set in motion deglaciation, a rapid rise in sea level, and an expansion of mesic tundra and then taiga into areas formerly occupied by arctic steppe. Bering and Anadyr Straits were submerged by at least 13,000, and possibly as early as 14,000, years ago but did not become formidable barriers to human dispersals until about 12,000 years ago. By that time, the Pleistocene large mammal fauna was rare, if not largely extinct, and caribou were the only remaining abundant ungulates. The human hunters increasingly depended upon these deer herds, which now ranged into the newly deglaciated mountains.

Rising sea levels set coastal glaciers afloat, leading to early deglaciation of the Koryak and Kamchatka coasts of Siberia and the Alaskan and British Columbian coasts of North America. If boat-using, coast-oriented people were on the scene, they could have dispersed from Beringia southwestward along the Pacific Coast to the Puget Sound area as early as 12,000 years ago.

In this chapter I attempt to summarize current knowledge of the changing landscape of Beringia (Figure 1.1) during times when the earliest aboriginal Americans were probably present, and I discuss some of the resources and environmental constraints with which early human populations would have had to deal. It is certain that men had reached the Lena River drainage in northeastern Siberia by 18,000 years ago (Mochanov 1975, 1976) and that men had crossed Beringia and were widely dispersed in the New World by 11,000 years ago (Haynes 1971, 1976). Men may well have been present in Beringia and the Americas 10,000 or 15,000 years earlier, but the evidence is fragmentary and remains debatable. Our knowledge of earlier paleogeography, climatic history, and vegetation patterns is also fragmentary, and so I shall begin my account with a description of Beringia and neighboring Arctic regions during the climax of the last glaciation, about 20,000 years ago.

Beringia during the Last Glaciation (20,000–14,000 years ago)

Extent of Glaciation

Twenty thousand years ago, the earth's northern regions were dominated by three gigantic continental glaciers: the Greenland, the Laurentide, and the northwest Eurasian ice sheets (Figure 1.2). The western North American Cordillera and northeastern Siberia also were heavily glacierized, as was the Brooks Range of northern Alaska, and it has been suggested that the Arctic Ocean

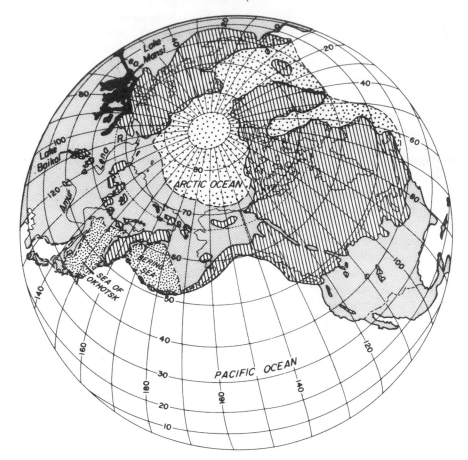

Figure 1.2. Paleogeography of the Arctic, 20,000 to 14,000 years ago. Distribution of glaciers after Degtyarenko (1963, northwest coast of Bering Sea), Prest *et al.* (1968, Canada), Ganeshin (1969, USSR), Flint (1971, North America and western Europe), Hopkins (1972, Beringia), Hoppe (1973, North Atlantic), Malekestsev *et al.* (1974, Kamchatka), Péwé (1975, Alaska), and Huges *et al.* (1977, Siberia). Distribution of year-round sea ice in Arctic and North Atlantic Oceans from CLIMAP (CLIMAP Project Members 1976). Inferred extent of winter sea ice in Bering and Okhotsk Seas is my own estimate. Extent of proglacial Lake Mansi in western Siberia from Volkov and Volkova (1975). Unglaciated land areas are shaded. Vertical patching, glaciers; light stippled areas, perennial sea ice; dark stippled areas, seasonal sea ice.

may have been covered by an ice shelf several hundred meters thick (Mercer 1970, Broecker 1975, Hughes *et al.* 1977).

At its maximum extent, the North American Cordilleran glacier system extended eastward into areas that at one time or another were also occupied by the westernmost part of the continental Laurentide ice, and it was formerly assumed that until about 13,500 years ago the two glacier systems actually merged along a suture some 1200 km long (Prest 1969), posing a certainly impassable barrier to human movements between Beringia and southern North America (Hopkins 1967, Müller–Beck 1967). Today, it is clear that the two ice sheets coalesced at least briefly along a front a hundred kilometers long in southern Alberta (Roed 1975), and the ice sheets may have coalesced for some time in northeastern British Columbia (Nat Rutter, University of Alberta, personal communication, May 10, 1977). Elsewhere, however, it seems that the Cordilleran piedmont glaciers reached their maximum extent early and then vacated areas later occupied by continuing expansion of the Laurentide ice (Reeves 1973, and earlier papers cited therein). Alaska and central North America may have been connected, then, by an ice-free corridor during most of the last glacial period, and the corridor may have been blocked in southern Alberta and northeastern British Columbia during an interval no longer than a couple of thousand years.

Mountainous coastal areas around the North Pacific were also heavily glacierized, with piedmont glaciers and small continental shelf ice caps fronting tidewater along much of the coast from Puget Sound to the western Aleutian Islands (Prest *et al.* 1968, Péwé 1975). Kamchatka and the mountainous Koryak coast of the western Bering Sea were also inundated by glaciers and ice fields, and glaciers extended to tidewater in many places there, too (Degtyarenko 1963, Melekestsev *et al.* 1974, Fig. 99).

Still, most of Asia and most of Beringia remained unglaciated (Figure 1.2). Vast, interconnected unglaciated plains and lowlands extended from southeastern Europe across southern Siberia, down the Lena River valley, along the exposed continental shelf, and across the dry floors of the Chukchi and Bering Seas into Alaska.

Sea Level, Position of the Shoreline, and Marine Food Resources

Enough water was stored in the world's glaciers to reduce sea level by at least 85 m during the last glaciation (CLIMAP Project Members 1976). However, because of local tectonic and isostatic effects, the relative lowering of sea level and the pattern of subsequent recovery differs from one coastal region to another (Curray and Shepard 1972). In Beringia, the tectonically active, heavily glaciated North Pacific coast has had quite a different history of relative sea level changes than have the tectonically quiet and mostly unglaciated continental shelf areas to the north. Figure 1.3 presents the history of relative sea level deduced from dated samples from the floors of the Bering and Chukchi Seas.

In the Bering Sea region, sea level was at its minimum position from about 20,000 until about 16,000 years ago. The shoreline lay somewhere inshore from the present 120-m isobath and probably near the 90-m isobath (Knebel 1972, Knebel *et al.* 1974). A zone nearly 300-km wide remained submerged on the present outer continental shelf, even when sea level was at its lowest (Figure 1.4). In a recent study of the areas most likely to contain submerged archaeologic

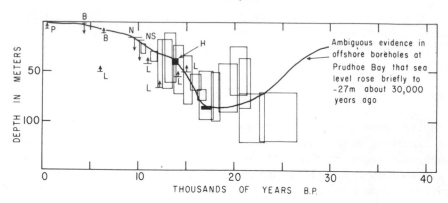

Figure 1.3. Reconstruction of sea level history based on data from the continental shelves off northern and western Alaska and northern Siberia. For data points indicated by boxes, width indicates standard deviation of radiocarbon age determination and height is range of uncertainty of position of sea level. Small boxes filled to emphasize good data points. For data points indicated by arrows, width of bar indicates standard deviation of radiocarbon age determination; upward pointing arrows are dates on subaerial peat beneath beach gravel, downward pointing arrows are wood or organic fibers in marine mud or sand. Data points B from Barrow, Alaska (Brown and Sellmann 1966); H, Hope Sea Valley (Creager and McManus 1967); L, Laptev Sea (Holmes and Creager 1974); N, Nome (Hopkins 1967); NS, Norton Sound, northern Bering Sea (Nelson and Creager 1977); P, Prudhoe Bay, Beaufort Sea (Hopkins 1977). All other data points are from the central Bering Sea (Knebel 1972, Knebel *et al.* 1974).

sites, S.W. Stoker (in Dixon *et al.* 1976) speculates that this remnant of the continental shelf " was then, as now, a region rich in marine life, fed by nutrition-rich upwelling, and supporting large populations of marine mammals, marine birds, marine and anadromous fish, and shellfish. For the most part . . . its margin was an open, storm-swept coast of sand-beaches and dunes in summer and of strong winds and extensive shore-fast and pack ice in winter time " (pp. 56–57). Stoker points out that mouths of large rivers whose drainage basins were largely unglaciated (e.g., the Yukon, the Kuskokwim, and the Anadyr, Figure 1.4) would have been the foci of especially high productivity, including runs of such anadromous fish as trout, salmon, and whitefish, and that rocky headlands on the site of the present-day Pribilof Islands and, a little later, St. Matthew Island, would then, as now, have been the homes of enormous populations of nesting sea birds. Ice-tolerant and ice-dependent pinnipeds, such as walrus, bearded seal, ringed seal, and harbor seal, would have been present, but ice conditions, in my opinion, probably rendered the Bering Sea coast unsuitable for the large breeding colonies of fur seal and Steller's sea lion that historically have occupied the Pribilof Islands. Scouring by sea ice impairs productivity of the intertidal zone in the southern Bering Sea north of the Aleutians, and I have observed that there is essentially no intertidal fauna along the present-day coasts north of the Pribilof Islands. During the interval 20,000 to 14,000 years ago, we must assume that the

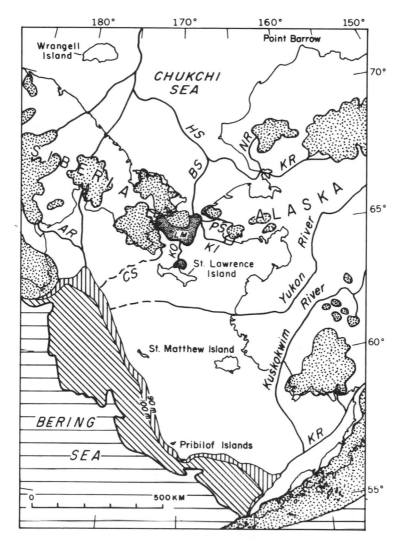

Figure 1.4. Paleogeography of central Beringia, 20,000 to 14,000 years ago (after Hopkins 1973). Glaciated areas stippled. Ancient lakes on continental shelves shaded. Northwest shore of Bering Sea lay somewhere within vertically hatched area between the 90- and 100-m isobaths, and a considerable part of the outer continental shelf remained submerged (diagonally hatched area). AR, Anadyr River; BS, Bering Strait sea valley; CS, Chaplin sea valley; HS, Hope sea valley; KR, Kobuk River; KI, King Island sea valley; KO, Kookoolik sea valley; KR, Kvichak River; LM, Lake Merklin; NR, Noatak River; PS, Port Clarence sea valley. Submerged drainage systems from Creager and McManus (1967), Kummer and Creager (1971), Hopkins (1972), Knebel (1972), Knebel *et al.* (1974), and Hopkins *et al.* (1976). [Reproduced with permission from Academic Press, Inc.]

intertidal zone along the south coast of the Bering Land Bridge had even less to offer human populations.

Most of the continental shelves of the Bering and Chukchi Seas lay exposed, joining Siberia and Alaska across a broad front. Farther north, the Arctic continental shelf was completely exposed. The shoreline lay only a few tens of kilometers north of the present Alaska coast, but it lay hundreds of kilometers north of the present Siberian coast (Figure 1.2).

The Arctic coast must have been a forbidding place. The stoney silts and clays that evidently represent full-glacial intervals in Arctic cores are devoid of organic remains (Herman 1970, 1974, 1977, Clark 1971), and the Arctic sea ice must have been thick and unbroken, if, indeed, it had not evolved into a floating ice shelf. If the entire continental shelf were emergent, there would have been no broad coastal zone where shallow water could have been warmed by the summer sun. Even if the Arctic Ocean had been covered by thick sea ice rather than by an ice shelf, therefore, summer shore leads would have been extremely narrow and the marine food resources available to man practically nil.

Hydrology

Major drainage changes were effected in eastern Europe and western Siberia by the presence of continental ice masses athwart north-flowing river valleys. The most significant of these was the blockage of the Ob' and Yenissei Rivers by the Kara Sea ice cap. A freshwater lake having the dimensions of a small sea, Pleistocene Mansi Lake accumulated in the lowlands of western Siberia and ultimately overflowed southward through the Turgai Channel at the head of the Tobol' River, draining to an enlarged Aral Sea, thence to the Caspian, and finally to the Black Sea (Volkov and Volkova 1972, 1975; Figure 1.2).

Major drainage changes were also effected on and near the exposed continental shelves of the Bering and Chukchi Seas (Figure 1.4). The Kuskokwim and Anadyr Rivers simply extended their courses, respectively, southwestward and southward but continued to enter the shrunken Bering Sea (Kummer and Creager 1971, Hopkins 1972). The Yukon River, during the interval 20,000 to 14,000 years ago, followed a more southern course across the present delta plain (Dupré 1977, written communication, April 1, 1977) and evidently reached the southwest shore of the land bridge at a point about 200 km northwest of St. Matthew Island (Knebel and Creager 1973).

The northern Bering shelf drained northwestward through the Kookoolik, King Island, and Port Clarence sea valleys (Hopkins et al. 1976) to a large lake, Lake Merklin[1] confined by a south-facing fault scarp (illustrated by Grim and

---

[1] The deposits of Lake Merklin, named here for the first time, were recognized by C.H. Nelson, R.B. Perry, and me in 1969 in the course of our study of high-resolution seismic reflection profile records collected aboard the R/V THOMAS G. THOMPSON in 1967 and 1968; some of these records are illustrated by Grim and McManus (1970). The lake is shown diagrammatically in Figure 5 of Hopkins (1972). The lake is named after the late Roman L. Merklin, distinguished and much loved molluscan paleontologist of the Paleontological Institute, in honor of his contributions to the study of the Pleistocene geology of the Bering Strait area.

McManus 1970) at the south entrance to Bering Strait. Lake Merklin fed a large stream that drained northward through the Bering Strait sea valley (Hopkins *et al.* 1976) to join the extended Noatak and Kobuk Rivers in the Hope sea valley, which leads northwest across the Chukchi Sea floor to enter the Arctic basin in a submarine canyon about 100 km east of Wrangell Island (Creager and McManus 1967) (Figure 1.4).

Climate

The periglacial climate of the unglaciated Arctic differed from the present climate as much in being drastically drier as in being drastically colder (CLIMAP Project Members 1976, Gates 1976). The causes reside in the state of the oceans and in the paleogeography of the time. The generally lowered surface temperatures of the oceans resulted in reduced evaporation and also tended to cool overlying air masses, reducing their moisture-carrying capacities. Meanwhile, the vast expanses of exposed continental shelf produced a much more continental climate, especially in Beringia. The drought in the Arctic was heightened by the rain shadow effect of ice caps on either side of the Atlantic Ocean.

Beringia, in particular, experienced a relatively modest reduction in mean annual temperatures, a substantial increase in summer temperatures, and a drastic reduction both in winter snow cover and in summer precipitation (Hopkins 1972, Streten 1974, Gates 1976, Matthews 1976, Sergin and Shcheglova 1976). Especially large sea areas were converted to exposed continental shelf, and eastern and western Beringia lay in the rain shadow of coastal mountains whose summits were raised and broadened by mountain ice fields and by ice caps on the adjoining continental shelves (Péwé 1975). Central Beringia, meanwhile, evidently replaced the present-day Atlantic Ocean as the main avenue through which relatively warm low-latitude air was advected, during summer, into the polar basin (Lamb 1970, Streten 1974).

In consequence, the seasonal round of weather in periglacial Beringia consisted of short summers, warmer and drier than at present, and long winters, evidently colder but with snow cover thinner than at present. Snowline was lowered (Péwé 1975) but not nearly as much as in more humid regions at middle latitudes (compare Heuberger 1974). Permafrost was thicker and more extensive, but the dry soils thawed more deeply each summer. One would predict, and the fossil evidence confirms, that the climate of unglaciated Beringia was dry enough so that moisture was a limiting factor governing the vegetation and the animal life. Furthermore, the vegetation cover was incomplete and the climate evidently windy, because fine sand was blown by northeast winds from the continental shelf southwestward onto the Pribilof Islands, the Seward Peninsula, and the Arctic coastal plain of Alaska and northeastern Siberia (Black 1951, Hopkins 1972, Péwé 1975, Tomirdiaro 1975). The paleoclimatic speculations of Lamb (1970) and Streten (1974) suggest that persistent winds out of a northern quadrant would have been winter winds, rather than summer winds as I had inferred in earlier discussions (Hopkins 1970, 1972). If so, then snow cover must have been thin indeed. Sand dunes and thick wind-blown silt dating from the last cold period also cover large areas in central Alaska, but these consist of sand and dust swept generally northward from former glacial outwash plains by katabatic winds draining from the glaciers (Péwé 1975).

Biota

The vegetation and the land mammal population that it supported differed drastically from the present land biota of Beringia (Colinvaux 1967, Guthrie 1968a, Yurtsev 1972, Matthews 1974, 1976, Ager 1975, Young 1976, many papers in Kontrimavichus 1976). The broad belt of coniferous forest (taiga)

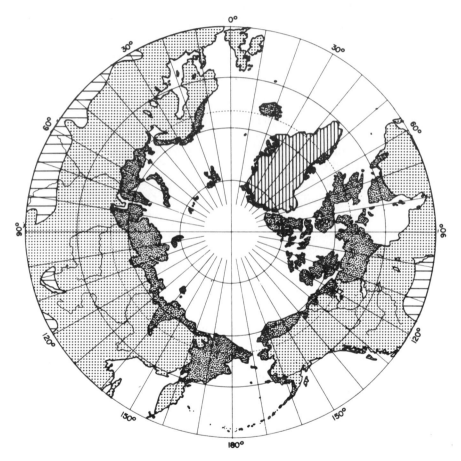

Figure 1.5. Distribution of major vegetation types in northern and arctic regions at the present time. Dark stippled, upland and alpine tundra; light stippled, taiga, wooded steppe, and (in the south) mixed deciduous and coniferous forest; horizontal hatching, cold steppe; vertical hatching, glaciers. After National Geographic Society Atlas of the World, 4th Ed. (1975) and Atlas of Lithologic and Paleogeographic Maps of the USSR (Ministry of Geology and Academy of Science, USSR, 1967).

which now separates the mesic arctic tundra from the arid steppes of the continental interior of Eurasia (Figure 1.5) was then reduced to a narrow and discontinuous strip (Figure 1.6). The midlatitude steppes merged northward in many places with a specialized and now nearly extinct cold climate xerophytic vegetation—tundra–steppe (Giterman and Golubeva 1967) or arctic steppe (Matthews 1976). Arctic steppe was a generally treeless vegetation dominated by grasses, sedges, and various species of *Artemisia* (sage). Ericaceous plants (heaths) were prominent. Prostrate dwarf birch and willow shrubs were present but not abundant. *Sphagnum* mosses and the tussock-forming sedges and grasses so conspicuous in modern low-Arctic tundra were almost lacking.

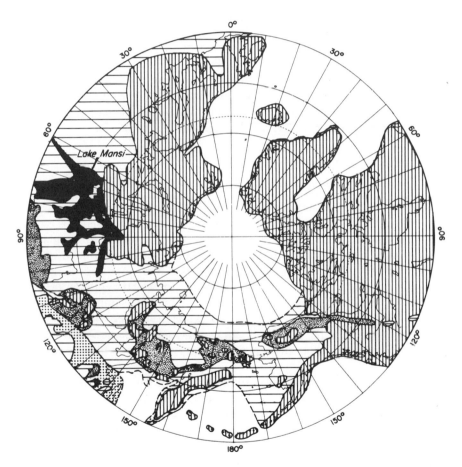

Figure 1.6. Distribution of major vegetation types in northern and arctic regions 20,000 to 14,000 years ago. Horizontal hatching, arctic and cold steppe; dark stippled, upland and arctic tundra; light stippled, taiga, wooded steppe, and, in southern Siberia, mixed deciduous and coniferous forest. After Frenzel (1968), Giterman *et al.* (1968), and Hopkins, (1972).

The lack of wood in deposits 20,000 to 14,000-years-old indicates that trees and large shrubs were extremely rare in Beringia. In an unpublished analysis of dated and identified wood from Alaska, I conclude that certainly spruce and probably tree birch and alder became extinct in Beringia, that aspen or cottonwood or both persisted, and that larch may well have persisted too. Aspen, if it did indeed persist, probably survived on sunny, south-facing slopes, and larch, cottonwood, and large willows, if present, probably persisted in sheltered valleys in the rolling upland regions that are now central Alaska and northern Chukotka.

The land mammal fauna was dominated by several species of large, gregarious herbivores, including bison, horse, mammoth, and reindeer; present in small numbers were musk ox, wapiti, and mountain sheep, and these were preyed upon by a predator fauna more diverse than the present one (Péwé 1975, Harington 1978). Although browsers were included in the fauna, grazing animals were numerically predominant (Guthrie 1968b). The presence in Beringia of *Camelops* (North American camel) and *Saiga* (Siberian steppe antelope) emphasizes the steppelike character of the large mammal fauna. The large herbivores must have interacted in a way that hastened early disappearance of the snow cover each spring, enhanced nutrient cycling, and maximized plant productivity, for it is clear that the standing crop of game animals was far larger than in present-day tundra and taiga (Matthews 1976, Schweger and Martin 1976).

Interactions with Human Populations

The vast expanse of arctic steppe extending from the Lena basin across the exposed Arctic shelf to Beringia comprised an extensive, homogeneous biome in which a nomadic big-game-hunting economy could be practiced. Unglaciated, steppe-clothed plains with generally similar game resources continued westward across southern Siberia into Europe, interrupted only by proglacial Lake Mansi (Figure 1.2). The hunters of the Arctic steppe, and those of the cold steppe as well, had available to them a meat supply far greater than can be found in present-day tundra or taiga. The smoother, drier ground and the thinner winter snow cover would have facilitated the driving of game and eased nomadic foot travel to an extent hardly imaginable to those of us familiar with today's hummocky Arctic tundra.

Dependence upon big game would have required a nomadic life during the summer season, when the prey was on the move and when warm temperatures precluded long-term preservation of surplus meat. However, storage of surplus meat became possible with the onset in autumn of predominantly subfreezing temperatures. Furthermore, as winter deepened, the large ungulates might be expected to minimize energy expenditures by moving as little as possible (C.E. Schweger, University of Alberta, oral communication, May 7, 1977). Thus, during winter, a more sedentary life based upon substantial shelter structures became both possible and necessary. Winter structures have not yet been recognized in Beringia except possibly in northeastern Siberia (Vereshchagin and Mochanov 1972, Vereshchagin 1974), but they have been found in many places in the cold-steppe regions of south-central Siberia and eastern Europe.

The rarity of trees and large shrubs, indicated in pollen spectra and confirmed by the lack of wood larger than matchsticks in alluvium, loess, and pond sediments between 20,000- and 14,000-years-old, would have forced reliance upon other materials for tent frames, spear shafts, and fuel. Paleolithic hunters evidently faced similar problems in the former cold-steppe regions of southern Siberia and eastern Europe. Winter habitation structures of Late Paleolithic age in southern Siberia and the Ukraine were framed with reindeer antler and the long bones and tusks of mammoths (Klein 1973, Chard 1974, Pidoplichko 1976) ; at Sungir', near Moscow, lances were made of straightened and sharpened mammoth tusks (O.N. Bahder, Institute of Ethnography, Academy of Sciences, Moscow, personal communications 1971, 1973) ; and the presence of quantities of burned bone in Late Paleolithic sites in the Ukraine seems to reflect reliance upon animal fat as an alternative to woody fuels (Klein 1973, Pidoplichko 1976).

Coastal occupation was possible only along the low-lying south shore of the land bridge, in areas that are now part of the submerged continental shelf of the Bering Sea. Certain areas, notably the river mouths and the cliffed headlands on the site of the present-day Pribilof Islands, are thought to have offered considerable potential food resources to human bands equipped with the appropriate technology (Dixon *et al.* 1976). Intense glacierization prevented occupation of the Pacific coast of Alaska to the southeast and the Koryak and Kamchatka coasts of Siberia to the west (Figure 1.2), and marine productivity along the Arctic coast must have been nearly nil.

The exposure of the continental shelf, the disruption in the continuity of the resource-hungry taiga, and the replacement of tundra by arctic steppe opened broad avenues for human dispersal. *Saiga*, the steppe antelope now native to the Ukraine and to Soviet Asia, expanded its range westward to Britain and eastward to the Arctic coast of Canada during the last glaciation, finally encompassing the entire region of the cold and arctic steppe (Sher 1967, Hopkins 1972). It is tempting to speculate that a single, interrelated human population also at some time might have occupied both the Eurasian cold steppe and the north Siberian–Beringian arctic steppe. However, a dichotomy between the lithic technology traditions of Siberian areas eastward and westward from the Yenissei River Valley (Mochanov 1975, 1976, Chard 1974) indicates that there were barriers to human communication and gene flow within the arctic steppe and the cold steppe regions during at least the later part of the period under discussion, and the one illustrated human burial at Sungir' near Moscow seems to be of European, not of Mongoloid or Paleo-Indian type (Debets 1967, W.S. Laughlin, University of Connecticut, written communication, May 1, 1977).

Although homogeneity of vegetation, game, landscape, and climate would have facilitated interchange of human populations between central Siberia and Beringia, the greatly expanded glacier systems impeded and possibly foreclosed human dispersals in other directions during the interval from 20,000 to 14,000 years ago. The wooded Amur Valley and the Mongolian steppe were isolated from the arctic steppe of northeastern Siberia by the alpine tundra of a partly glaciated mountain chain (Figures 1.2 and 1.6), forming an ecologic barrier that is likely to have limited human contact between these two regions. Intense highland glaciation in Kamchatka and in the Koryak Range and extensive outlet glaciers reaching tidewater along the adjoining coasts made foot travel difficult, food resources severely limited, and boat travel unattractive in the region between the Amur Valley and the south coast of the land bridge. Heavily glaciated coastal mountains and the presence of ice caps on the exposed continental shelf foreclosed the possibility of human dispersals southeastward along the Pacific coasts of present-day Alaska and British Columbia prior to 14,000 years ago. The merged or nearly merged North American Laurentide and Cordilleran ice sheets tended to isolate Beringia from interior central North America; yet we cannot ignore the possibility that ancestral Paleo-Indian bands may have been able to disperse from Beringia into central America at a time near the maximum of the last glaciation.

Beringia during Deglaciation (14,000–10,000 years ago)

Chronology of Deglaciation

About 14,000 years ago, glaciers began a sudden, drastic thinning and an oscillating retreat. Many thousands of years were required to complete the destruction of the Laurentide and Fennoscandian ice sheets, but the smaller mountain glaciers were almost completely eliminated within the next 2000 years. Most mountain passes in northeastern Siberia and Alaska were free of ice soon after 12,000 years ago (Kind 1972, Funk 1973, Denton 1974, Hamilton and Porter 1975), and by 11,500 years ago glaciers had ceased to be a barrier to the dispersals of plants, animals, and humans along the inland route between Beringia and central North America (Prest 1969), if, indeed, they had been earlier.

Coastal areas were deglaciated quickly. Rising sea levels flooded the continental shelves, accelerating the breakup of glaciers and ice caps that had been grounded there. Fjords and inlets along the North Pacific coast were cleared of ice, and from 13,000 to 10,000 years ago they were submerged more deeply than at present (Easterbrook 1969, Schmoll *et al.* 1972, Miller 1973, Clague 1975); coastal Kamchatka, the Koryak coast, and the shores of the Aleutian Islands were probably deglaciated just as early.

Sea Level History

The sea level had begun to rise about 16,000 years ago (Figure 1.3), but the Bering Land Bridge was not breached until 2000 or 3000 years later. Submerged shorelines at −38 m were occupied no later than 13,000, and possibly as early as 14,000 years ago (Creager and McManus 1967, Hopkins 1973), and Alaska was then separated from Siberia by a shallow and sinuous seaway extending from the southern Bering Sea through the Anadyr and Bering Straits, and thence northward across the partly flooded continental shelf of the Chukchi Sea to the Arctic Ocean (Figure 1.7A). This early seaway should not have been a formidable barrier to dispersals between the continents. Currents were weak, permitting smooth, landfast ice to form each winter, and land was and still is visible across the seaway at its two narrowest points between Cape Chaplin, Siberia, and the northwest corner of St. Lawrence Island (then the St. Lawrence Peninsula of Alaska) and between Cape Dezhnev, Siberia, and westernmost Seward Peninsula, Alaska.

The inception of the swift currents and formidable winter sea-ice conditions that characterize the modern northern Bering Sea, Bering Strait, and Chukchi Sea took place about 12,000 years ago, when sea level rose to −30 m, flooding Shpanberg Strait and isolating St. Lawrence Island from Alaska (Knebel 1972, Hopkins 1973) (Figure 1.7B). Areas of exposed continental shelf were rapidly shrinking, and by the time that sea level had risen to ancient shorelines at −20 m, shortly after 10,000 years ago, only the shallowest embayments along the Alaskan and Siberian coasts remained exposed (Figure 1.7C).

Mountainous coastal regions had been isostatically depressed during glaciation and consequently were drowned more deeply than at present, immediately after deglaciation. As the ice load lightened, these coasts rebounded and began to emerge so that shortly after 10,000 years ago, the shoreline stood near or below its present position.

Climate

The almost catastrophic deglaciation that began about 14,000 years ago reflected a sudden, drastic, worldwide climatic warming. Feedback effects caused by the presence, at first, of still extensive continental glaciers and by the continuing presence of large areas of exposed continental shelf retarded the full impact of the climatic change. Nevertheless, the sudden stagnation of valley glaciers in Alaska in itself indicates an abrupt lifting of the snowline and a sudden increase in mean annual temperatures (Denton 1974, Hamilton and Porter 1975).

Vegetation changes, discussed below, indicate that the warming was accompanied by an increase in total precipitation and suggest that winter snow was deeper than it had been previously, probably as a consequence of more effective penetration of moist air masses into Beringia following the warming of the North Pacific Ocean and the deglaciation of the coastal mountains. Deeper snow on pond ice is also suggested by a change in freshwater mollusk faunas; thaw-lake sediments deposited during this interval in the Kotzebue Sound area contain richer freshwater mollusk faunas than do older lacustrine deposits, perhaps because a deeper, softer, and more insulative snow cover prevented lakes from

30

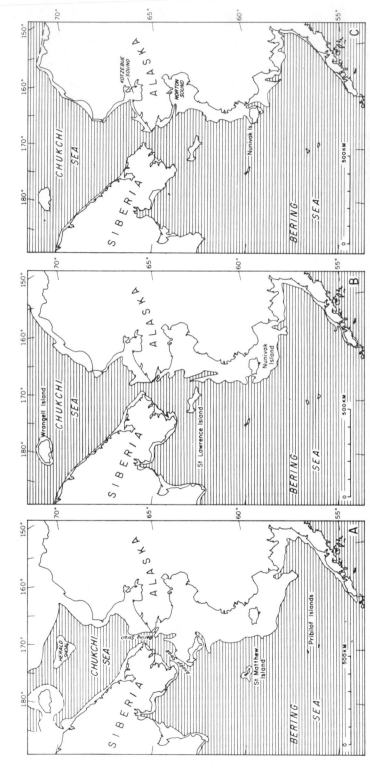

Figure 1.7. Extent of submerged areas in the Bering Sea region when sea level stood (A) near −38 m, 13,000 or 14,000 years ago; (B) near −30 m, about 12,000 years ago; and (C) near −20 m, shortly after 10,000 years ago (Hopkins 1973).

freezing to the bottom.[2] When late-glacial times drew to a close 10,000 years ago, the climate of Beringia had become much like its modern climate.

Biota

A dramatic vegetation change is expressed in the late-glacial segment of every Beringian pollen profile that extends back to full-glacial times (Colinvaux 1967, Rampton 1971, Ritchie and Hare 1971, Ager 1975, Lozhkin 1975). Arctic steppe vanished, to be replaced by mesic arctic tundra. Dwarf birch shrubs, previously only a minor constituent of the vegetation, proliferated and became dominant. *Sphagnum* moss, previously so rare that its spores are generally not reported in pollen spectra of full-glacial age, began to recolonize the tundra vegetation, clogging drainage and heightening the unsuitability of silty soils for xerophytic herbs and grasses. A slowing down of the loess rain and of the movement of aeolian sand may have resulted in gradually lowered soil fertility and thus enhanced the effects of an increasingly moist climate. Peat, wood fragments, and other detrital organic debris became more abundant in sediments dating from late-glacial time in western Alaska, supporting the inference that soils were becoming increasingly moist and waterlogged. Frost heaving increased, and the ground became hummocky and uneven.

Forest trees and associated shrubs and herbs began to expand from their glacial refugia, but some of them required several thousand years to reach their potential climatic–geographic limits. Poplar and perhaps aspen seem to have persisted in Beringia and appear as fossils at and beyond their present limits in western and northern Alaska as early as 11,300 years ago (Hopkins 1972 and unpublished data). Spruce took much longer to recolonize eastern Beringia, evidently spreading from a refugium south of the former Laurentide ice sheet. Spruce trees appeared at the Arctic coast in northwestern Canada as early as 11,500 years ago (Ritchie and Hare 1971), in central Alaska shortly after 10,000 years ago (Ager 1975), and near their present altitudinal limits in the foothills of the Alaska Range and their latitudinal limits in northwestern Alaska about 5000 years ago (Schweger 1975, Thorson and Hamilton 1977).

From the point of view of the large herbivores, a serious range deterioration was in progress, as mesic tundra expanded and the arctic steppe contracted. The once interconnected arctic steppe eventually fragmented into isolated relict grasslands on the few remaining active glacial outwash plains and small patches of relict steppe vegetation on a few steep south-facing valley slopes and river bluffs scattered through areas of continuing strongly continental climate in central Chukotka, central and eastern Alaska, and Yukon Territory, Canada (Yurtsev

---

[2] A contrast between freshwater mollusk faunas in pond sediments of full-glacial and of early Holocene age in the Kotzebue Sound area was partly documented by McCulloch *et al.* (1965), but field work by R.W. Rowland and myself in 1970 established that the change in freshwater mollusk faunas took place earlier than 11,500 years ago rather than about 10,000 years ago as suggested by McCulloch and Hopkins (1966). The 1970 collections were identified by Len V. Kalas, Environment Canada (written communication 1972).

1972, Young 1976). As suitable range areas became constricted and compart-
mented, the large herbivores became increasingly vulnerable to human predators.
Mammoth and horse were quickly exterminated, and bison persisted only in
eastern Beringia and only in very small numbers. By 10,000 years ago, the rich
arctic-steppe fauna had been replaced by the impoverished modern ungulate
fauna, consisting of caribou and a few musk oxen in the tundra, moose in the
taiga, and a few bison in the relict steppe areas.[3]

Interactions with Human Populations

The rapid paleogeographic and vegetational changes of late glacial time
drastically altered the options available to human bands. The possibility of an
economy based upon nomadic hunting of a broad spectrum of gregarious, plains-
dwelling herbivores was rapidly eliminated by the progressive submergence of
the continental shelf and the almost complete disappearance of the arctic-steppe
biome. As the Dry Creek (Alaska) site shows, bison, horse, and probably mammoth hunting continued in local relict arctic-steppe areas as late as 10,600 years
ago (Thorson and Hamilton 1977), but within the next few centuries the hunters'
focus must of necessity have shifted to caribou, the only remaining abundant
arctic ungulate. As shrub tundra and taiga expanded in the lowlands, the caribou
increasingly ranged into the newly deglaciated mountains. Alaskan occupation
sites on the order of 10,000-years-old tend to be concentrated in mountain passes,
on glacial moraines and eskers, and at river crossings still frequented by caribou
(Dixon 1976, West 1975, 1976).

Communication by an interior route between eastern Beringia and central
North America was possible through a steadily broadening ice-free inland corridor
during part and perhaps all of the interval from 14,000 to 10,000 years ago. The
ice-free corridor may have had a steppe landscape during part of this interval,
but the early appearance of spruce at the mouth of the Mackenzie River suggests
that the corridor became extensively clothed in taiga forest by 11,000 years ago.

Communication on foot between Siberia and Alaska was a continuing but
increasingly narrowly focused possibility during the interval 14,000 to 12,000
years ago, as the land bridge became constricted by rising sea level and then
flooded by a shallow seaway which was smoothly frozen during much of the year
(Figure 1.7A). After 11,800 years ago, the breaching of Shpanberg Strait isolated
St. Lawrence Island from the Alaskan mainland (Figure 1.7B). The strengthened

---

[3] The youngest mammoth remains directly dated so far in Alaska are 15,380 ± 300 years old
(SI-456; Stuckenrath and Mielke 1970), but the Berelyakh mammoth in northeastern Siberia is
12,240 ± 160 years old (LU-149, Mochanov 1975). Both occurrences are associated with evidence
of human butchering. The youngest directly dated horse remains reported so far from Beringia are
13,640 ± 410 years old (I-9422). Horse, bison, and probable proboscidian bones occur in component
II of the Dry Creek archaeologic site, which is evidently about 10,600 years old and may have
adjoined a relict steppe area in the northern foothills of the Alaska Range (Thorson and Hamilton
1977). Bison remains as young as 5340 ± 110 years old (SI-845) have been recovered in the Fair-
banks area of central Alaska and two dates on a single bison bone from near Anchorage in southern
Alaska are 470 ± 90 years (SI-851) and less than 180 years (I-9277). (The dates for horse and bison
bones are partly unpublished and were kindly furnished by R.D. Guthrie, University of Alaska,
personal communication, May 27, 1977).

currents and turbulent ice conditions that prevailed thereafter would have made the Bering Strait a formidable barrier to any except skilled boatmen.

Perhaps skilled boatmen were already present. If communities of sea hunters had already evolved, the early deglaciation of coastal areas would have permitted them to disperse along the shores of the Bering Sea, the Sea of Okhotsk, and the North Pacific Ocean as early as 12,000 years ago. The time and place at which a pelagic economy was first developed in the North Pacific region remains unknown, but that this adaptation appeared rather early is attested by evidence of trade in obsidian through the fjords and channels of southeastern Alaska (R. Ackerman, personal communication 1976) and of dolphin hunting off the coast of British Columbia as early as 9000 years ago (Conover 1972, Fladmark 1975) and by the presence of a large, permanent community of sea hunters on Anangula Island off Umnak Island in the eastern Aleutians as early as 8700 years ago (Laughlin 1975, Black 1976). Water barriers at the Bering Strait impeded movements between Siberia and North America only briefly, if at all.

Dispersal of men along the south coast of Beringia was accompanied by a wave of extinctions comparable to the Late Pleistocene extinctions in which man seems to be implicated (Martin 1973, 1974). Steller's sea cow (*Hydrodamalis gigas*) and the flightless spectacled cormorant (*Phalacocorax perspicillatus*) were historically confined to unoccupied and undiscovered northern islands but must once have ranged widely around the North Pacific. Steller's sea cow, for example, is represented by fossils of Late Pleistocene age on the continental shelf of central California (Jones 1968). Both are extremely vulnerable to human predation. Steller's sea cow was exterminated within a couple of decades and the spectacled cormorant within a little more than a century after the discovery of relict populations on the Commander Islands by the Bering Expedition in 1741 (Laughlin 1967).

It seems certain that former, more widely ranging populations of Steller's sea cow and spectacled cormorant were exterminated by human hunters. Walrus, often thought of as obligatory dwellers on floating ice, also once ranged south to the latitudes of central California, North Carolina, and Paris (Repenning and Redford, 1977) and they, too, have probably contracted their range in response to human predation. Any occupation sites around the shores of the North Pacific Ocean discovered to contain large quantities of Steller's sea cow, walrus, or spectacled cormorant remains should probably be assumed to record the first appearance of coastal hunters at that locality.

Beringia during Holocene Time
(10,000 years ago to the present)

Climatic, vegetational, and paleogeographic changes have been slight in most Arctic regions and minimal in Beringia during the last 10,000 years (Figure 1.5). The sea level reached almost its present position by 4000 years ago and has risen only a few meters since (Figure 1.3). The present interglacial peaked about 6000 to 4000 years ago (Deevey and Flint 1957), but this maximum of summer warmth was muted within Beringia by the cooling influence of the nearby seas.

More notable was an episode of greater summer warmth recorded in north-western Alaska early in Holocene time, 10,000 to 8000 years ago (McCulloch and Hopkins 1966). The thermal maximum of 6000 to 4000 years ago was strongly felt, however, outside the limits of Beringia in northern Canada (Nichols 1974, 1975) and in parts of western Siberia (Kind 1967). Lighter sea ice in the channels of the Canadian Arctic Archipelago at that time (Blake 1975) may have facilitated the dispersal of early Eskimos across the Canadian Arctic to Greenland.

Small cold fluctuations during the last 3000 years have been accompanied by small glacial advances, heightened rates of peat accumulation in some areas, and reactivation of frost cracking and ice-wedge growth in areas of discontinuous permafrost. The limits of poplar, aspen, and birch trees may have fluctuated, but the limits of spruce and alder have not. Instead, pollen and macrofossil records seem to indicate a steady advance, uninterrupted but varying in rate, of spruce trees and alder shrubs into former shrub–tundra areas (Heusser 1960, D.M. Hopkins, manuscript). The impact of these landscape changes on aboriginal populations must have been minute compared to the dramatic consequences of the climatic and paleogeographic events of late glacial time.

Discussion

I have not attempted to reconstruct early human history in Beringia and Arctic regions to the west, or even to review our sketchy present knowledge. However, the growing understanding of the climatic and landscape history of Beringia and other northern regions permits some assumptions and provides some constraints that can be used as limiting factors in future attempts to reconstruct early human history:

1. The Arctic steppe was not a hostile environment to man and, once occupied by humans, it probably was never vacated. Scenarios for human dispersals that call for an abandonment of Beringia between 25,000 and 12,000 years ago are not likely to be correct.

2. The pan-Arctic steppe region seems to have been so homogeneous that I would expect colonization of the entire region soon after human populations first developed the necessary survival and hunting techniques.

3. The presence of Lake Mansi in the Ob' and Yenissei River Valleys severely limited the width of the cold-steppe belt in western Siberia, forming a constriction that could have limited gene flow between populations residing in the cold-steppe regions of eastern Europe and in central Siberia during major glacial intervals. The geochronology of Lake Mansi is not yet adequately known, but it probably created a filter barrier in southwestern Siberia during most of the interval 20,000 to 14,000 years ago.

4. The Bering Strait has never remained a barrier to human dispersals for periods longer than a millenium or two. During intervals when the sea level lay within 40 m of its present position, however, the Bering Strait area has been a funnel limiting the avenues of human dispersal and communication and poten-

tially limiting gene flow between populations resident in Alaska and northeastern Siberia. The Bering Strait area was probably a filter barrier of this type prior to 25,000 years ago and again after 14,000 years ago.

5. Merger of the Laurentide and Cordilleran ice sheets has probably never been a barrier to human dispersals for periods longer than a millenium or two. During major glacial intervals, however, the great length and small width of the ice-free corridor has made it a funnel narrowly limiting avenues of human dispersal and severely limiting potential gene flow between human populations resident in Alaska and in central North America. The narrow ice-free corridor was probably a filter barrier of this type during much of the interval from 25,000 to 12,000 years ago.

6. Coastal dispersals between Beringia and the Soviet Far East and between Beringia and Puget Sound were probably prevented by the presence of ice caps and piedmont glaciers on the continental shelves during major glacial intervals. However, the Alaskan and the British Columbia coasts and probably also the Koryak and Kamchatka coasts were deglaciated early, so that coastal dispersals from Beringia to Puget Sound or between Beringia and the Sea of Okhotsk could have taken place as early as 12,000 years ago.

7. As S.W. Stoker's unpublished study emphasizes (Dixon *et al.* 1976), concentrated food resources were probably available to man locally at the south shore of the Bering Land Bridge, even during full-glacial times. The early dates for North American coastal settlements relative to those postulated for coastal settlements on the shores of the Okhotsk Sea (Chard 1974) require that we give serious consideration to the possibility that sea mammal hunting originated somewhere in southern Beringia before 10,000 years ago.

8. The drastic changes in vegetation and in the ungulate population during the interval 14,000 to 10,000 years ago must have placed great stress upon nomadic arctic hunting populations. This seems to me to be the interval during which major ethnic or demographic population changes would be most likely to occur.

Acknowledgments

I am grateful to the Wenner-Gren Foundation and especially to Lita Osmundsen for the opportunity to participate in the Wenner-Gren Burg Wartenstein Symposium, on which this volume is based, and to the other participants for stimulating discussion that clarified my thinking on many anthropologic issues. This chapter is a product of discussions with these and many other colleagues, among whom R. Dale Guthrie, William S. Laughlin, John V. Matthews, Charles E. Schweger, Stephen Young, and Boris Yurtsev were especially helpful. The chapter was greatly improved by reviews of an early version by David Adam and Thomas D. Hamilton (US Geological Survey, Menlo Park, California), Robert Ackerman (Washington State University, Pullman, Washington), Douglas Anderson (Brown University, Providence, Rhode Island), and Igor Volkov and the late Sergei L. Troitskiy (Institute of Geology and Geophysics, Novosibirsk, USSR).

References

Ager, T.A. (1975). Late Quaternary Environmental History of the Tanana Valley, Alaska. Ohio State Univ. Inst. Polar Studies, Rept. 54, 117 p.

Black, R.F. (1951). Aeolian Deposits of Alaska. *Arctic* 4:89–111.

Black, R.F. (1976). Geology of Umnak Island, Eastern Aleutian Islands, as Related to the Aleuts. *Arctic Alpine Res.* 8:7–35.

Blake, W. Jr. (1975). Climatic Implications of Radiocarbon-Dated Driftwood in the Queen Elizabeth Islands, Arctic Canada. *In* Y. Vasari, H. Hyvärinen, and S. Hicks, Eds., *Climatic Changes in Arctic Areas during the Last 10,000 Years.* Acta Univ. Ouluensi, Ser. A., No. 3, Geol. No. 1. pp. 78–104.

Broecker, W.S. (1975). Floating Glacial Ice Caps in the Arctic Ocean. *Science* 188:1116–1118.

Brown J., and P.V. Sellmann (1966). Radiocarbon Dating of a Buried Coastal Peat, Barrow, Alaska. *Science* 153:299–300.

Chard, C.S. (1974). *Northeast Asia in Prehistory.* Madison: Univ. Wisconsin Press, 214 p.

Clague, J.J. (1975). Late Quaternary Sea Level Fluctuations, Pacific Coast of Canada and Adjacent Areas. *Geol. Surv. Can. Pap.* 75-1C:17–20.

Clark, D.L. (1971). Arctic Ice Cover and Its Late Cenozoic History. *Geol. Soc. Am. Bull.* 82:3313–3324.

CLIMAP Project Members (1976). The Surface of the Ice-Age Earth. *Science* 191:1131–1137.

Colinvaux, P.A. (1967). Quaternary Vegetation History of Alaska. *In* D.M. Hopkins, Ed., *The Bering Land Bridge.* Stanford, Calif.: Stanford Univ. Press, pp. 207–231.

Conover, K. (1972). *Archaeological Sampling at Namu. A Problem in Settlement Reconstruction.* Ph.D. Thesis, University of Colorado, Boulder, Colorado.

Creager, J.S., and D.A. McManus (1967). Geology of the Floor of Bering and Chukchi Seas—America Studies. *In* D.M. Hopkins, Ed., *The Bering Land Bridge.* Stanford, Calif.: Stanford Univ. Press, pp. 7–31.

Curray, J.R., and F.P. Shepard. (1972). Some Major Problems of Holocene Sea Levels. *Abs. Am. Quatern. Assoc. (AMQUA)* 2nd Bien. Mtg., Miami, Florida, pp. 16–18.

Debets, G.F. (1967). Skelet Pozdnepaleoliticheskogo Cheloveka iz Pogrebeniia na Sungirskoi Stoyanke [Skeleton of Late Pleistocene Man in Burial at Sungir' Site]. *Soviets. Arkheol.* 3:160–164.

Degtyarenko, Yu.P. (1963). Osnovnye Cherty Geomorfologicheskogo Stroeniya Koryakskoi Gornoi Systemy [Basic Features of the Geomorphic Structure of the Koryak Ridge System]. *Geol. Koryakskogo Nagorya,* pp. 169–184.

Deevey, E.S., and R.F. Flint. (1957). Post-glacial Hypsithermal Interval. *Science* 125:182–184.

Denton, G.F. (1974). Quaternary Glaciations of the White River Valley, Alaska, with a Regional Synthesis for the Northern St. Elias Mountains, Alaska and Yukon Territory. *Geol. Soc. Am. Bull.* 85:871–872.

Dixon, E.J. (1976). The Gallagher Flint Station, an Early Man Site on the North Slope of Brooks Range, Arctic Alaska, and Its Relationship to the Bering Land Bridge. *In* V.L. Kontrimavichis, Ed., *Beringia in Cenozoic.* Vladivostok: Acad. Sci. USSR, Far-East. Sci. Ctr., pp. 467–475. [In Russian with English abs.]

Dixon, E.J., G.D. Sharma, R.D. Guthrie, and S.W. Stoker. (1976). Unpublished Report to the US Dept. Interior, Bureau of Land Management, Contract 08550-CTS-45.

Dupré, W.R. (1977). Late Quaternary History of the Yukon–Kuskokwim Delta Complex, Alaska. *Geol. Soc. Am. Abs. Programs* 9:413.

Easterbrook, D.J. (1969). Pleistocene Chronology of the Puget Sound Lowland and San Juan Islands, Washington. *Geol. Soc. Am. Bull.* 80:2273–2286.

Fladmark, K. K. (1975). A Paleoecological Model for Northwest Coast Prehistory. *Can. Archeol. Surv. Pap.* 43, 328 p.

Flint, R.F. (1971). *Glacial and Quaternary Geology.* New York: John Wiley & Sons, 892 p.

Frenzel, B. (1968). The Pleistocene Vegetation of Northern Eurasia. *Science* 161:637–649.

Funk, J.M. (1973). *Late Quaternary Geology of Cold Bay, Alaska, and Vicinity.* M.S. Thesis, University of Connecticut, Storrs, Conn. 45 p.

Ganeshin, G.S. (Ed.). (1969). Karta Chetvetichnykh Otlozheniy SSSR [Map of the Quaternary Deposits of the USSR], 1:5,000,000. Vses. Nauchno-Issled. Geol. Inst., VSEGEI and Vses. Aerogeologicheskiy Trest., 4 sheets.

Gates, W.L. (1976). Modeling the Ice-Age Climate. *Science* 191:1138–1144.

Giterman, R.E., and L.V. Golubeva. (1967). Vegetation of Eastern Siberia During the Anthropogene Period. *In* D.M. Hopkins, Ed., *The Bering Land Bridge.* Stanford, Calif.: Stanford Univ. Press, pp. 232–244..

Giterman, R.E., L.V. Golubeva, E.D. Zaklinskaya, E.V. Koreneva, O.V. Matveeva, and L.A. Skiba. (1968). The Main Development Stages of the Vegetation of North Asia in Anthropogene. *Acad. Sci. USSR, Geol. Inst., Trans.* 177:1–269.

Grim, M.S., and D.A. McManus. (1970). A Shallow Seismic-profiling Survey of the Northern Bering Sea. *Mar. Geol.* 8:293–320.

Guthrie, R.D. (1968a). Paleoecology of the Large Mammal Community in Interior Alaska. *Am. Mid. Naturalist* 79:346–363.

Guthrie, R.D. (1968b). Paleoecology of a Late Pleistocene Small Mammal Community from Central Alaska. *Arctic* 21:223–244.

Hamilton, T.D., and S.C. Porter. (1975). Itkillik Glaciation in the Brooks Range, Northern Alaska. *Quatern. Res.* 5:471–498.

Harington, C.R. (1978). Quaternary Vertebrate Faunas of Canada and Alaska and Their Suggested Chronological Sequence. *Natl. Mus. Nat. Sci., Syllogeus*, No. 15. Ottawa: National Museum of Canada, 105 p.

Haynes, C.V. (1964). Fluted Projectile Points. Their Age and Dispersion. *Science* 145:1408–1413.

Haynes, C.V. (1971). Time, Environment, and Early Man. *Arct. Anthropol.* 8(2):3–14.

Haynes, C.V. (1976). Mammoth Hunters of the USA and the USSR. *In* V. Kontrimavichis, Ed., *Beringia in Cenozoic.* Vladivostok: Acad. Sci. USSR, Far-East. Sci. Ctr., pp. 427–438. [In Russian with English abs.]

Herman, Y. (1970). Arctic Paleo-oceanography in Late Cenozoic Time. *Science* 169:474–477.

Herman, Y. (1974). Arctic Ocean Sediments, Microfauna, and the Climatic Record in Late Cenozoic Time. *In* Y. Herman, Ed., *Marine Geology and Oceanography of the Arctic Seas.* New York: Springer-Verlag, pp. 283–348.

Herman, Y. (1977). Faunal Composition, Eolian Dust, and Ice Rafting as Indicators of Arctic Ice Extent and Thickness in Late Cenozoic Time. *Proc. Intl. Quatern. Assoc. (INQUA)*, X Cong., Birmingham, UK, p. 203.

Heuberger, H. (1974). Alpine Quaternary Glaciations. *In* J.D. Ives and R.G. Barry, Eds., *Arctic and Alpine Environments.* London: Methuen, pp. 319–338.

Heusser, C.J. (1960). Late-Pleistocene Environments of North Pacific North America. *Am. Geog. Soc. Spec. Pub.* 35, 308 p.

Holmes, M.L., and J.S. Creager. (1974). Holocene History of the Laptev Sea Continental Shelf. *In* Y. Herman, Ed., *Marine Geology and Oceanography of the Arctic Seas.* New York: Springer-Verlag, p. 229.

Hopkins, D.M. (1967). The Cenozoic History of Beringia—A Synthesis. *In* D.M. Hopkins, Ed., *The Bering Land Bridge.* Stanford, Calif.: Stanford Univ. Press, pp. 451–481.

Hopkins, D.M. (1970). Paleoclimatic Speculations Suggested by New Data on the Location of the Spruce Refugium in Alaska During the Last Glaciation. *Abs. Am. Quaternary Assoc. (AMQUA)*, 1st Bien. Mtg., Bozeman, Montana, p. 67.

Hopkins, D.M. (1972). The Paleogeography and Climatic History of Beringia During Late Cenozoic Time. *Internord* 12, pp. 121–150.

Hopkins, D.M. (1973). Sea Level History in Beringia During the last 250,000 Years. *Quatern. Res.* 3:520–540.

Hopkins, D.M. (1977). Offshore Permafrost Studies, Beaufort Sea. Environmental Assessment of Alaskan Continental Shelf, Principal Investigator's Repts. for Year Ending March 31, 1977. Washington, D.C.: Natl. Oceanog. and Atmospheric Adm., pp. 396–518.

Hopkins, D.M. (manuscript). Dated Wood from Alaska: Implications for Forest Refugia in Beringia.

Hopkins, D.M., C.H. Nelson, R.B. Perry, and T.R. Alpha. (1976). Physiographic Subdivisions of the Chirikov Basin, Northern Bering Sea. Washington, D.C.: *US Geol. Survey Prof. Pap.* 759-B, 7 p.

Hoppe, G. (1973). Ice Sheets Around the Norwegian Sea During the Würm Glaciation: *Ambio Spec. Rept.* No. 2, pp. 25–29.

Hughes, T., G.H. Denton, and M.G. Groswald. (1977). Was There a Late-Würm Ice Sheet. *Nature* 266:596–602.

Jones, R.E. (1968). A *Hydrodamalis* Skull from Monterey Bay, California. *J. Mammal.* 48: 143–144.

Kind, N.V. (1967). Radiocarbon Chronology in Siberia. *In* D.M. Hopkins, Ed., *The Bering Land Bridge.* Stanford, Calif.: Stanford Univ. Press, pp. 172–192.

Kind, N.V. (1972). Late Quaternary Climatic Changes and Glacial Events in the Old and New World—Radiocarbon Chronology. *24th Intl. Geol. Cong., Repts.*, Sec. 12, pp. 55–61.

Klein, R.G. (1973). *Ice-age Hunters of the Ukraine.* Chicago: Univ. of Chicago Press, 140 p.

Knebel, H.J. (1972). *Holocene Sedimentary Framework of the East-Central Bering Sea Continental Shelf.* Ph.D. Thesis, University of Washington, Seattle, 186 p.

Knebel, H.J., and J.S. Creager. (1973). Yukon River. Evidence for Extensive Migration During the Holocene Transgression. *Science* 179:1230–1232.

Knebel, H.J., J.S. Creager, and R.J. Echols. (1974). Holocene Sedimentary Framework, East-Central Bering Sea Continental Shelf. *In* Y. Herman, Ed., *Marine Geology and Oceanography of the Arctic Seas.* New York: Springer-Verlag, pp. 157–172.

Kontrimavichus, V.L. (Ed.). (1976). *Beringia in Cenozoic.* Vladivostok: Acad. Sci. USSR, Far-East. Sci. Ctr. 594 p. [In Russian with English abs.]

Kummer, J.T., and J.S. Creager. (1971). Marine Geology and Cenozoic History of the Gulf of Anadyr. *Mar. Geol.* 10:257–280.

Lamb, H. H. (1970). Atmospheric Circulation During the Last Ice Age: *Quatern. Res.* 1:29–58.

Laughlin, W.S. (1967). Human Migration and Permanent Occupation in the Bering Sea Area. *In* D.M. Hopkins, Ed., *The Bering Land Bridge*. Stanford, Calif.: Stanford Univ. Press, pp. 409–450.

Laughlin, W.S. (1975). Aleuts. Ecosystem, Holocene History and Siberian Origin. *Science* 189:507–515.

Lozhkin, A.V. (1975). Absolyutnaya Geochronologiya i Sobbitiya Pleistotsena na Territorii Severo-Vostoka SSSR [Absolute Geochronology and Occurrences in Northeastern USSR], *In Geologicheskie Issledovaniya na Severo-Vostoke SSSR*. Acad. Sci. USSR, Far-East. Sci. Ctr., *Northeastern Complex Inst., Bull.* 68:127–129. [In Russian with English abs.]

Martin, P.S. (1973). The Discovery of America. *Science* 179:969–974.

Martin, P.S. (1974). Paleolithic Players on the American Stage. Man's Impact on the Late Pleistocene Megafauna. *In* J.D. Ives and R.G. Barry, Eds., *Arctic and Alpine Environments*. London: Methuen, pp. 669–702.

Matthews, J.V. (1974). Quaternary Environments at Cape Deceit (Seward Peninsula, Alaska): Evolution of a Tundra Ecosystem. *Geol. Soc. Am. Bull.* 85:1353–1384.

Matthews, J.V. (1976). Arctic Steppe: An Extinct Biome. *Abs. Am. Quatern. Assoc.* *(AMQUA)*, 4th Bien. Mtg., Tempe, Arizona, pp. 73–77.

McCulloch, D.S., and D.M. Hopkins. (1966). Evidence for an Early Recent Warm Interval in Northwestern Alaska. *Geol. Soc. Am. Bull.* 77:1089–1108.

McCulloch, D.S., D.W. Taylor, and M. Rubin. (1965). Stratigraphy, Non-marine Mollusks, and Radiometric Dates from Quaternary Deposits in the Kotzebue Sound Area, Western Alaska. *J. Geol.* 73:442–453.

Melekestev, I.V., O.A. Braitseva, E.N. Erlich, A.E. Shanster, A.I. Chelebaeva, E.G. Lulikina, I.A. Egoraeva, and N.N. Kzhemyaka. (1974). Istoriya Razvitiya Rel'efa Sibiri i Dal'nego Vostoka. *Kamchatka, Kuril'skie, i Kmoandorskie Ostrova, Izd-vo "Nauka"*, Moscow, 438 p.

Mercer, J.H. (1970). A Former Ice Sheet in the Arctic Ocean? *Paleogeog. Paleoclimatol. Paleoecol.* 8:19–27.

Miller, R.D. (1973). Gastineau Channel Formation, a Composite Glaciomarine Deposit Near Juneau, Alaska. *U.S. Geol. Surv. Bull.* 1394-C:1 20.

Mochanov, Yu.A. (1975). Stratigrafiya i absolyutnaya Khronologiya Paleolita Severo-Vostochnoi Azii [Stratigraphy and Absolute Chronology of the Paleolithic of North East Asia], *in Yakutya i éé Sosedi v Dvernosti [Yakutia and Neighboring Regions in Antiquity]*. Izd-vo Yakutsk. Fil., Acad. Sci. USSR, pp. 9–30.

Mochanov, Yu.A. (1976). The Paleolithic of Siberia. *In* V.L. Kontrimavichis, Ed., *Beringia in Cenozoic*. Vladivostok: Acad Sci. USSR., Far-East Sci. Ctr., pp. 540–565. [In Russian with English abs.]

Müller-Beck, H. (1967). On Migrations of Hunters across the Bering Land Bridge in the Upper Pleistocene. *In* D.M. Hopkins, Ed., *The Bering Land Bridge*. Stanford, Calif.: Stanford Univ. Press, pp. 373–408.

Nelson, C.H., and J.S. Creager. (1977). Displacement of Yukon-derived Sediment from Bering Sea to Chukchi Sea During Holocene Time. *Geology* 5:141–146.

Nichols, H. (1974). Arctic North American Paleoecology. The Recent History of Vegetation and Climate Deduced from Pollen Analysis. *In* J.D. Ives and R.G. Barry, Eds., *Arctic and Alpine Environments*. London: Methuen, pp. 637–668.

Nichols, H. (1975). Palynological and Paleoclimatic Study of the Late Quaternary Displacement of the Boreal Forest–Tundra Ecotone in Keewatin and MacKenzie, N.W.T., Canada. Univ. of Colorado: *Inst. Arctic and Alpine Research, Occas. Pap.* 15, 87 p.

Péwé, T.L. (1975). Quaternary Geology of Alaska. Washington, D.C.: *US Geol. Surv. Prof. Pap.* 835.

Prest, V.K. (1969). Retreat of Wisconsin and Recent Ice in North America. Ottawa: *Geol. Surv. Can., Map* 1257A.

Prest, V.K., D.R. Grant, and V.N. Rampton. (1968). Glacial Map of Canada (1:5,000,000). Ottawa: *Geol. Surv. Can., Map* 1253A.

Rampton, V. (1971). Late Quaternary Vegetational and Climatic History of the Snag–Klutlan area, Southwestern Yukon Territory, Canada. *Geol. Soc. Am. Bull.* 82:959–978.

Reeves, B. (1973). The Nature and Age of the Contact Between the Laurentide and Cordilleran Ice Sheets in the Western Interior of North America. *Arctic Alpine Res.* 5:1–16.

Repenning, C.A., and R.H. Redford. (1977). Otaroid Seals of the Neogene. Washington, D.C.: *US Geol. Surv. Prof. Pap.* 992, 93 pp.

Ritchie, J.C., and F.K. Hare. (1971). Late Quaternary Vegetation and Climate Near the Arctic Treeline of Northwestern North America. *Quatern. Res.* 1:331–342.

Roed, M.A. (1975). Cordilleran and Laurentide Multiple Glaciation, West-Central Alberta, Canada. *Can. J. Earth Sci.* 12:1493–1515.

Schmoll, J.R., B.J. Szabo, M. Rubin, and E. Dobrovolny. (1972). Radiometric Dating of Marine Shells from the Bootlegger Cove Clay, Anchorage Area, Alaska. *Geol. Soc. Am. Bull.* 83:1107–1114.

Schweger, C.E. (1976). *Late Quaternary Paleoecology of the Onion Portage Region, Northwestern Alaska.* Ph.D. Thesis, University of Alberta, Alberta. 183 p.

Schweger, C.E., and J. Martin. (1976). Grazing Strategies of the Pleistocene Steppe–Tundra Fauna. *Abs. Am. Quatern, Assoc. (AMQUA),* 4th Bien. Mtg., Tempe, Arizona, p. 157.4.

Sergin, S.Ya., and M.S. Shcheglova. (1976). The Climate of Beringia During Glacial Epochs: A Result of the Influence of Local and Global Factors. *In* V.L. Kontrimavichus, Ed., *Beringia in Cenozoic.* Vladivostok: Acad. Sci. USSR, Far-East. Sci. Ctr., pp. 171–176. [In Russian with English abs.]

Sher, A.V. (1967). Iskopaemaya Saiga na Severe Vostochnoi Sibiri i Alyeske [Fossil Saiga in North East Siberia and Alaska]. *USSR Acad. Sci., Kommiss. Izuch. Chetvertichn. Perioda, Bull.* 33:97–112. [In Russian.]

Streten, N.A. (1974). Some Features of the Summer Climate of Beringia. *Arctic* 27:273–286.

Stuckenrath, R. Jr., and J.E. Mielke. (1970). Smithsonian Institution Radiocarbon Measurements VI. *Radiocarbon* 12:193–204.

Thorson, R.M., and T.D. Hamilton. (1977). Geology of the Dry Creek Archaeological Site, Alaska. *Quatern. Res.* 7:149–176.

Tomirdiaro, S.V. (1975). Lessovo-delovaya Formatsiya Verkhnepleistotsenovoi Giperzony v Severovom Polusharii [Loess-ice Formation of Upper Pleistocene Hyperzone in the Northern Hemisphere]. *In Geologicheskie Issledovaniya na Severo-Vostoke SSSR.* Acad. Sci. USSR, Far-East. Sci. Ctr. Northeastern Complex Inst., Magadan, *Bull.* 68:170–197. [In Russian with English abs.]

Vereshchagin, N.K. (1974). The Mammoth "Cemeteries" of North-East Siberia. *Polar Rec.* 17:3–12.

Vereshchagin, N.K., and Yu.A. Mochanov. (1972). Samyye Severnyye v Mire Sledy

Verkhnego Paleolita: Berelëkhskoye Mestonakhozhdeniye v Nizov'yakh r. Indigirki [The Northernmost Traces of the Late Paleolithic in the World: the Berelyokh Site on the Lower Indigirka]. *Soviets Arkheol.* 2:332–336.

Volkov, I.A., and V.S. Volkova. (1972). Pleistocene Mansi Lake in the South of West Siberia. *V. Intl. Verein. Limnol.* 18:1083–1085.

Volkov, I.A., and V.S. Volkova. (1975). Istoriya Ozer v Pleistocene. Leningrad: Acad. Sci. USSR, Inst. Limnology. *IU Vsesoyuzniy Symposium Istorii Ozer*, Abs., 2:133–141.

West, F.H. (1975). Dating the Denali Complex. *Arctic Anthropol.* 12:76–81.

West, F.H. (1976). Old World Affinity of Archaeological Complexes from Tangle Lakes, Central Alaska. *In* V.L. Kontrimavichis, Ed., *Beringia in Cenozoic.* Vladivostok: Acad. Sci. USSR, Far-East. Sci. Ctr., pp. 439–458. [In Russian with English abs.]

Young, S. (1976). Is Steppe Tundra Alive and Well in Alaska? *Abs. Am. Quatern. Assoc.* (*AMQUA*), 4th Bien. Mtg., Tempe, Arizona, pp. 84–88.

Yurtsev, B.A. (1972). Phytogeography of Northeastern Asia and the Problem of Trans-beringian Floristic Interrelations. *In* A. Graham, Ed., *Floristics and Paleofloristics of Asia and Eastern North America.* Amsterdam: Elsevier, pp. 19–54.

Chapter 2

The Origin and Dispersion of American Indians in North America[1]

James B. Griffin

Introduction

The historically effective discovery of the New World by Europeans in 1492 opened the Americas to large-scale colonization and to a long period of speculation on the ultimate origin of the New World peoples. For almost 400 years the chronologic framework was based on biblical sources and was not adequate to explain the cultural diversity of New World culture. Prominent among the speculations on origins were the multiple-migration hypotheses, which provided civilized Mediterranean, European, and even Southeast Asian groups to bring a level of higher culture into the Americas and to produce the more complex civilizations of Mesoamerica and the Andes. Such explanations were also employed to explain a presumed multiracial origin of the American Indian.

As early as the sixteenth century some writers recognized the predominantly Asiatic relationships of the Native American populations and postulated that the movement of people took place from northeast Asia to northwestern North America. This is, of course, the only route of entry seriously considered by contemporary scholars, but there is considerable room for disagreement in many phases of the origin and dispersion of American Indian populations. I have chosen to present a view that is on the conservative side, but not as conservative as some of my colleagues, such as C. Vance Haynes (1964, 1967, 1969) or Paul S. Martin (1963, 1967, 1973), normally are. There are, however, some evaluations of the antiquity of man in the New World measured in the high tens of thousands

[1] This chapter is a revised version of a paper by the same title published in *Biomedical Challenges Presented by the American Indian*, Scientific Publication No. 165, Pan American Health Organization, Washington, D.C., 1968, pp. 2–10.

to hundreds of thousands of years or more on what I regard as either provocative or very slim evidence. These are simply not regarded as demonstrated by a large number of competent authorities. A site that is proposed in the New World as a definite "Early Man" site could be readily acceptable by most archaeologists if it had the following characteristics:

1. A clearly identifiable geologic context, either in an open location or in shelters or caves. This context examined by a number of competent geologists specializing in the type of formation in which the site is located. Reasonable agreement on the context, manner of primary deposition, and relative antiquity of the deposits by almost all of the geologists, with no possibility of intrusion or secondary deposition.

2. Recovery of an adequate sample of material culture items representing activity areas and a range of tool forms and debris adequate to characterize a significant portion of a society's material culture.

3. Well-preserved animal remains to aid in interpreting environment as well as hunting practices, protein sources, and probable clothing sources.

4. Pollen studies from site and environment to aid in interpreting climate, food, and shelter supplies. Macrobotanical materials identified for environmental, food, or utilitarian interpretations.

5. Human skeletal materials to aid in recognizing biologic relations and rough temporal estimates as well as to yield data on group nutrition and other social patterns.

An adequate series of C^{14} dates should be obtained from each of one or more levels at the site. Other independent age assessments should be made if possible by other radiometric time clocks or other means of dating.

All of the data should agree on the age, season, environment, and cultural level of the occupants of the site. Admittedly, this is an ideal model for an "Early Man" site to have, but insofar as finds that have been proposed fail to satisfy such criteria they are inevitably open to question, rejection, or suspended judgment. In the Eastern Hemisphere on its major continents there are finds of *Homo sapiens* not only measured in the tens of thousands but of hominids and their cultural products measured in the hundreds of thousands of years. When sites in the Americas produce the kind of evidence known from the Old World there will be little reluctance by scientists to accept the evidence.

The Temporal Framework for Early Indian Groups

The time of arrival of the first human groups is far from settled and is likely to remain a research problem for some time. For the purposes of this chapter, I shall refer to the period of time from the first arrival of man to about 8000 B.C. as the Paleo-Indian period, and all cultural complexes and skeletal material of this age, if any, will be included within this period. A rapid review of evidence from the United States and Canada includes only one fairly homogeneous and widely dispersed complex, a few examples of other early Paleo-Indian period peoples, and a number of purported sites that are not regarded in this chapter as soundly established.

The Fluted Point Hunters are the sole recognizable group with a continent-wide occupancy in the latter part of the Paleo-Indian period from about 10,000 to 8000 B.C. The earliest radiocarbon dates in association with an established cultural context are about 9300 to 9000 B.C., from sites in Arizona, New Mexico, Colorado, and Oklahoma. This is referred to as the Llano complex, with the Clovis fluted point as a major diagnostic tool. The Sandia finds are regarded as a potential predecessor, with an uncertain age—probably within a thousand years of Clovis. In the western plains the later Folsom assemblages at several sites have been dated between 9000 and 8000 B.C. In the eastern United States and Canada, fluted points are known from the southern Atlantic and Gulf coasts to the Great Lakes and Ontario and into New England, and as far northeast as Nova Scotia. At the Debert site in Nova Scotia, a series of radiocarbon dates suggests that man occupied the area between 9000 and 8500 B.C. (MacDonald 1968). Provisional correlations of the distribution of fluted points in the Great Lakes area have proposed an age of from 10,000 to 8000 B.C. for that area (Griffin 1965), and the evidence from the southeast also suggests an antiquity for fluted points of about the same order.

In the area west of the Rocky Mountains there are a number of complexes that may begin before 8000 B.C., but as yet this has not been adequately demonstrated. Among these are the Desert culture, Lake Mohave, and the San Dieguito culture of southern California, and a number of sites and complexes from Oregon to British Columbia and Washington compressed by some archaeologists into an Old Cordilleran culture. In extreme northwest Canada, the British Mountain complex is assigned considerable antiquity, but its age, except on typologic grounds, is not known. Some of the oldest radiocarbon dates believed to be in association with human habitation in the western United States are from Wilson Butte Cave in south-central Idaho, where from the lower zone of stratum C there is a date of 12,550 ± 500 B.C. (M-1409) and from stratum E a date of 13,050 ± 800 B.C. (M-1410). The context of the material in the cave indicates the presence of man with extinct fauna, such as camel and horse and some boreal zone microfauna, but the few artifacts found are not particularly diagnostic (Gruhn 1965).

Very recent excavations at the Meadowcroft rock shelter about 40 miles southwest of Pittsburgh have been producing a series of radiocarbon dates in what is called stratum IIa running from 11,000 to 14,000 B.C. This location has possibilities but until an analysis of the deposits, of the cultural complex, and of the composition and matrix of the material dated is provided it is better to suspend judgment on the meaning of the material (Adovasio *et al.* 1977, 1978).

Fluted points are known from Mexico and as far south as Guatemala and perhaps Costa Rica. In none of these instances, however, is there an associated industry. Most of the few fluted points are found in northern Mexico, where they are on the southern fringe of the Llano and Folsom concentrations in the south-west and western Texas. There are no soundly established cultural assemblages in Mexico and Central America directly dated by radiocarbon before 8000 B.C.

There is a large number of locations in North America for which considerable antiquity has been claimed as places inhabited by early Indians. Even whole books have been published on nonsites. The reasons it is now difficult or impossible to include such "finds" here varies from location to location; a detailed dissent is not within the scope of this chapter. In this category I would include the claims for occupation of Tule Springs, Nevada, on the order of 20,000 to 30,000 years ago; on Santa Rosa Island, much before 8000 B.C.; at the Scripps Institute bluff at La Jolla, California, slightly over 20,000 B.C.; at the Del Mar site and others recently receiving aspartic acid racemization or amino acid dating, providing age estimates in the 40,000 years ago range; at Lewisville, Texas, more than 37,000 B.C.; at Sheguiandah, Manitoulin Island, Ontario, for a cultural complex much before 7000 B.C.; for a completely pebble-tool culture in northern Alabama of extreme antiquity; for a chopper/chopping-tool complex of an interglacial or interstadial period, or of a simple bone-tool tradition of any age; or for Pleistocene man in the Trenton, New Jersey, gravels.

Various types of tool assemblages are fairly widespread in Alaska by about 8000 B.C. From the Kobuk River–Akmak complex to the Mt. McKinley–Dry Creek site there are indications of early hunting groups and at the latter site occupation may be back considerably earlier, perhaps as much as 12,000 B.C. The earliest occupations in the Aleutian chain are between 6000 and 7000 B.C., although it is likely that earlier sites are now below sea level. On the Seward Peninsula the lowest level of Trail Creek Cave dates about 12,000 B.C., but there is little cultural evidence (Larsen 1968). Claims for significantly earlier finds attributable to man in the western Yukon or Alaskan area of 20,000 to 30,000 years ago are not yet accepted by many archaeologists. The presence of fluted points in Alberta, the Yukon, and Alaska are best regarded as the result of a south to north movement in the western high plains (Dixon 1976).

In Mexico particularly, and also in Central America, there are indications of the presence of man before 8000 B.C., but many of these identifications have been made quite a number of years ago and suffer from a lack of sound dates or are isolated artifacts inadequate for the reconstruction of a cultural assemblage. Some of these finds from the Late Pleistocene Upper Becerra formation may well record human occupation, and continuing excavations in Mexico will eventually place the temporal position and industrial activities during the Paleo-Indian period on a firmer basis. The excavations conducted in the Valsequillo gravels at Puebla and at the Tlapacoya site in the Valley of Mexico are some of the more recent contributions. As usual, there is some uncertainty about the temporal correlation of gravel deposits between one area and another, about radiocarbon dates of 35,000 to 24,000 years ago of high antiquity but not directly associated with adequate artifacts, or about the precise age of a cultural complex on a buried living surface.

The South American data have recently been admirably reviewed by Lynch

(1976). He believes that when the early American hunter–gatherers had reached the northwestern part of the continent they could have moved south fairly rapidly along both the east and west sides of the Andes to reach the southern part of the continent. There is considerable evidence of occupancy over a wide area between 10,000 and 7000 B.C., including southern Patagonia and Tierra del Fuego. One of the best of the oldest South American dates is at the Los Toldos site in Santa Cruz, Argentina where human occupation associated with guanaco bones has a radiocarbon date of 10,650 ± 650 B.C. MacNeish (1976) and his associates from excavations in highland Peru believe they have evidence of a series of early hunting cultures back as early as 16,000 to 15,000 B.C. On present evidence the wide distribution of "man" in South America by at least 10,000 to 11,000 B.C. would seem to require significantly greater age in Mexico and Central America and in North America. The position in this chapter is that, following these early population movements, there were no significant group movements into the southern continents. Some population spread in small numbers may well have occurred from Mesoamerica to South America, and vice versa, for there is reasonably good archaeologic and other evidence to support it. The viewpoint adopted in this chapter does not recognize as convincing the several arguments for the intrusion of African, Phoenician, outer space, or Eastern Asian peoples or artifactual complexes that have been proposed. The varieties of Indian populations of 1500 A.D. developed in South America from the early migrations well before 8000 B.C.; after that it is a different story.

The South American data, with their wide geographic spread of early man around or before 9000 to 8000 B.C., imply the arrival of man on that continent substantially before the known dated complexes. Similarly, in Mexico and Central America the wide distribution of human occupations just before or after 8000 B.C. implies that the arrival of the first human groups was substantially before this date. If the proposed dates for man in the Valley of Puebla and the Valley of Mexico are supported in the coming years by sound evidence of an age between 40,000 and 24,000 years ago, there will be much work for archaeologists to do in the future to find substantiating evidence in the rest of the world. The North American geographic spread of dated evidence and the considerable diversity of assemblages shortly after 8000 B.C. imply an antiquity of man in North America considerably greater than the known age of the Fluted Point Hunters or of the occupants of Wilson Butte Cave. In summary, an age of about 20,000 years for man in the New World is viewed as not unreasonable but, also, as not yet satisfactorily proved.

The Temporal Framework for Northeast Asia

Sound dating of the Late Pleistocene occupations of Siberia is just beginning; in fact, adequate investigation of ancient man in northeastern Siberia has only recently been initiated. Most of the sites with an age of more than 4000 or 5000 years are along the southern borders of Siberia from Russia to the Japanese islands. A radiocarbon date of 12,900 ± 120 B.C. (GIN-97) has recently been obtained on a fossil bone from the lower cultural level of Mal'ta near Irkutsk, often attributed to the older Upper Paleolithic of the Irkutsk area. A date of 18,950 ± 300 B.C. was obtained on charcoal from the lower cultural level of Afontova Gora II in the Upper Yenesei Valley near Krasnoyarsk (Klein 1967). From Kamchatka there is a date of close to 18,000 B.C. on charcoal from the Ushki I site. At the Ninth Congress of the International Union of Prehistoric and Protohistoric Sciences in Nice, France in September, 1976, Dr. V. Larichev reported on two newly discovered sites in southwest Siberia. The oldest, of about 33,000 years ago, was a forerunner of the Mal'ta complex and was occupied during the interstadial before the last major glaciation of the area, whereas the second was slightly later than the now known oldest levels of the Mal'ta site.

The southern Siberian complexes, of about 30,000 to 12,000 B.C., have a strong early relationship to the Late Mousterian stone industries of eastern Europe. As the Upper Paleolithic developed, there was corresponding modification in Siberia, but the patterns of change in Siberia are sufficiently different from those of the better known areas to the west that an easy alignment has not been possible. The industrial development of these Siberian populations was the result of the long Eurasian cultural development, which became adapted to the Late Pleistocene. The level of occupation is attributed to the lower section of deposits on Terrace II of the Yenesei. The faunal composition represents cold periglacial conditions, and the date corresponds to about the maximum of the last major Siberian glaciation, called the Sartan. Also in the Middle Yenesei, at the Kokorevo sites, radiocarbon dates range from about 14,000 to 11,000 B.C. In the same area at the Mal'ta site, a date of about 7000 B.C. was obtained on a Mesolithiclike complex located on the higher areas of the Terrace I floodplain deposits.

In these Siberian sites there are crude heavy chopping tools; a variety of flake implements, including scrapers and knives; discoidal cores; and some bifacially flaked points or knives. There is a trend toward greater use of true blades made from prepared cores and toward the fashioning of end and side scrapers, piercers, perforators, gravers, and burins and an increase in bone tools and ornaments. The animals on which the people fed are those from arctic to subarctic and cold arid-steppe environments. They made skin clothing and had substantial houses, in the construction of which they used the bones of large mammals, such as woolly rhinocerus and mammoth. Probably the most important animal was the reindeer in the tundra area. It might be said that the spread of man into North America awaited the presence of arctic–alpine tundra species, during the Late Pleistocene, on which man could live as he hunted his way across northeastern Siberia into North America.

Although the extent of the mountain glaciers in eastern Siberia is not satisfactorily known, it is certain that only a small part of the land mass was glaciated and that most of the area was occupied by xerophytic arctic tundra or alpine tundra. A long tongue of steppe or periglacial steppe extended from southwest Siberia eastward between the Central Siberian Plateau and Lake Baikal as far as Yakutsk on the Middle Lena. From this area the best access route to the north was down the Lena Valley to the Arctic Ocean.

The fall of the sea level during the last major glacial advance of the Wisconsin–Würm is estimated to have produced a land bridge at the Bering Strait from about 24,000 to 8000 B.C., with two periods of submersion of the highest part of the shelf corresponding to major ice-melting phases of the retreat of the Wisconsin ice. The size of the exposed land was considerable. Most of it was not forested but was occupied by tundra vegetation similar to that of the Siberian arid-steppe tundra. During the last glacial dominance, between 23,000 and 10,000 B.C., the arctic trees and shrubs were more restricted in their distribution than they are today and the climate was colder than it is now (Hopkins 1967).

The Movement of Early Man into North America

Keeping in mind the lack of direct evidence for the presence of Early Man in northwestern North America and northeastern Siberia, we can still present an acceptable hypothesis for a spread of hunting bands from west to east. Their way of life was developed from southern Russia to south-central Siberia during the latter part of the Pleistocene; it was based on a Late Mousterian industry, modified by the initial elements of Upper Paleolithic emphasis on blade tools and the beginnings of a bone industry that was an aid in the production of skin clothing and shelter. This gradual expansion northward into new territories from northeastern Siberia to Alaska would have taken place without resistance from resident hunters. If it took place in the time period suggested, then a substantial number of hunting camps must now be under ocean water, but some will eventually be found in favorable areas, such as elevations overlooking passes followed by game animals in moving from one feeding ground to another.

This early population spread is believed to have been diverted south along the west side of the McKenzie Valley. Several recent papers have emphasized the difficulty of passing from the Lower McKenzie Valley into the eastern Rocky Mountain slopes of the United States because of the presence of the coalesced continental and Cordilleran ice from Montana to the Yukon Territory along the eastern margin of the Canadian Rockies. The evidence for the closing and opening of the corridor between these sheets is not so firmly established that sound datings for these events are available. The position adopted in this chapter is that the corridor would have been closed only at the maximum of the Wisconsin glaciation for a few thousand years, about 19,000 to 15,000 years ago.

If the early hunters came into the United States before 17,000 B.C., then archaeologists in the United States either have been unlucky or have not been able to correctly evaluate the evidence for their occupancy before the 13,000 B.C. date mentioned at Wilson Butte Cave. Entry shortly before 17,000 B.C. would allow ample time for penetration into extreme southern South America for the known occupancy there but would not accommodate the proposed Valsequillo and Tlapacoya occupations in Central Mexico. If the corridor was closed between 21,000 and 10,000 B.C., archaeologists are faced with at least as impressive dilemmas in the form of an absence of sound data representing man in North

America before 21,000 B.C. and the long period from then to 13,000 B.C., or in accepting the speed with which man moved from Alberta to Tierra del Fuego.

The environmental changes in North America as a result of the retreat of the Wisconsin ice would have had an effect upon the way of life of ancient man through the shift of climatic zones, vegetation, and animal life. The expansions of the Canadian continental ice sheets effectively obliterated the vegetation and animal life from much of Canada. The expansion of the ice into the northern sections of the United States markedly altered the biota and compressed and interdigitated elements of previous periods into assemblages distinctive to Late Wisconsin times. The climatic conditions during the life of the western mountain glaciers lowered the tree line, changed the faunal associations and distributions, and produced thousands of lakes in the now dry basins of the western plains, in the southwest, in the intermontane plateau region, and in the Pacific coast states and Mexico. The lowered forest zones and more extensive and effective grass-lands supported the large grazing and browsing animals of the Late Pleistocene fauna. There were more streams, with corridors of pine and spruce crossing the grassland.

The changes in climatic regime accompanying the withdrawal of the Wisconsin ice had already produced notable shifts from the full-glacial environments by the time of the early Fluted Point Hunters of 10,000 to 9000 B.C. The shift in vegetation and accompanying animal life took place on a large scale over North America, causing some shifts in hunting and collecting areas, and assisted in the displacement or disappearance of a small number of game animals. The north-ward movement of musk ox and mammoth is thought to have been in a park–tundra vegetation zone that initially occupied the soils left free of glacial ice in the Great Lakes area. It has been suggested that there was an early post-Valders invasion of the east by animal forms now associated with western prairie environments, and this correlates with recent similar hypotheses of prairie vegetation movement eastward at an early period. The park–tundra and cool prairie would be suitable for barren-ground caribou, which have been identified in Michigan and New York. Did they penetrate this far south before the last Wisconsin advance, or did they arrive with the "reopening" of the corridor?

The Fluted Point Hunters of North America

By 10,000 to 8000 B.C. the people of the Late Paleo-Indian period had occu-pied sparsely most of the area south of the present Canadian boreal forest to South America and from the Pacific to the Atlantic. Most of these populations were strongly dependent on hunting, as we know from the spear and dart points, knives, and scrapers of various kinds to work skins and from the fact that these tools have been found in association with a small number of large game animals. Data from such sites as Lindenmeier in Colorado, Graham Cave in central Missouri, and others prove that the meat diet was quite varied, and at least in the east there is very little evidence of early man killing the mammoth and mastodon. In addition, the early Fluted Point Hunters would have recognized a large variety of the plant foods available from nuts to berries and tubers. The diet of early man was not likely to stay restricted to a few classes of foods when he entered environments with a wide variety of them. It should be possible to discover sites that reflect varieties of food gathering and processing activities that were part of the life of the Fluted Point Hunters. They should have had some seasonal activity patterns. A number of students of the Paleo-Indian cultures are beginning to recognize regional tool and behavior complexes that will aid our understanding of this earliest known complex.

The wide distribution of Fluted Point bands and the relative homogeneity of the implements recovered implies a rather rapid spread of these early hunting people, and apparently into areas not hitherto occupied. There is also the implication that there would have been continuing contact between neighboring bands, perhaps for group hunting or other food-procuring activities at favorable seasonal locations or at locations favorable for shelter during the winter seasons. Such collective activity would have permitted exchanges of new cultural developments--in terms of sources of food, raw materials, manufacturing techniques, hunting technology—and of people. It is doubtful that individual bands would have been isolated from other groups for extensive periods or that peoples moving into new regional environments would have been cut off by those environments from culture sharing with peoples in their former territory.

Archaic Period Adaptations in North America

In the long Archaic period in the United States between 8000 B.C. and the effective introduction of agriculture around 1 B.C., many regional cultural developments occurred as the Indian groups became more familiar with local resources and developed the knowledge for successfully exploiting them. As they did so, successful adaptations to particular environments tended to restrict band and group activity to these environments and to produce a higher level of exchange of culture and people within these areas than between them. This is reflected archaeologically by the growth of distinguishable regional cultural traditions.

One of the best documented cultural continuities from the Paleo-Indian period Fluted Point Hunters to later complexes is in the western plains states. The production of fluted points was gradually abandoned and nonfluted points and knives of essentially the same basic form continued in use along with the rest of the stone tools. New tools appear, such as the specialized Cody knife, and new techniques, such as the parallel flaking of the Scottsbluff and Eden forms. All the evidence from sites in this region, from the Rio Grande north into the Canadian prairie provinces, continues to reflect the existence of a hunting economy, with bison as an important supplier of food, tools, and clothes. At some locations there is clear evidence of mass killings, evidently the result of communal drives. Indications of variability of animal food come from sites where giant beaver, pronghorn antelope, elk, deer, raccoon, coyote, and smaller mammals as well as bison were part of the food supply. At other sites there are indications of grinding and milling stones, burins, sandstone abraders, and whetstones.

West of the Rocky Mountains from around 8000 B.C., and continuing for many millenia, archaeologists recognize the Desert culture, which has a number of named regional variants from Mexico into Oregon and Washington. These variants emphasize the gathering and preparation of small seeds by hand and milling stones; the hunting of a wide variety of animals; and the extensive utilization of wood, bone, hide, and vegetational sources for tools, ornaments, and containers. It was a gradually developing adjustment to the essentially desert environment, which supported only small bands and in which population density remained low up to the historic period. It was not, however, a static complex, for many significant changes took place in the technology, some of them representing almost continent-wide shifts in tool forms, and new weapons, and shifts in the techniques of manufacturing baskets. Important variants are recognized in areas along streams and lakes, in upland forested or alpine areas, and where minor shifting climatic patterns allowed the expansion of desert bands into sometimes better watered areas or the penetration of foreign groups into the Desert culture region.

Between 8000 and 6000 B.C. along the northwest coast and into the interior along the major rivers, at least some part of the year was spent in obtaining food from the spawning runs and in otherwise exploiting the food supply associated with the streams and coastal areas. The latter is a reasonable inference, for some parts of what was then the coast are now under water. It was to be a long time, however, before the striking northwest coast sea-adapted complex would develop between Washington and the Alaskan peninsula.

In the interior, from the Yukon territory to Idaho, there is an Old Cordilleran complex that may be viewed as an Asiatic-derived parent to the Fluted Point, as a collateral contemporaneous variant, or as the result of the northern and western expansion of the Llano to Plano tradition. At present there would seem to be a basic relationship, and current radiocarbon dates indicate the time period of Old Cordilleran as not over 8000 B.C.

In southern California the San Dieguito hunting culture, with percussion-flaked lanceolate points, knives, scrapers, and choppers, is known from about 8000 to 6000 B.C. Shortly afterward there is a development of a number of areal specializations in coastal, desert, and forest environments, which during the last few thousand years resulted in an unusually dense population for a hunting–gathering culture in the oak-forest area of central California.

Southwestern Alaska was sparsely populated by around 8000 to 6000 B.C. by people with a unifacial core and blade industry whose movement into Alaska is likely to have been from the Pacific side of the Chukchi peninsula into coastal Alaska and south to the eastern Aleutian area. If these early coastal-adapted groups were the first Aleuts, as is implied, it would suggest that the Bering Strait area was occupied by Eskimoan-speaking peoples longer ago than has been thought. In interior and northern Alaska there is a considerable variety of assemblages reflecting inland and coastal developments with continuing ties with Siberia, and also influences from northward spreading groups primarily moving with the expansion of bison and other game animals.

Between 3000 and 2000 B.C. the Denbigh Flint complex, with a marked coastal adaptation, spread with surprising speed eastward to provide the first successful occupation of the eastern Arctic. The first Eskimo bands reached northern Greenland by 2000 B.C. Eskimo cultural traditions have a considerable time depth, and a main hearth area was the Bering Strait on both sides of the International Date Line, where people have been moving across in both directions for many millenia.

In the woodland area of the eastern United States, a gradual transition is recognized in several areas from the Fluted Point Hunters complex to assemblages maintaining the same basic manufacturing stone-working techniques and tools, but with the development or adoption of nonfluted projectile forms very similar to some of the early Plano forms of the plains. By 7000 to 6000 B.C. stemmed and notched projectile forms become common, and an increasing diversity of regional areas through time reflects the increasing specialization of groups learning to exploit the resources of these local regions. Many of the changes are related in form and function to those of areas to the west, but the east acquires a distinctive flavor of its own through the development of a series of woodworking tools and ground and polished stone forms. Whereas regional specializations are present in the eastern Archaic complexes, there is also evidence of increasing exchange of raw materials or manufactured objects as travel or trade routes become established.

Many archaeologists view most of the cultural complexes of Mexico from about 8000 to 4000 B.C. as a southern extension of the Desert culture. Certainly the general hunting–gathering pattern is similar, and there are some tool forms that are held in common with early Archaic and Desert culture groups from California to Texas. The most important features of the Mesoamerican scene is the early domestication of plants, from Tamaulipas to Chiapas, which have been demonstrated where systematic efforts have been expended to search for such evidence. This area shows a very gradual increase in the number of plants domesticated in the several regions and in the proportion of domesticates consumed. Marked population increase in some areas, such as the Valley of Mexico, the Valley of Oaxaca, the coastal lowland of southern Vera Cruz and Tabasco, and the Pacific Coast of adjacent Chiapas and Guatemala, is observed by 1000 B.C., when agricultural practices were well developed and the Early Formative cultures were becoming established.

Summary and Conclusions

All the supportable evidence available indicates that the first human occupants of North America came from northeastern Asia. Some archaeologists support the view that this first occurred from 30,000 to 40,000 or more years ago,

others believe it was from about 25,000 to 20,000 years ago, and some have con-
tended that it could not have been until about 12,000 years ago. The time of
arrival has not been settled.

Some archaeologists emphasize the Mousterian origins of the first emigrants,
believing that the earliest American cultural complexes indicate a spread into
North America before elements of Upper Paleolithic origin had reached eastern
Asia. Many archaeologists, however, believe that Upper Paleolithic develop-
ments were a part of the cultural mechanisms that allowed man to move into
North America and spread throughout the New World.

The main access route into interior North America was east of the Rocky
Mountains, and dispersion into most of the Americas was by this route. Population increase and any physical differentiation of human groups south of the
Arctic area is derived primarily from the population of the Paleo-Indian period.
There are not indications in the prehistoric record of any later substantial
migrating groups influencing the cultural life of the residents of North America
south of the Alaskan and Canadian Arctic region.

The archaeologic evidence, except in rare instances, supports the view that,
in spite of regional adaptations to food supplies and raw materials, there was a
continuous exchange of new developments between regions, with the additional
implication of population interaction as well.

References

Adovasio, J.M., J.D. Gunn, J. Donahue, and R. Stuckenrath. (1977). Meadowcroft
    Rockshelter: A 16,000 Year Chronicle. *Ann. N.Y. Acad. Sci.* 288:137–159.
Adovasio, J.M., J.D. Gunn, J. Donahue, and R. Stuckenrath. (1978). Meadowcroft
    Rockshelter, 1977: An Overview. *Am. Antiq.* 43:632–651.
Dixon, E.J. (1976). The Pleistocene Prehistory of Arctic North America. *In* J.B. Griffin,
    Assembler, *Habitats Humains Anterieur a L'Holocene en Amerique.* Colloque XVII. 9th
    Congr. Un. Intl. Sci. Prehist. Protohist. Nice. pp. 168–198.
Griffin, J.B. (1965). Late Quaternary Prehistory in the Eastern Woodlands. *In* H.E.
    Wright, Jr., and D.G. Frey, Eds., *The Quaternary of the United States.* Princeton, N.J.:
    Princeton Univ. Press, pp. 665–667.
Gruhn, R. (1965). Two Early Dates from the Lower Levels of Wilson Butte Cave, South
    Central Idaho. *Tebiwa* 8:57.
Haynes, C.V. (1964). Fluted Projectile Points: Their Age and Dispersal. *Science* 145:1408–
    1413.
Haynes, C.V. (1967). Muestras de C-14, de Tlapacoya, Estado de Mexico. Mexico, D. F.:
    Inst. Nacc. Antropol. Hist. Bol. 29:49–52.
Haynes, C.V. (1969). The Earliest Americans. *Science* 166:709–715.
Hopkins, D.M. (1967). *The Bering Land Bridge.* Stanford, Calif.: Stanford Univ. Press.
Klein, R.G. (1967). Radiocarbon Dates on Occupation Sites of Pleistocene Age in the
    USSR. *Arctic Anthropol.* 4:224–226.
Larichev, V.E. (1976). Discovery of Hand-axes in China and the Problem of Local
    Cultures of Lower Paleolithic of East Asia. In A.K. Ghosh, Assembler, *Le Paléolithique
    Inférieur et Moyen en Inde, en Asie Centrale, en Chíne et dans le sud-est Asiatique.* Colloque IX.
    9th Congr. Un. Int. Sci. Prehist. Protohist. Nice, pp. 154–178.

Larsen, H. (1968). Trail Creek, Final Report on the Excavation of Two Caves on Seward Peninsula, Alaska. *Acta Arctia* (Copenhagen), Fasc XV.

Lynch, T.F. (1976). The Entry and Postglacial Adaptation of Man in Andean South America. *In* J.B. Griffin, Assembler, *Habitats Humains Anteriéurs a L'Holocene en Amerique.* Colloque XVII. 9th Congr. Un. Intl. Sci. Prehist. Protohist. Nice. pp. 69–98.

Martin, P.S. (1963). *The Last 10,000 Years.* Tucson: Univ. Arizona Press.

Martin, P.S. (1967). Prehistoric Overkill. *In* P.S. Martin and H.E. Wright, Eds. *Pleistocene Extinctions: The Search for a Cause.* New Haren: Yale Univ. Press, pp. 75–120.

Martin, P.S. (1973). The Discovery of America. *Science* 179:969–974.

MacDonald, G.F. (1968). *Debert: A Paleo-Indian Site in Central Nova Scotia.* Anthropol. Pap. 16. Ottawa: National Museum of Canada.

MacNeish, R.S. (1976). Early Man in the New World, *Am. Sci.* 64:316–327.

Chapter 3

Anthropometry of Siberian Peoples

Valeri P. Alexseev

Introduction

Geography

Geographers and anthropologists use different definitions of the word "Siberia." In the geographic literature, Siberia refers only to the internal continental regions of northern Asia limited in the east by the territory of Yakutia. Siberia, in the anthropologic sense, includes this area plus the Soviet Far East, the Soviet part of the Asiatic coast of the Pacific Ocean, and the Amur Valley. So used, the term refers to the whole Asiatic part of the Soviet Union with the exceptions of the Caucasus, Middle Asia, and Kazahkstan. In this paper, "Siberia" will be used in the anthropologic sense. Such usage flows from a long anthropologic tradition based on the historical and cultural unity of the peoples of the area. Also, it is far more convenient to use a single term to denote the whole territory under consideration.

Data

Obviously, all somatologic data on living populations collected during the last three decades reflect the composition of the populations during that period of time. Time depth in this study is determined by the availability of crania with which to characterize the physical features of modern populations. Despite the fact that these features no doubt change over time, the amount of change in the last two or three centuries has probably not been great, and so cranial material from the eighteenth and nineteenth centuries can be included with modern material in the anthropologic analysis of populations. Cranial series from the second half of the eighteenth century and from the nineteenth century are treated as synchronic with contemporary data. Paleoanthropologic collections from Siberia and surrounding areas also exist but because of lack of space they cannot be presented systematically here.

Head and Facial Anthropometry

Historical Perspective

The first attempts to take anthropometric measurements of the aboriginal peoples of Siberia were begun in the latter half of the nineteenth century and continued into the first decades of the twentieth. Hair, eye, and, in some cases, skin pigmentation were described, but measurements of the soft parts of the face were not included in those studies. Each investigator generally concentrated on only one group of people, and results obtained by different workers using different scales of measurement were not comparable. Such limitations make these early reports of limited value. They have been reviewed by Levin (1958) and are not discussed in detail here.

In the decade following the revolution, strenuous efforts were made to create more satisfactory methods of describing facial variation and of measuring pigmentation. In addition to improvements in methodology, basic principles of somatologic studies, such as the choice of representative groups and the determination of adequate sample sizes, were established. One important step in the collection of anthropometric data was taken at the end of the 1920s. This was a study of Altay–Sayan peoples conducted by Yarkho (1929, 1947). He measured Soyots, as well as all tribal groups among the Altay and Khakass peoples.

In the 1930s, G.F. Debets and T.A. Trofimova conducted studies in western Siberia where they measured Samoyed peoples—Nenets and Selkups (Debets 1947)—as well as western Siberian Tatars (Trofimova 1947). S.A. Shluger conducted an extensive investigation of the Nenets but unfortunately his findings were not published at that time and have only recently appeared (see Alexseev 1971). Debets later continued his studies of the peoples of northeastern Siberia. During his work on Chukotka and Kamchatka, he measured nearly all ethnic groups in these regions except Aleuts (Debets 1951). On the basis of these data, Debets was able to classify the local anthropologic variants of the groups and to propose hypotheses about their genealogic relationships.

Paralleling the work of Debets on Chukotka and Kamchatka, Levin conducted studies in the Amur Valley. It is important to note that Debets also participated in the study of the Amur peoples, and this personal contact between Debets and Levin gave them an opportunity to compare methodologies and to eliminate some of the subjectivity in their techniques.

Levin (1958) measured Nivkhs, Orochs, Negidals, Ulchs, and Evens from the coast of the Sea of Okhotsk. He had previously studied the Yakuts (Levin 1947a, b). The results of these anthropometric investigations could be compared in a broad way with ethnologic and archaeologic data and used in ethnogenetic reconstructions. This approach was also used in studies of the Altay and Sayan Mountain regions where Levin (1954) gathered extensive data on the Tofalars, Reindeer Soyots, and the Soyots of the central regions.

The contributions of Yarkho, Debets, and Levin made it possible to describe the anthropometric characteristics of most major Siberian groups and to consider geographic variations of cranial and facial dimensions and of the soft parts of the face.

However, of the major Siberian groups, the Buryats had not been well studied. Debets and Levin had taken measurements on only two local groups of Buryats. Also, data were lacking for some small groups in western and central Siberia, as well as for several local subgroups among the peoples investigated earlier. This gap was filled by Zolotareva (1960, 1962, 1965, 1968, 1974, 1975a, b), whose work completed the general anthropologic survey of Siberia.

To complete this review of the literature, some papers on individual Siberian groups should be mentioned. These are articles by Rychkov (1961) on the Evenks of Podkamennaya Tunguska, Alexseev (1971) on Forest Nenets, Rosov (1961a, b) on Chulym Tatars and Selkups, Alexseev *et al.* (1968) on southern groups of Yakuts, and Aksyanova (1975) on some groups of Nenets.

I began collecting anthropometric data in 1969, but it has remained unpublished. In 1969 the main tribal groups of the Altay people were measured. In 1920 and 1971, the Eskimos and Chukchis were measured. During the years 1972 to 1976, Koryaks, Itelmens, the Evens of Kamchatka, and the Aleuts of the Commander Islands were studied. All data were collected and classified by population. Demographic and genetic data were gathered along with anthropometric measurements. The investigation therefore has generated a complex body of information, of which only the anthropometric material is discussed in this chapter.

Materials

The foregoing survey of the literature allows at least a preliminary review of all anthropometric data gathered so far. Virtually every Siberian group has been investigated by a caliper-wielding anthropologist. Large groups, such as the Buryats, have been examined down to the level of the local population. Unfortunately, the number and composition of the groups studied have not been uniform from study to study, and the aims of the investigators have also varied. These circumstances have made the general study of geographic variation of the anthropometric characteristics of Siberians quite different, and in many cases extrapolation from subgroups is required. Nevertheless, this chapter on geographic variation is based on data from about 100 population samples which, unfortunately, are not distributed quite evenly throughout Siberia. As previously mentioned, the different samples are not necessarily equivalent to one another. In some cases, the sample represents a territorial group, in others an administrative region, but only rarely does the sample represent a true population. However, these are the best data available.

Methods: Measurements and Descriptions

Russian anthropology has traditionally sought to standardize methods of data collection. Chepurkovskiy (1913) appears to be the first to have pointed out the incomparability of measurements made by different investigators and to clearly formulate the problems inherent in the subjective evaluation of descriptive features. Later, this same problem attracted the attention of Mahalanobis (1928), who confirmed the incomparability of different sets of data in many cases. This still remains one of the most vexing anthropometric problems.

The solution proposed to the problem of subjectivity was collaboration among different investigators in field work where they could carefully study one another's techniques. In the Soviet Union, this effort was based, with some modifications, on the techniques of Martin (1928). One of the most important changes was to measure facial and nasal height from the lower border of the eyebrows and the deepest point on the bridge of the nose rather than from the nasion as Martin had recommended. V.V. Bunak (1941) and other workers after him stressed that it is virtually impossible to fix the nasion accurately on living subjects and, therefore, any attempt to use the nasion to measure facial and nasal heights would result in wide variation. Facial and nasal heights, therefore, as well as indices based on them, for Siberian peoples may not be exactly comparable to data from other parts of the world.

Consistent evaluation of the soft parts of the face could not be achieved, even by personal contacts among investigators. Therefore, scales for the most typical variants of soft facial parts were developed. Plaster models of these variants were then made so that the features of living subjects could be compared to the models and scored accordingly (Yarkho 1932a, Bunak 1941). The use of models did not fully solve the problem, but it did significantly reduce observational differences among investigators. Similar scales were also developed for the measurement of hair, eye, and skin pigmentation. The 12-grade scale developed by Bunak was used for the evaluation of eye color.

Geographic Variation of Metric Characters

The complex of metric characters that characterizes Siberian Mongoloids includes: (1) large cranial and facial dimensions; (2) considerable nasal breadth, both absolute and relative; and (3) great mandibular breadth. Within this general complex, local combinations of metric characters clearly exist, but they are not necessarily confined to definite geographic areas.

Some decrease in facial dimensions is characteristic of Ugrian peoples and some groups of Nenets and Selkups in western Siberia as well as of the northern Altay groups of southern Siberia.

Definite decreases in facial height combined with considerable facial breadth and highly developed Mongoloid features are observed among certain peoples of western Siberia (Chulym Tatars), southern Siberia (Tofalars and Reindeer Soyots), and central Siberia (Evenks of Podkamennaya Tunguska). The latter two characteristics (facial breadth and Mongoloid features) make it impossible to explain the decreased facial height by Caucasian admixture. Many anthropologists have separated this complex as a special local variant among Siberian Mongoloids. Debets (1951) designated it as the "Katanga Type." Rychkov (1961) considered this type to be widely distributed in western and central Siberia.

However, the low-faced forms do not have a continuous distribution. They differ from one another in many other characters and may well have different origins. For example, the main features of this combination are expressed much more strongly in the Evenks of Podkamennaya Tunguska than in the Chulym Tatars.

Clearly, facial dimensions among Siberians vary regularly, reaching high values in areas of Tungus–Manchu and Turko–Mongolic peoples. Yakuts, Soyots, and Buryats, as well as many groups of the Altay and Khakass peoples, have faces larger than the Evenks and the Tungus–Manchu peoples of the Amur Valley.

In northeastern Siberia, increasing facial size corresponds to a definite but unexplained increase in mandibular breadth. Debets (1951), who was the first to note this, attributed it to the extensive use of raw fish in the diet and the stress this would place on the mandible during mastication, but no concrete evidence exists to support this hypothesis.

Early publications on Siberian Mongoloids proposed that dolichocephaly characterized the Tungus–Manchu peoples, whereas other peoples of northern Asia were brachycephalic. In fact, the first anthropologic classifications of Siberian Mongoloids were based on the cephalic index (see, for example, Debets 1934). However, further investigation revealed a different distribution of cephalic indices. Some Tungus–Manchu peoples, for example the Negidals, were found to be brachycephalic, whereas dolichocephaly was found in some tribal groups of the Northern Altay. Therefore, cranial dimensions and indices were shown not to be characters of high taxonomic value but could only be considered as local variants of limited distribution.

Maps of the geographic distributions of bizygomatic breadth, upper face height, and cephalic index in Siberia are shown in Figures 3.1, 3.2, and 3.3. Such maps are complicated by the fact that the distribution of Siberian peoples has changed considerably from its original pattern over the last three centuries. The seventeenth century distribution was reconstructed by Dolgikh (1960) on the basis of historical sources and ethnologic data, and his map is used as a basis for drawing maps of anthropologic variation. An additional complication arises because data exist only for modern peoples. Anthropometric variation in the seventeenth century could have been different, and the extrapolation of modern data to populations of the seventeenth century may produce some errors. However, these are the only data available, and the maps shown here represent modern anthropometric data superimposed on the reconstructed distribution of Siberian peoples in the seventeenth century.

Figures 3.1, 3.2, and 3.3 demonstrate the trends discussed above: comparatively compact areas of facial breadth and height and a discontinuous distribution of the cephalic index. Other metric traits have geographic distributions similar to the cephalic index.

Geographic Variation of Descriptive Characters

Because of the widespread use of different measurement scales, only partial uniformity in the evaluation of descriptive characters has been achieved, and it

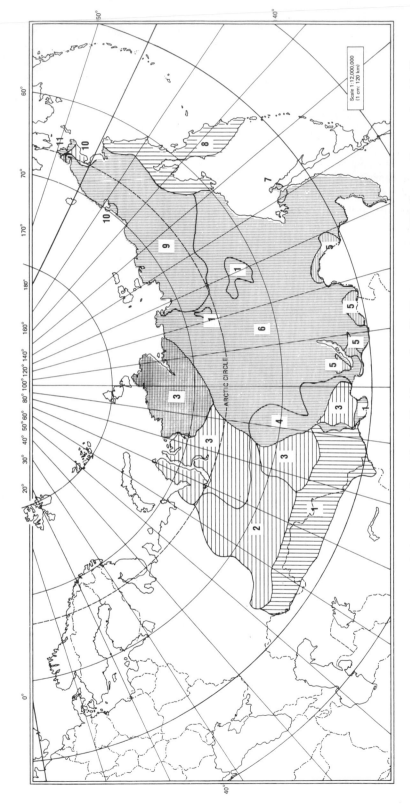

Figure 3.1. Geographic distribution of bizygomatic breadth. (1) Turkic peoples, (2) Ugrian peoples, (3) Samoyedian peoples, (4) Kets, (5) Buryats, (6) Tungus peoples, (7) Nivkhs, (8) Kamchatka peoples, (9) Yukagirs, (10) Chukchis, (11) Eskimos. Broad vertical hatching, 143.0–145.0 mm; broad horizontal hatching, 145.1–147.1 mm; narrow vertical hatching, 147.2–149.2 mm; narrow horizontal hatching, 149.3–151.3 mm; cross-hatching, 151.4–153.4 mm.

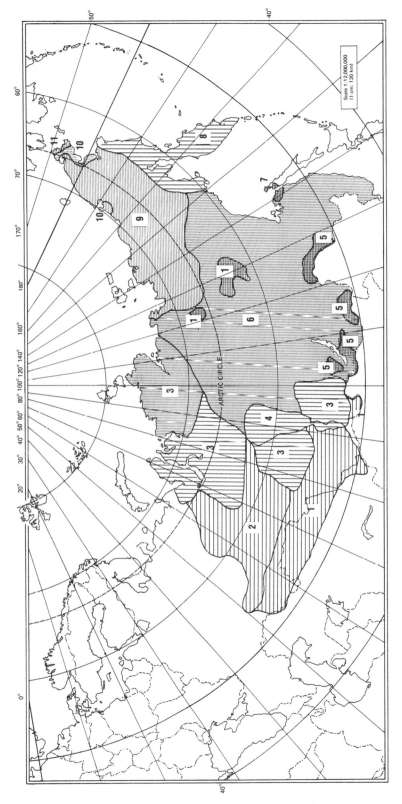

Figure 3.2. Geographic distribution of upper facial height. Numbers refer to the same groups as in Figure 3.1. Broad vertical hatching, 126.7–129.0 mm; broad horizontal hatching, 129.1–131.4 mm; narrow vertical hatching, 131.5–133.8 mm; narrow horizontal hatching, 133.9–136.2 mm; cross-hatching, 136.3–138.6 mm.

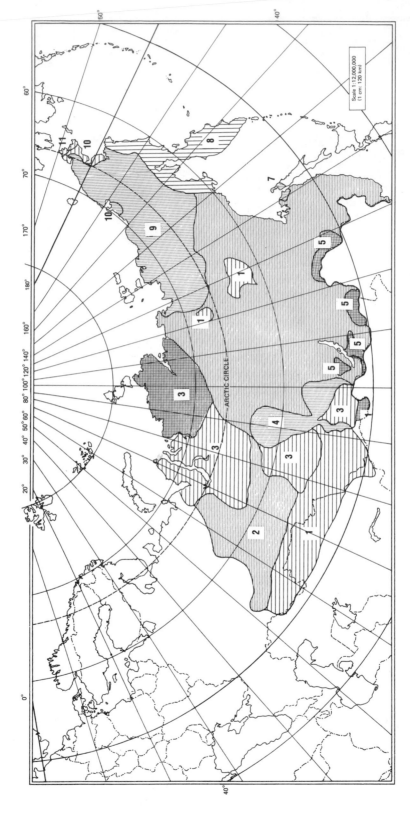

Figure 3.3. Geographic distribution of cephalic index. Numbers refer to the same groups as in Figure 3.1. Broad vertical hatching, 78.9–80.2; broad horizontal hatching, 80.3–81.6; narrow vertical hatching, 81.7–83.0; narrow horizontal hatching, 83.1–84.4; cross-hatching 84.5–85.8.

is impossible to neglect this fact in tracing their distribution. Therefore, some important but local studies which embrace groups inhabiting only one small territory will not be considered. Preference is given to data on peoples of different origins and languages collected by one investigator. The data themselves are grouped and discussed by investigator. Then general tendencies in variation between ethnic and territorial groups in Siberia are compared with the results obtained by each investigator whose material has been included in the analysis.

Data of G.F. Debets

Intensive investigations carried out by Debets are reported in two publications—one devoted to the peoples of western Siberia (Debets 1947) and one dealing chiefly with the peoples of Chukotka and Kamchatka (Debets 1951) but also including some data on Buryats and Evenks of Podkammenaya Tunguska. Variation in descriptive characters is shown in Table 3.1.

Consideration of these data, gathered by Debets during 8 years of continuous field work between 1939 and 1947 when he had no opportunity to alter his methods of evaluating descriptive characters, demonstrates that variation in these characters can differentiate Siberian peoples. All characters except the position of the nasal tip show significant variation. Evens and Evenks are similar in many respects and tend to show extreme development of many specifically Mongoloid features, including minimal growth of beard, maximal development of the upper eyelid fold, nearly maximal development of the epicanthus, the lowest nasal bridge, and the flattest face. Hair, however, is relatively light, which seems odd when combined with the intense development of other Mongoloid features. Other investigators have established that this combination is a peculiarity not only of Evens and Evenks but also of other Siberian representatives of the Tungus–Manchu linguistic family. In other descriptive features the Evens and Evenks do not occupy an extreme position among Siberian peoples.

All peoples of western Siberia are characterized by decreased development of Mongoloid features. Beard growth is heavier, hair is lighter even in comparison with the Evens and Evenks, and development of the epicanthus is weaker, as is the development of the upper eyelid fold. These features are combined with a relatively high nasal bridge, a less flat face, thinner lips, and a less pronounced prochelon (prominence of the center of the upper lip). However one may regard this complex, the combination is similar in some respects to morphologic variants that predominate among populations of western and eastern Europe. The western Siberian populations and representatives of the Tungus–Manchu peoples are opposite poles of anthropologic variation among Siberian Mongoloids. Debets' material can also be used to identify other morphologic complexes specific for populations of central Siberia and the northern part of the Soviet Far East. The Buryats, for example, have relatively dark hair and very dark eyes. They also have strongly developed Mongoloid features, including a highly developed epicanthus and upper eyelid fold, flat faces and nasal bridges, maximally thick lips, nearly

Table 3.1. Variation in Descriptive Characters among Siberian Peoples[a]

Characters[b]						Peoples					
	Khanty	Nenets	Chulym Tatars	Selkups	Kets	Buryats	Evenks	Evens	Itelmens	Reindeer Chukchis	Eskimos
Number of subjects	128	52	144	71	79	250	75	131	89	96	191
Beard growth, after 25 years	1.92	1.61	1.23	2.10	1.42	1.51	1.04	1.04	1.56	1.15	1.62
Hair color (% of N27)	2.80	8.20	10.10	0.00	21.40	20.80	8.70	16.10	42.20	25.00	73.80
Eye color	1.28	1.42	1.58	1.25	1.66	1.74	1.90	1.56	1.76	1.93	1.95
Epicanthus (% of existence)	23.70	37.40	45.10	25.40	24.00	54.40	52.00	74.90	31.50	60.40	33.90
Upper eyelid fold, p	1.31	1.32	1.21	1.20	1.33	1.99	2.25	2.53	1.00	2.08	1.39
Upper eyelid fold, m	1.86	1.63	1.95	1.82	1.61	2.25	2.36	2.57	1.40	2.27	1.49
Upper eyelid fold, d	1.86	1.63	1.87	1.83	1.58	2.22	2.32	2.55	1.40	2.23	1.45
Nasal bridge height	1.65	1.44	1.56	1.67	1.81	1.42	1.22	1.14	1.76	1.39	1.60
Horizontal nose profile	2.09	2.00	1.99	1.97	2.10	1.71	1.72	1.59	2.10	1.69	2.08
Vertical nose profile (% concave forms)	30.50	21.20	25.70	41.40	17.70	6.40	18.90	18.80	22.50	26.00	10.60
Nasal tip	1.55	1.58	1.79	1.40	1.52	1.53	1.66	1.52	1.63	1.70	1.78
Facial flatness	1.20	1.31	1.17	1.47	1.41	1.05	1.01	1.01	1.20	1.05	1.07
Upper lip height	2.18	2.02	2.58	2.33	2.20	2.36	2.27	2.42	2.27	2.32	2.40
Prominence of upper lip	1.47	1.37	1.20	1.45	1.42	1.13	1.06	1.07	1.46	1.20	1.27
Thickness of upper lip	1.44	1.84	1.59	1.52	1.50	2.29	1.75	1.75	2.18	2.26	2.19
Thickness of lower lip	1.98	2.02	2.04	1.85	1.82	2.51	2.01	1.95	2.30	2.53	2.60

[a] From the data of G.F. Debets. Males. All measurements taken as described by Levin (1958).

[b] All values are central values, except as noted.

maximal upper lip heights, and a strongly prochelous upper lip. This combination of features plus the greatest facial dimensions in Siberia yield a morphologic complex as distinctive as that previously described for the Evens and Evenks.

A distinctive complex also exists among the Eskimos of northeastern Siberia. They have beard development similar to that found among western Siberians but are darkly pigmented and differ from other peoples in having a relatively weak epicanthus and upper eyelid fold, a convex nose with a low tip, and a maximally high and prochelous upper lip. Debets, who described this complex, emphasized its difference from classic Siberian Mongoloids but also considered it to be nearer to northern representatives of Asiatic Mongoloids than to North American Indians. Although these resemblances have never been quantified, there is no doubt that Eskimos tend to resemble North American Indians. Debets included Chukchis among the Eskimos, but his own data contradict such a classification. In several features—beard growth, hair color, the upper eyelid fold, and the nasal bridge—the Chukchis clearly fall near the Mongoloids of central Siberia. Taking into account that as Eskimos they have a prochelous upper lip, relatively great upper lip height, and considerable thickness of both lips, which characters are also fixed among Mongoloids of internal Siberia, the differences between Chukchis and Eskimos in overall morphologic variation are still more impressive.

Debets' data show the existence of four geographically delineated morphologic complexes among Siberian peoples. The first occurs among groups in western Siberia (Khanty, Nenets, Chulym Tatars, Selkups, and Kets) and is characterized by decreased development of Mongoloid features. The second complex occurs in southern Siberia and is typified by the Buryats, who have dark pigmentation and maximally great facial dimensions and, who approach the Siberian maximum in the development of Mongoloid descriptive characters in the eye region and in facial flatness. The third complex is found in central Siberia and the Amur Valley among the Tungus–Manchu peoples (Evens and Evenks) who are characterized by maximal facial flatness and the most extreme development of Mongoloid features of the eye region. Hair, however, is relatively light in color. The fourth complex occurs in northeastern Siberia. The Eskimos resemble North American Indians despite some morphologic differences. The Chukchi complex obviously resembles the southern and central Siberian, but it is impossible to determine whether the Chukchis are closer to the Buryats or to the Evenks. The Itelmens resemble Eskimos in beard growth, eye color, the percentage of individuals having an epicanthus, development of the upper eyelid fold, and the nasal bridge. The Itelmens show lesser development of typical Eskimo features than do Eskimos themselves.

If the morphologic similarities of Eskimos and Chukchis to North American Indians are regarded as a reflection of their ancient genetic relationships, it may be concluded that the population which is ancestral to the Indians inhabits the coastal regions of northern Asia. This leads to the further conclusion that ancient peoples moved from Asia to America along the Asiatic coast of the Pacific Ocean from south to north. Hence the populations from the interior of Siberia probably had little or nothing to do with the peopling of the New World.

Data of M.G. Levin

The data collected by Levin appear in Table 3.2. The material shows that the morphologic complex found among the Evens and the Evenks also occurs among the Negidals and, to a lesser extent, among the Oroks, Ulchs, and Nanays. Therefore, great morphologic similarities can be seen among all Tungus–Manchu peoples of Siberia who occupy the immense area from the Yenisei in the west to the mouth of the Amur River in the southeast. Local variation occurs at the boundaries of this area, such as an increase in beard growth among the Amur peoples and darker pigmentation among the Nanays, but they are of little taxonomic importance.

According to Levin's data, the Tofalars, as well as the Reindeer Soyots, have rather low faces and differ from surrounding groups in having even less beard growth and rather light hair but very dark eyes. Levin, with good reason, has noted the similarity between this group of Soyots and the Evenks of Podkem-manaya Tunguska, and he regards both groups as representatives of a single local morphologic complex. This conclusion is supported by the existence of moderate lip thickness in both groups and a lower nasal tip than is found among the Buryats.

It is easy to see that the Soyots fall in the same group as the Buryats. Beard growth among the Soyots is only slightly greater than the average for the Tungus–Manchu peoples. They have a high percentage of black hair, very dark eyes, and a nasal bridge that is slightly higher than among Tungus–Manchu groups. The upper lip is prochelous and both upper and lower lips are thick. Soyot facial dimensions are as large as among the Buryats. Yakuts belong to the same local complex, but their facial dimensions are greater than the Buryats'. Levin's data therefore confirm the existence of two local complexes among Mongoloids of central Siberia and the Soviet Far East who speak Turko–Mongolic and Tungu–Manchu languages. Both of these complexes undoubtedly constitute a single taxonomic variety contrasting with western Siberia, on the one hand, and northeastern Siberia, on the other, and occupying a vast uninterrupted area in southern and central Siberia as well as in the southern part of the Soviet Far East.

The anthropologic features of the Nivkhs present a special case. They show the most intensive beard development of all the Amur peoples studied by Levin. Their hair color resembles that of the Turko–Mongolic rather than the Tungus–Manchu peoples, even though it is lighter than hair of the Nanays. Levin pointed out the possibility of a systematic bias in the determination of hair color among the Nanays, but it is not clear why bias should appear in this case only. Very dark hair occurs among the surrounding Ainu, Japanese, and Korean populations. The percentages of No. 27 (on the scale of Fischer) were 81.8 percent, 76.8 percent, and 76.9 percent in these three groups, respectively. N.N. Cheboksarov found 79.3 percent No. 27 among the Chinese (see Levin 1958). The very dark hair observed among the Nivkhs as contrasted with their Tungus–Manchu neighbors can be explained as admixture with Ainu, Japanese, or Koreans. Beard growth among Japanese and Koreans is sparser than among the Nivkhs, whereas it is about equal to the Ainu. Long contact between the ancestors of the Nivkhs

Table 3.2. *Variation in Descriptive Characters among Siberian Peoples[a]*

Characters[b]	Peoples								
	Soyots	Tofalars	Buryats	Yakuts	Nivkhs	Oroks	Negidals	Ulchs	Nanays
Number of subjects	101	41	120	597	245	19	52	125	50
Beard growth, after 25 years	1.62	1.37	1.59	1.18	2.01	1.60	1.28	1.76	1.30
Hair color (% of N27)	52.40	16.30	55.30	—	42.30	36.40	2.30	18.50	63.90
Eye color	1.90	1.95	1.84	1.85	1.84	1.44	1.60	1.74	1.52
Epicanthus (% of existence)	44.60	52.50	50.40	65.40	53.50	31.60	52.00	64.80	44.00
Upper eyelid fold, p	2.27	2.15	2.31	2.18	2.36	2.47	2.48	2.36	1.98
Upper eyelid fold, m	2.59	2.42	2.60	2.44	2.45	2.79	2.44	2.39	2.14
Upper eyelid fold, d	2.65	2.45	2.63	2.42	2.45	2.79	2.40	2.39	2.12
Nasal bridge height	1.35	1.43	1.38	1.61	1.23	1.05	1.19	1.14	1.12
Horizontal nose profile	2.05	1.98	1.92	—	1.85	1.68	1.69	1.65	1.76
Vertical nose profile (% concave forms)	20.80	12.50	10.80	13.00	12.00	26.30	9.60	27.20	12.00
Nasal tip	1.70	1.95	1.81	1.71	1.47	1.74	1.58	1.62	1.46
Facial flatness	1.03	1.00	1.04	1.12	1.04	1.00	1.02	1.02	1.00
Prominence of upper lip	1.15	1.27	1.37	1.45	1.06	1.16	1.06	1.06	1.02
Thickness of upper lip	2.05	1.78	2.10	—	2.21	1.42	1.71	2.04	2.00
Thickness of lower lip	2.19	2.00	2.42	—	2.60	2.00	2.13	2.30	2.47

[a] From the data of M.G. Levin. Males.
[b] All values are central values, except as noted.

and of the Ainu is well documented historically and the morphologic peculiarities of the Nivkhs may be attributed to it.

Levin emphasized the uniqueness of the Nivkhs by giving them a separate taxonomic status equal in systematic rank to other Siberian morphologic complexes. In his opinion, the anthropologic peculiarities of the Nivkhs arose not as a product of direct admixture between Tungus–Manchu people of the Amur and the Ainu, but as a much earlier event between an ancient group from internal Siberia and ancient coastal Mongoloids. This appears to be a viable hypothesis, but it has not yet been confirmed by paleoanthropologic data.

Data of I.M. Zolotareva

Some groups studied by Zolotareva show unique anthropologic traits. First among these are the Yukagirs. Zolotareva considered them one of the local variants of the Tungus–Manchu morphologic complex, which is most intensively expressed among the Evens and Evenks and has been characterized by Debets and Levin. Zolotareva's conclusion supports Levin's (1958) hypothesis that the Yukagirs belong to the so-called Baikal race of the Tungus–Manchu complex, a point which had been argued before the Yukagirs were thoroughly studied. However, comparison of the Yukagirs to the Chukchis (Table 3.3) casts doubt on this conclusion. Significant differences between Yukagirs and Chukchis occur only in the degree of facial flatness, which is greater among the Yukagirs, and upper lip thickness, which is less among the Yukagirs. Values for other traits are nearly equal.

How can this similarity be explained? As mentioned above, Chukchis differ from Eskimos and show some resemblance to central Siberian Mongoloids. However, both the Yukagirs and the Chukchis have hair too dark to consider them representatives of the Baikal complex.

Zolotareva formulated another hypothesis on the basis of data from the Nganasans. She proposed that among modern Siberian populations we can find surviving examples of ancient morphologic complexes. This seems a fruitful idea, even though paleoanthropologic data demonstrating such continuity are lacking.

The problem of similarity between the Nganasans and the Yukagirs as postulated by Zolotareva cannot be solved definitively. The two groups resemble one another, but the Yukagirs stay nearer to the Chukchis in upper lip height and prochelon development than to the Nganasans. However, the Yukagirs resemble the Nganasans more than the Chukchis in facial flatness. On the basis of the hypothesis of the survival of ancestral complexes, the slight differences among the three groups may be interpreted as local variations within a protomorphic complex from which either all or some Siberian Mongoloids are derived.

Zolotareva's data on the morphologic differences among peoples of western and southern Siberia and between them as a whole and the peoples of central Siberia as a whole confirm previous observations. The Nenets show decreased development of all Mongoloid features in comparison with other Siberian groups.

Table 3.3 *Variation in Descriptive Characters among Siberian Peoples[a]*

Characters[b]	Peoples							
	Nenets	Ngarasans	Dolgans	Buryats	Yakuts	Evenks	Yukagirs	Chukchis
Number of subjects	70	122	106	498	63	40	44	36
Beard growth, after 25 years	1.67	1.11	1.15	1.51	1.39	1.07	1.05	1.10
Hair color (% of N27)	13.20	43.50	18.70	36.00	53.00	—	44.40	44.80
Eye color	1.54	1.84	1.61	1.69	1.65	1.50	1.75	1.78
Epicanthus (% of existence)	32.90	43.40	49.50	61.90	58.70	47.50	59.10	58.30
Upper eyelid fold, p	1.23	1.88	1.81	2.17	2.41	—	2.48	2.53
Upper eyelid fold, m	2.19	2.72	2.56	2.59	2.78	—	2.86	2.78
Upper eyelid fold, d	2.19	2.70	2.49	2.40	2.69	—	2.86	2.69
Nasal bridge height	1.61	1.16	1.29	1.41	1.30	1.37	1.09	1.25
Horizontal nose profile	2.10	2.11	2.57	2.16	2.33	1.78	1.98	2.03
Vertical nose profile (% concave forms)	57.70	30.60	17.90	24.10	17.20	60.00	34.10	33.40
Nasal tip	1.22	1.44	1.31	1.28	1.47	—	1.48	1.36
Facial flatness	1.61	1.29	1.38	1.25	1.23	1.08	1.11	1.64
Upper lip height	2.15	2.46	2.23	2.15	2.39	—	2.76	2.65
Prominence of upper lip	1.48	1.28	1.32	1.23	1.22	—	1.05	1.15
Thickness of upper lip	1.52	1.31	1.78	1.90	1.75	—	1.59	1.97
Thickness of lower lip	1.84	1.88	2.21	2.41	2.19	—	2.34	2.41

[a] From the data of I.M. Zolotarava. Males.
[b] All values are central values, except as noted.

The differences between Nenets and Nganasans are important because they reveal (1) that the western Siberian complex, having, as mentioned above, some common features despite relative morphological heterogeneity, is not geographically limited to western Siberia, and (2) that among Samoyed speakers, a morphologic complex exists that definitely resembles complexes found in the Siberian interior.

The Buryats of southern Siberia possess a combination of traits characterized by intense development of Mongoloid features but distinguished from the Tungus–Manchu peoples by very dark pigmentation. The Yakuts also represent the southern Siberian complex, but this can be explained by their rather late and historically well-documented northward migration to their present territory from the northern regions of the Baikal Valley, where their ancestors were in close contact with those of the Buryats and with whom they may well have formed a single population.

The importance of Zolotareva's data lies in their combination to racial taxonomy and to the understanding of the genealogic relationships among Siberian trait complexes. Before publication of her material, anthropologists assumed that Mongoloids of the Siberian interior constituted two morphologic variants represented by the Tungus–Manchu peoples, on the one hand, and the Turko–Mongolic groups, on the other. Both types were considered to be local variants of a single higher taxonomic level. However, three additional complexes have been identified—the Western Siberian, the Eskimo, and the Nivkh. Zolotareva's investigations among small relic populations in Siberia, including the Yukagirs and the Nganasans (the Dolgans belong to the same group) have shown similarities between them, as well as between them and the Chukchis. The discontinuous and geographically isolated positions of these relic populations may indicate that they are survivors of a single ancestral population. This raises the problem of the taxonomic status of this sixth or ancestral complex and its genetic and systematic relations to other complexes. This problem can only be solved by paleoanthropologic studies in Siberia. The relationship of this ancestral complex to the morphology of the North American Indian is also of great interest.

Data of V.P. Alexseev

The Forest Nenets (Alexseev 1971) are the most Mongoloid of all local groups of Nenets investigated, but they still differ from other Siberians in the development of some Mongoloid features, including nasal bridge height, beard growth, horizontal facial flatness, and the existence of the epicanthus. Nenet values for all these traits resemble European (Table 3.4). This illustrates the peculiar Europoid character of the western Siberian morphologic complex.

Altay groups studied in southern Siberia have very dark pigmentation and markedly developed Mongoloid features of the same grade as Turkic-speaking southern Siberians. A small group of Yakuts from southern Siberia (not shown in Table 3.4.) resembles the Turkic groups and, probably as a result of its isolation

Table 3.4   *Variation in Descriptive Characters among Siberian Peoples*[a]

Characters[b]	Peoples					
	Nenets	Altays	Telengets	Chukchis	Eskimos	Aleuts
Number of subjects	28	102	31	114	67	14
Beard growth, after 25 years	1.52	1.40	1.52	1.13	1.15	1.50
Hair color (% of N27)	4.00	38.10	35.50	86.10	82.00	28.40
Eye color	1.20	1.59	1.63	1.95	1.94	1.93
Epicanthus (% of existence)	42.80	58.40	48.30	73.90	66.70	42.60
Upper eyelid fold, p	2.21	2.36	2.50	2.04	1.84	0.71
Upper eyelid fold, m	2.50	2.57	2.57	2.17	2.14	1.00
Upper eyelid fold, d	2.50	2.56	2.57	2.06	2.14	1.00
Nasal bridge height	1.50	1.49	1.35	1.65	1.72	2.29
Horizontal nose profile	1.57	1.54	2.32	1.58	1.68	1.86
Vertical nose profile (% concave forms)	53.50	30.90	36.70	46.00	29.90	57.40
Nasal tip	1.21	1.68	1.32	1.47	1.76	1.57
Facial flatness	1.46	1.51	1.42	1.33	1.21	1.71
Upper lip height	1.93	2.08	2.06	2.16	2.12	2.43
Prominence of upper lip	1.30	1.35	1.71	1.54	1.48	1.71
Thickness of upper lip	1.43	2.18	2.16	2.31	2.36	2.29
Thickness of lower lip	1.82	2.26	2.42	3.11	3.03	2.79

[a] From the data of V.P. Alexseev. Males.
[b] All values are central values, except as noted.

from other Yakuts, shows depigmentation. Generally, however, Yakuts and Altays constitute a single morphologic complex (especially when the large facial dimensions of both are considered) which is also found, with some modification, among the Buryats.

Evens from central Kamchatka (not shown in Table 3.4.) have the same complex of features as other Tungus–Manchu peoples- -intensively developed Mongoloid features with relatively light pigmentation.

The Eskimo complex stands out and appears very similar to the Chukchi. This results from the fact that most data come from Coastal Chukchis who have had long contact with Eskimos. Debets (1951) pointed out that Coastal Chukchis resemble Eskimos more than do Reindeer Chukchis. In general, these materials fully confirm the existence of four main morphologic complexes among Siberian peoples: (1) Western Siberian, (2) Southern Siberian, including the Yakuts, (3) Tungus–Manchu (or Baikal), and (4) Eskimo (or Eskimo–Chukchi).

The anthropologic characteristics of the Aleuts pose a special problem. A hypothesis based on the temporal dynamics of cranial traits places them, with some reservations, among the Arctic peoples along with Eskimos and Chukchis. However, Aleuts differ markedly from Eskimos in somatologic traits, even when the influence of Russian, American, and European admixture is excluded. Aleuts show some resemblance to certain Japanese groups, but the problem demands further investigation.

Some Systematic Conclusions

The distinctive features of the various morphologic complexes can be seen to a greater or lesser degree in the data of each investigator reviewed here. Disregarding data on local variation, it is possible to define the morphologic characteristics and geographic distribution of six complexes from the data presented in this chapter:

1. The first complex is characterized by decreased development of Mongoloid features and is expressed by the Ugrian and Samoyed peoples excepting the Nganasans and some Nenets.

2. The second complex, typically Mongoloid and characterized by dark hair and eyes, is found among the Turko–Mongolic peoples of southern Siberia and the Yakuts.

3. The third complex shows maximum development of Mongoloid features together with relatively light pigmentation and is widespread among the Tungus–Manchu peoples of central Siberia, Kamchatka, and the lower part of the Amur Valley.

4. The fourth complex differs from all others in some respects, thus showing its southern origin. Skin pigmentation is relatively dark, and there is decreased development of some Mongoloid features along with strong expression of others. Typical representatives are the Eskimos and Coastal Chukchis and to some extent the Itelmens and Koryaks. Further investigation may place the Aleuts in this complex as well.

5. The fifth complex is represented by the Nivkhs. They differ from the Tungus–Manchu people in having darker pigmentation and heavier beard growth.

6. The sixth complex is characterized by pigmentation intermediate between that found in the second and third complexes along with strong development of other Mongoloid features. It occurs among the Nganasans, Dolgans, Yukagirs, Western Chukchis, and in part among the Reindeer Chukchis of the Chukotka Peninsula. These groups may be survivals of an ancestral morphologic complex which gave rise to the modern complexes found today in internal Siberia.

Nomenclature and Some Genealogic Considerations

In the Soviet literature during the last few decades, a geographic nomenclature for racial complexes has been adopted. The racial distinctiveness of the enumerated complexes stands out because they occupy definite areas. Geographic nomenclature is convenient; it avoids the use of a mixture of ethnic and racial names, but it also appears to correctly describe a biologic reality. Racial complexes in Siberia, as in other parts of the world, do not necessarily coincide with ethnic entities. Each is found not just among one people but among groups of related peoples. Therefore, ethnic names cannot be used for racial complexes. Using the geographic approach, the first complex is called Uralian (Cheboksarov 1951), the second Central Asiatic (Yarkho 1932b), the third Baikal (Roginsky 1934), the fourth Arctic (Debets 1951), and the fifth Amuro–Sakhalinian (Levin 1958). The sixth and final complex may be called Protomorphic Siberian or Paleosiberian (see, for example, Debets 1929a).

The problem of the origin of the Paleosiberian complex has been mentioned above, but the salient paleoanthropologic data are not available at the present time. A great deal of paleoanthropologic data comes from the steppe regions of the Altay–Sayan Mountain area, but these data are irrelevant to the formation of the Mongoloid race because during the late Bronze and early Iron Ages the area was occupied by a Europoid race. There is a small collection of specimens from the forested part of the Altay–Sayan Mountain area but it is of mixed origin, clearly containing Europoid and Mongoloid combinations. The paleoanthropologic material relevant to the origins of the Mongoloid race therefore is limited to (1) the numerous data from the Baikal region (Debets 1929a, 1948, 1951, Levin 1956, 1958, Mamonova 1973), (2) some finds from Yakutia (Yakimov 1950, Debets 1955), and (3) a small collection from the area between the Altay and Sayan Mountains and Lake Baikal (Gerasimova 1964).

Direct comparison of modern and ancient anthropometric data is difficult. Some possible comparisons are discussed later in the context of modern craniologic material. The groups retaining the ancestral morphologic complex (Nganasans, Dolgans, and Yukagirs) either are not represented in the craniologic series or are represented by single skulls. It is therefore hard to decide whether they really belong to the complex which, because it is regarded as protomorphic, must have been widely distributed in ancient populations. As was mentioned previously, there is no paleoanthropologic information on the formation of the Amuro–Sakhalinian complex.

The two morphologic complexes found among Siberians of the interior are undoubtedly more Mongoloid in appearance than any paleoanthropologic series assembled up to the present time. This supports Roginsky's (1937) hypothesis that the Mongoloid complex has undergone specialization from a less distinctive form since ancient times. However, this creates difficulties in the interpretation of similarities between ancient and modern populations. If the Baikal Neolithic series shows Europoid admixture, as many authors believe, then the Mongoloid component of the series may be considered as a basis for the formation of both the Baikal and Central Asiatic complexes. If, however, the relative similarity to Europoid groups reflects an ancient less specialized Mongoloid template, then the Baikal and Central Asiatic complexes arose relatively late in time, and they cannot be considered as ancestral morphological combinations. If this is the case, the distinction between the complexes is of low taxonomic value, and both should be lumped together in a single category the taxonomic rank of which is the same as the Uralian, Arctic, and other complexes. This seems the most logical solution to the problem on the basis of available information.

A large series of skulls, only recently described (Levin 1964, Debets 1975), comes from the northeastern part of the Chukotka Peninsula and from the Uelen and Ekven cemeteries. Archaeologic evidence connects them with the ancestors of Eskimos in the region and dates them to the first century A.D. On the basis of these archaeologic observations and the morphology of the skulls themselves, it is obvious that the Arctic complex clearly existed at the beginning of our era and, consequently, had begun to form at a much earlier time. Brachycephaly, which now characterizes Asiatic Eskimos, is not found in the ancient populations of the Asiatic coast, and, as Levin (1947c) states, must have developed at a later time.

The origin of the Uralian complex is in dispute. From comparisons of cranio-logic data representing ancient and modern populations, several authors conclude that the Uralian complex represents the survival of the undifferentiated ancestral Mongoloid template rather than Europoid admixture. However, the majority of workers hypothesize that the Uralian complex arose from long-standing inter-breeding between Europoids and Mongoloids (see Alexseev 1961, Debets 1961). This means that modern Uralian Mongoloids possess a significant number of Europoid genes, a contention fully supported by ethnologic evidence. The ad-mixture hypothesis rests on the geographic distribution of racial features in eastern Europe and western Siberia. The development of Mongoloid features intensifies along a west to east gradient, and the association of morphologic characteristics with particular areas is documented historically. The hypothesis of an undifferentiated origin of the Uralian race cannot account for either of these facts.

Body Anthropometry

Materials

Studies of body measurements and somatic features are much less numerous than studies of measurements and descriptive features of the head. Within the last few decades, data have been presented by Rudenko (1914) and S.N. Shluger (see Alexseeva et al. 1972) on Khanty and Nenets and by Sergeev (1932) and Petrov (1933) on Buryats from several different areas. All these publications include a rather limited number of different measurements and therefore do not adequately describe the physical characteristics of the investigated groups.

More useful somatologic data have been collected among the Forest Nenets Buryats, Chukchis, and Eskimos who have been studied according to an extensive morphophysiologic and genetic protocol during a project led by T.I. Alexseeva and V.P. Alexseev (Alexseeva et al. 1971, 1972, Smirnova 1973, Klevtsova and Smirnova 1974, Klevtsova 1976). Smirnova and Klevtsova took all body measure-ments using exactly the same protocol. Although all the data have not yet been published, they yield a general picture of territorial variation in many characters.

Geographic Variations and Their Interpretations

Siberian Mongoloids in general are short in stature and have relatively short legs and a massive body build. However, local variations do occur within this overall complex (Table 3.5.). The Nenets are the shortest and most gracile of the groups studied, but they have relatively long legs in comparison with the Buryats.

Table 3.5. *Variation in Body Measurement among Siberian Peoples*[a]

| | Peoples | | | |
Dimensions	Nenets	Buryats	Chukchis	Eskimos
Number of subjects	46	136	88	57
Stature (cm)	159.9	165.4	164.5	162.8
Weight (kg)	57.7	63.8	63.2	64.2
Chest circumference (cm)	88.7	87.7	92.2	91.0
Biacromial diameter (cm)	37.7	37.4	38.3	38.7
Biiliac diameter (cm)	27.3	27.5	28.7	28.4
Trunk length (cm)	49.4	—	50.4	49.2
Leg length (cm)	84.4	85.3	88.0	87.4
Arm length (cm)	70.8	72.0	71.7	70.2
Arm circumference (cm)	26.3	28.6	28.3	29.3
Forearm circumference (cm)	26.1	—	27.4	27.0
Thigh circumference (cm)	48.4	52.5	52.2	52.7
Lower leg circumference (cm)	31.4	34.0	33.9	33.8
Waist circumference (cm)	78.4	79.4	80.4	79.5
Athletic body constitutions (%)	32.1	15.0	37.6	56.1

[a] Males.

They have little fat tissue and well-developed musculature. A substantial percentage of Nenets have an athletic body constitution. The Buryats are heavier than the Nenets and have more fat, which results in increased limb circumferences. Only half as many Buryats as Nenets have the athletic body constitution. Chukchis and Eskimos form a separate group. They have relatively long legs (as do the Nenets), strongly developed musculature, and very little fat. Table 3.5 shows that 56.1 percent of Eskimo males but only 37.6 percent of Chukchi males have athletic body constitutions.

Marked variation in many physiologic traits has been used to establish hypotheses concerning adaptation and population differentiation. Consideration of these hypotheses lies beyond the scope of this chapter, but a few remarks are relevant.

The concept of adaptive types forms the basis of these hypotheses (Alexseeva 1972, 1975a, b). As distinct from racial units, to which a single origin is usually ascribed, adaptive types are combinations of physical traits that arise in response to certain environmental factors and may develop in many different populations exposed to similar circumstances. Alexseeva distinguished two adaptive types in Siberia—the Arctic type, found among peoples of the Asian Arctic, and the Continental type, found among the Buryats. The two types differ from one another in numerous physiologic features (hemoglobin, total protein, and blood cholesterol levels) as well as in somatologic features, such as the relative proportions of muscle and fat. The Arctic type has more muscle and the Continental type more fat. Resource utilization, as well as environmental and nutritional selection, may have played a role in creating adaptive complexes. Whale and walrus hunting among Arctic peoples may be physically much more taxing work than nomadic cattle breeding.

Craniology

Historical Perspective

There is no need to enumerate papers published at the turn of the century that deal with cranial data from Siberia; these works have been discussed by Levin (1958). Such studies are numerous, but their methodology is largely outmoded and they shed little light on the origins and ethnohistory of Siberian peoples. They have played only a limited role in the history of anthropology, therefore, and in the study of the anthropologic composition of Siberian populations.

This situation changed dramatically in the 1930s, when Levin's papers began to appear. They were devoted to the description and comparison of craniologic series from various Siberian peoples. Levin described the cranial features of Coastal Orochs (Levin 1936), the Ulchs (Levin 1937), the Khanty and the Mansi (Levin 1941), and the Chukchis and Eskimos (Levin 1949). Levin not only used much larger samples sizes than earlier workers, who had usually described single skulls, but he also provided complete metric and anatomic descriptions of his specimens. Furthermore, he endeavored to discriminate, by the most up to date somatologic principles, between primary and secondary features, and he laid great stress on the importance of facial dimensions. Finally, Levin tried to coordinate cranial data with ethnologic reports and to interpret his findings not by formal morphology alone but also in an historicogenealogic framework.

During the same period, works were published by Debets (1929a) on the Soyots, Trofimova (1932) on the Oroks, Tokareva (1937) on the Aleuts, and Yuzefovich (1937) on the Yakuts. These studies all incorporated the methodologic innovations introduced by Levin and along with his works from the basis of all subsequent investigations of cranial variation in Siberia.

Hrdlička (1924) published an important description of Buryat skulls in United States collections and also works on the craniology of the Eskimos and Aleuts (Hrdlička 1942a, 1942b, 1944, 1945).

Cheboksarov (1947) dealt with the systematics of Asiatic Mongoloids in general and based his classifications primarily on cranial data. Cheboksarov presented the first cranial data on the Nanays and illustrated some craniologic peculiarities of Southern, Eastern, and Northern Mongoloids. He also used cranial data to distinguish among Arctic, Central Asiatic, and Baikal groups of Northern Mongoloids. The review article by Hrdlička (1942b) was mostly devoted to the demonstration of analogies between modern populations and the morphologic complexes of the Neolithic populations of the Baikal region. The article also contained information on the cranial collections from modern populations which had been measured in the museums of the USSR, including Evenks, Chukchis, Eskimos, Nenets, Buryats, Yakuts, Khanty, Mansi, Ulchs, Orochs, Nivkhs, and Yukagirs.

Hrdlička's publication contained only indices and gave detailed descriptions only of Neolithic and Late Neolithic skulls.

The importance of these works was surpassed by the publication of Debets'

(1951) anthropologic investigations in Kamchatka. This work also contained a very complete summary of cranial materials from Siberia. He also studied nearly all the collections housed in Soviet museums and introduced methods of measuring facial and nasal flatness that had been proposed by members of the English biometrical school and by Soviet scholar N.A. Abinder, whose manuscript, unfortunately, remained unpublished.

All these studies have contributed to the understanding of Siberian craniology. They provide a detailed picture of geographic variation in cranial features and also reveal that variation occurs in patterns which generally coincide with the Arctic, Baikal, Central Asiatic, and Uralian complexes identified on somatologic evidence.

Many established ideas have been altered by later findings. The affiliation of the Aleuts is a case in point. Cheboksarov placed the Aleuts near the Central Asiatic groups on the basis of cranial height. Debets, in contrast, used a genealogic argument based on the temporal dynamics of cranial characters to place the Aleuts near the Eskimos.

The continuing study of Siberian craniology has included both the collection of new data and the restudy of museum collections. Much new information has come from the excavation of recent cemeteries. This includes data on the Selkups and Chulym Tatars (Rozov 1956), Khakass (Alexseev 1960), Yakuts (Tomtosova 1974), Eskimos (Alexseev and Baluyeva 1976), and the vast body of unpublished data on the Trans-Baikal Buryats collected by N.N. Mamonova. Some small new museum collections of Eskimos (Alexseev 1961a) and Orochs (Alexseev 1964b) have recently been described. Finally, in line with recent advances in craniologic methodology and the new focus on features showing discrete variation, many Siberian collections have been reexamined using a much more inclusive protocol (Rychkov and Movsesyan 1972, Movsesyan *et al.* 1975).

Materials

The craniology of different Siberian peoples has been studied to varying extents. The Evenks, for example, occupy a huge territory stretching from the Ob' in the west to the Lena in the east. The individual specimens of a series of several dozen Evenk skulls were collected from sites that might lie 1000 km apart. The series is therefore a pooled sample from different areas. Pooling all available specimens is typical of cranial series representing widely distributed population groups.

A great deal of data come from the excavation of one or more cemeteries in an area. This type of sampling yields information not only on a population group as a whole but also on its local variants. Data of this type exist for the Khakass and the Buryats. In some cases, data are available on small ethnographic subgroups of a population, for example the Telengets or Shors among the Altay peoples, and lacking on the major part of the population.

The geographic locations of cemeteries yielding cranial materials appear in Figure 3.4. The series of Yukagirs measured by Debets has been excluded because

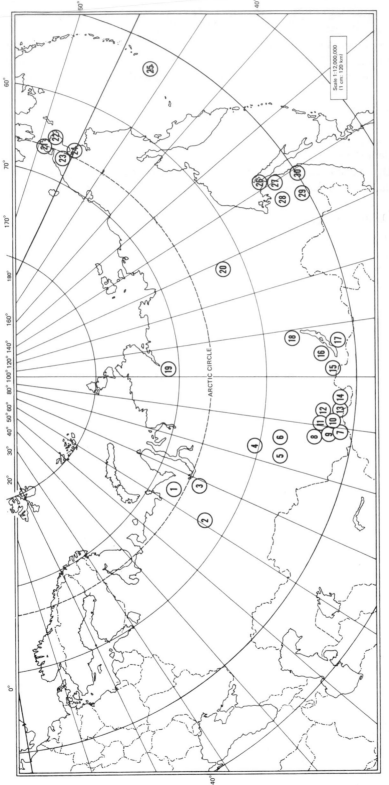

Figure 3.4. Probable geographic locations of the cemeteries from which recent craniologic materials are considered. (1) Nenets, (2) Mansi, (3) Khanty, (4) Northern Selkups, (5) Southern Selkups, (6) Chulym Tatars, (7) Telengets, (8) Shors, (9) Sagayts, (10) Beltirs, (11) Khatshints, (12) Koybals, (13) Western Soyots, (14) Western Soyots, (15) Soyots of the central regions, (16) Tunkin Buryats, (16) Western Buryats, (17) Eastern Buryats, (18) Evenks, (19) Dolgans, (20) Yakuts, (21) Naukan Eskimos, (22) Southeastern Eskimos, (23) Reindeer Chukchis, (24) Coastal Chukchis, (25) Aleuts, (26) Nivkhs, (27) Ulchs, (28) Negidals, (29) Nanays, (30) Orochs.

its ethnic origin is uncertain. Major groups and their subdivisions are shown on the map, but they are pooled in Tables 3.6 through 3.9 in order to obtain an overall picture of cranial variation in Siberia. Each value for the Khakass (Table 3.7.) therefore represents the unweighted mean of series from four local ethnographic groups. The Buryat values are derived from three local groups. The values for Selkups (Table 3.6), Soyots (Table 3.7), Eskimos (Table 3.9), and Chukchis (Table 3.9) come from two subgroups within each population.

A glance at the map shows that large gaps exist in the data. Even in southern Siberia, where the data are very numerous, there is no information on northern groups of the Altay peoples, on the Udegeits of the Amur Valley, on the Evenks, or from Kamchatka. Therefore, all conclusions must be tentative.

Methods

The materials discussed above were obtained with measurement protocols varying in completeness from the highly detailed to the comparatively limited. Measurements were taken as recommended in the handbook by Martin (1928)

Table 3.6. *Measurements of Male Crania from Western Siberia*

		Peoples[b]				
No.[a]	Measurements	Nenets	Mansi	Khanty	Selkups	Chulym Tatars
1	Cranial length (mm)	179.2 (38)	183.9 (28)	181.1 (111)	178.9 (54)	178.0 (34)
8	Cranial breadth (mm)	146.6 (38)	139.5 (28)	143.7 (111)	144.5 (54)	139.5 (34)
17	Cranial height, basion–bregma (mm)	129.0 (38)	126.1 (28)	127.2 (116)	132.6 (51)	132.8 (32)
8:1	Cranial index	81.9 (38)	76.1 (28)	79.5 (111)	80.8 (54)	78.3 (34)
45	Bizygomatic breadth (mm)	139.1 (38)	135.3 (25)	139.4 (114)	140.2 (54)	136.1 (31)
48	Upper facial height (mm)	73.9 (35)	70.8 (27)	73.9 (113)	70.8 (51)	68.5 (33)
48:45	Upper facial index	53.1[c]	52.3[c]	53.0[c]	50.5[c]	51.3 (31)
48:17	Upper craniofacial index	57.4 (35)	56.2 (27)	58.2 (113)	53.4[c]	51.6[c]
51	Orbital breadth from mf, left (mm)	42.7 (37)	41.8 (27)	43.7 (116)	42.6 (52)	39.8 (32)[d]
52	Orbital height, left (mm)	34.4 (37)	34.3 (27)	35.4 (116)	33.8 (52)	33.4 (33)
54	Nasal breadth (mm)	25.0 (37)	26.3 (28)	25.9 (114)	26.2 (51)	26.1 (33)
55	Nasal height (mm)	53.0 (37)	52.5 (28)	53.8 (114)	52.6 (52)	50.4 (33)
DS	Dacrial subtense (mm)	9.4 (36)	10.1 (27)	9.9 (113)	10.6 (49)	10.6 (30)
DS:DC	Dacrial index	45.6 (36)	47.9 (27)	47.4 (113)	46.9 (49)	50.9 (30)
SS	Simotic subtense (mm)	2.7 (36)	2.8 (27)	2.8 (114)	8.3 (49)	3.1 (32)
SS:SC	Simotic index	37.3 (36)	40.2 (27)	36.8 (114)	38.0 (49)	38.7 (32)
72	General facial angle (°)	86.1 (37)	86.5 (27)	84.7 (114)	84.4 (45)	82.4 (30)
75 (1)	Nasal prominence angle (°)	23.3 (27)	20.2 (20)	20.7 (100)	18.7 (37)	16.8 (28)
77	Nasomolar angle (°)	146.4 (36)	142.1 (26)	143.9 (112)	145.9 (49)	143.9 (28)
	Zygomaxillary angle, zm′-ss-zm′ (°)	135.8 (35)	135.0 (24)	132.6 (113)	134.3 (43)	133.0 (26)

[a] Number given to measurement in the work of Martin (1928).

[b] Numbers in parentheses are sample sizes.

[c] Index of average means.

[d] 1.5 mm has been added to orbital breadth from d.

V.P. Alexseev

Table 3.7. *Measurements of Male Crania from Southern Siberia*

No.[a]	Measurements	Telengets	Shors	Khakass	Soyots	Buryats
				Peoples[b]		
1	Cranial length (mm)	176.7 (51)	175.8 (30)	179.0 (144)	182.2 (53)	182.1 (137)
8	Cranial breadth (mm)	151.5 (51)	143.5 (30)	146.6 (146)	147.2 (53)	150.1 (137)
17	Cranial height, basion–bregma (mm)	130.6 (51)	132.9 (31)	133.1 (143)	129.2 (51)	133.2 (133)
8:1	Cranial index	85.9 (51)	81.9 (30)	82.0 (143)	80.8 (53)	82.6 (137)
45	Bizygomatic breadth (mm)	142.1 (49)	136.6 (32)	140.5 (143)	139.5 (44)	142.6 (135)
48	Upper facial height (mm)	74.2 (41)	73.0 (30)	75.2 (136)	77.2 (39)	77.6 (121)
48:45	Upper facial index	52.2[c]	53.6 (30)	53.4 (130)	54.9 (39)	54.2[c]
48:17	Upper craniofacial index	56.7 (41)	55.2 (29)	56.3 (128)	55.3[c]	58.3[c]
51	Orbital breadth from mf, left (mm)	42.4 (49)	42.1 (32)	42.9 (147)	43.9 (44)	42.3 (133)
52	Orbital height, left (mm)	34.2 (49)	33.8 (32)	34.9 (149)	35.4 (44)	35.5 (133)
54	Nasal breadth (mm)	27.0 (49)	25.0 (32)	25.8 (146)	29.5 (44)	26.8 (131)
55	Nasal height (mm)	53.9 (49)	52.4 (32)	54.2 (148)	58.5 (44)	55.8 (131)
DS	Dacrial subtense (mm)	9.2 (49)	9.6 (32)	10.2 (147)	8.8 (44)	9.2 (110)
DS:DC	Dacrial index	41.5 (49)	50.2 (32)	50.7 (147)	41.0 (44)	42.8 (109)
SS	Simotic subtense (mm)	3.2 (50)	3.0 (31)	3.6 (147)	2.8 (46)	2.9 (115)
SS:SC	Simotic index	40.4 (50)	44.1 (31)	48.3 (147)	34.3 (46)	35.9 (115)
72	General facial angle (°)	87.2 (45)	88.1 (31)	87.0 (130)	87.5 (41)	87.5 (106)
75 (1)	Nasal prominence angle (°)	23.3 (37)	24.5 (31)	23.3 (126)	19.7 (21)	19.9 (91)
77	Nasomolar angle (°)	149.3 (49)	146.5 (32)	145.5 (147)	146.8 (47)	145.8 (116)
	Zygomaxillary angle, zm′-ss-zm′ (°)	134.5 (47)	132.9 (32)	134.4 (145)	138.2 (43)	140.8 (111)

[a] Number given to measurement in the work of Martin (1928).
[b] Numbers in parentheses are sample sizes.
[c] Index of average means.

and in the works of English biometricians. Modifications of methods are described in Alexseev and Debets (1964).

The major innovation is in the measurement of facial flatness. Facial flatness was determined from horizontal angles measured at the level of the nasomolar points and nasion and of the zygomaxillary points and subspinale. Nasal flatness was measured from the angle of the nasal bones to the line of the facial profile [measurement 75(1) in Martin (1928)] as well as with the help of the dacryal and simotic heights and indices. These measurements cannot only discriminate Mongoloid crania, characterized by large facial flatness angles and small dacryal and simotic heights, from Europoid crania but can also distinguish among local Mongoloid variants. A methodologic peculiarity common to most papers on Siberian craniology is the use of the vertical cranio-facial index (Martin's measurement 48:17) which is the ratio of facial height to cranial height. This ratio also discriminates among local groups of Siberian Mongoloids.

Geographic Distribution of Traits

The general geographic distribution of traits is shown in Tables 3.6 to 3.9. Siberian Mongoloids differ from other Mongoloids in having the most extreme

Table 3.8. *Measurements of Male Crania from Central Siberia*

	Measurements	Peoples[b]		
		Evenks	Dolgans	Yakuts
1	Cranial length (mm)	185.5 (28)	176.3 (11)	185.9 (19)
8	Cranial breadth (mm)	145.7 (28)	147.3 (11)	148.5 (19)
17	Cranial height, basion–bregma (mm)	126.3 (27)	125.0 (11)	137.2 (19)
8:1	Cranial index	78.7 (28)	83.6 (11)	80.0 (19)
45	Bizygomatic breadth (mm)	141.6 (28)	139.6 (11)	146.3 (19)
48	Upper facial height (mm)	75.4 (28)	76.3 (10)	82.5 (19)
48:45	Upper facial index	53.2[c]	54.7[c]	55.9 (19)
48:17	Upper craniofacial index	60.0 (27)	61.2 (10)	59.9 (19)
51	Orbital breadth from mf. left (mm)	43.0 (27)	43.1 (10)	44.3 (19)
52	Orbital height, left (mm)	35.0 (27)	34.4 (10)	35.2 (19)
54	Nasal breadth (mm)	27.1 (28)	25.8 (10)	27.0 (19)
55	Nasal height (mm)	55.3 (28)	54.9 (10)	57.7 (19)
DS	Dacrial subtense (mm)	8.7 (28)	7.6 (10)	9.5 (19)
DS:DC	Dacrial index	40.1 (28)	39.4 (10)	44.2 (19)
SS	Simotic subtense (mm)	2.4 (28)	2.3 (10)	2.8 (19)
SS:SC	Simotic index	32.3 (28)	41.0 (10)	39.4 (19)
72	General facial angle (°)	86.6 (28)	87.5 (10)	88.8 (19)
75 (1)	Nasal prominence angle (°)	18.7 (22)	19.4 (8)	20.7 (19)
77	Nasomolar angle (°)	149.1 (28)	149.8 (11)	147.1 (19)
	Zygomaxillary angle, zm'-ss-zm' (°)	141.6 (28)	136.8 (10)	136.5 (19)

[a] Number given to measurement in the work of Martin (1928).
[b] Numbers in parentheses are sample sizes.
[c] Index of average means.

development of characteristically Mongoloid features, including very large facial dimensions, an orthognathic and much flattened face, high orbits, and flattened nasal bones and nasal bridge. Some individual features of this complex are found outside Siberia, but the complex as a whole is specific for all Siberian populations.

Local variations show up clearly within the general Siberian complex. The peoples of Western Siberia are characterized by decreased facial, and in some cases nasal, flatness and by some reduction in facial dimensions. These tendencies appear among the Shors and Khakass of southern Siberia.

Eskimos, Aleuts, and Chukchis have a more vertically profiled facial skeleton and the most pronounced nasal bones found in Siberia. This combination occurs together with the narrowest nasal openings of any Siberian group. Aleuts differ markedly from Eskimos in cranial height, but the difference probably developed late in time. Hrdlička's (1945) data from the Aleutian Islands showed that the so-called Pre-Aleuts had a much greater cranial height than modern Aleuts. Also Pre-Aleuts were dolichocephalic as were Eskimos and Chukchis.

Some marked local variation occurs among groups in central Siberia, parts of southern Siberia, and the Amur Valley. Buryats, Yakuts, and Soyots have very high and broad facial skeletons. The Tungus–Manchu peoples and the Dolgans have somewhat reduced facial skeletons, but they have the greatest facial and nasal flatness in Siberia.

V.P. Alexseev

Table 3.9. Measurements of Male Crania from the Far East

No.^a	Measurements	Peoples^b							
		Eskimos	Chukchis	Aleuts	Nivkhs	Ulchs	Negidals	Nanays	Orochs
1	Cranial length (mm)	183.1 (152)	183.7 (57)	179.8 (133)	179.9 (15)	183.3 (31)	180.9 (16)	184.6 (11)	177.0 (12)
8	Cranial breadth (mm)	141.7 (153)	142.2 (57)	147.6 (133)	147.8 (15)	142.3 (31)	145.6 (16)	142.3 (11)	148.9 (12)
17	Cranial height, basion–bregma (mm)	136.2 (146)	135.4 (55)	127.9 (128)	133.5 (15)	134.4 (30)	130.4 (16)	137.8 (11)	130.8 (12)
8:1	Cranial index	77.6 (152)	77.5 (57)	82.1 (133)	82.2 (15)	78.3 (31)	80.6 (16)	77.1 (11)	84.2 (12)
45	Bizygomatic breadth (mm)	139.7 (148)	140.8 (53)	143.1 (116)	142.5 (13)	139.9 (30)	141.3 (16)	139.7 (10)	139.4 (12)
48	Upper facial height (mm)	77.9 (141)	78.4 (54)	74.6 (122)	77.3 (13)	77.6 (30)	76.8 (14)	73.0 (10)	73.3 (12)
48:45	Upper facial index	55.8^c	55.7^c	52.1^c	54.2^c	55.5^c	54.4^c	52.3^c	52.6^c
48:17	Upper craniofacial index	57.2^c	58.1 (53)	58.3^c	58.0^c	57.8 (29)	58.9 (14)	57.1 (9)	56.1 (12)
51	Orbital breadth from mf, left (mm)	43.8 (153)	43.9 (55)	43.3 (121)	43.7 (13)	43.4 (31)	42.4 (16)	43.4 (10)	42.8 (12)
52	Orbital height, left (mm)	36.3 (153)	36.6 (55)	36.2 (121)	34.8 (13)	35.7 (31)	35.3 (16)	35.3 (10)	35.0 (12)
54	Nasal breadth (mm)	24.6 (152)	24.7 (55)	25.4 (127)	26.5 (13)	26.7 (30)	27.3 (15)	27.1 (10)	25.6 (12)
55	Nasal height (mm)	55.2 (152)	55.9 (55)	52.1 (127)	53.4 (13)	55.4 (30)	56.5 (15)	55.4 (10)	53.2 (12)
DS	Dacrial subtense (mm)	10.5 (98)	10.3 (51)	10.2 (35)	9.9 (13)	8.7 (31)	8.3 (15)	8.7 (8)	8.4 (12)
DS:DC	Dacrial index	54.0 (98)	53.4 (51)	49.3 (35)	51.4 (13)	44.3 (31)	39.4 (15)	42.3 (8)	43.5 (12)
SS	Simotic subtense (mm)	3.0 (106)	2.8 (52)	2.9 (35)	3.0 (13)	2.2 (31)	2.3 (16)	2.5 (10)	1.8 (12)
SS:SC	Simotic index	44.4 (106)	41.5 (52)	34.7 (35)	43.5 (13)	32.8 (31)	32.1 (16)	35.0 (10)	26.7 (12)
72	General facial angle (°)	83.0 (128)	83.2 (54)	82.3 (36)	84.5 (13)	86.0 (31)	85.8 (16)	85.9 (10)	86.0 (12)
75 (1)	Nasal prominence angle (°)	23.4 (82)	22.5 (33)	25.8 (24)	16.9 (8)	17.0 (21)	15.3 (16)	15.1 (7)	17.0 (11)
77	Nasomolar angle (°)	146.8 (134)	147.3 (55)	145.5 (30)	146.0 (13)	146.2 (30)	148.6 (16)	147.7 (11)	148.7 (12)
	Zygomaxillary angle, zm'-ss-zm' (°)	135.6 (129)	138.1 (53)	138.1 (28)	135.9 (13)	137.5 (30)	142.3 (16)	137.3 (10)	138.3 (12)

^a Number given to measurement in the work of Martin (1928).
^b Numbers in parentheses are sample sizes.
^c Index of average means.

In each case of local variation discussed above, the results of cranial studies discriminate the same complexes—Uralian, Arctic, Central Asiatic, and Baikal —identified by using anthropometric and descriptive characteristics. The Amuro–Sakhalinian complex is distinct from Siberian complexes in descriptive characteristics and its independent position therefore cannot be confirmed by cranial data. The Paleo-Siberian complex cannot be studied as yet because of the lack of cranial data; the Dolgans are the only Paleo-Siberian groups for which some data are available.

Some Genealogic Hypotheses

The genetic relationships of the various complexes can only be determined using both cranial and anthropometric data. Cheboksarov's (1949) hypothesis of an ancient division of Asiatic Mongoloids into two stocks—Pacific Coast and Internal Asiatic —was confirmed on the basis of Siberian cranial material (Debets 1951). Arctic Mongoloids were considered to be a northern branch of the Pacific Coast stock, while all Siberian Mongoloids of the interior were treated as a northern group of the Internal Asiatic stock. According to Roginsky's (1937) hypothesis, a relatively undifferentiated group of Mongoloid ancestors, morphologically similar to North American Indians, formed the basis of both stocks. The skulls from the upper cave at Choukoutien and Lutsian provide good paleoanthropologic evidence for this hypothesis because they have less developed Mongoloid features than do modern Asiatic populations. The Arctic complex resembles the North American more closely than any other Siberian morphologic combination.

Some additional information on the origin of the Uralian complex can be extracted from cranial data. Deniker (1889, 1900) first described this complex. It has often been considered as protomorphic in the evolution of Siberian Mongoloids and to have formed without any Europoid admixture in spite of its decreased development of Mongoloid features (Bunak 1924, 1956, Yakimov 1956, 1960, Schwidetzky 1914). We have stated elsewhere that the intergroup correlations of features among Uralian peoples differ markedly from intragroup correlations bearing in mind that such differences can only have resulted from a long period of intensive racial differentiation. The same differences in correlation coefficients within groups and between groups were obtained from cranial materials of Western Siberian Ugrian and Eastern European Finnic origin (Alexseev 1973). At the same time, craniologic analysis shows regularities in the geographic distribution of traits that distinguish Europoid and Mongoloid populations; definite gradients of increase and decrease run from west to east. Both the differences between intergroup and intragroup correlation coefficients and the regular geographic distribution of traits can be better explained by the admixture hypothesis than by the hypothesis of *in situ* development from an undifferentiated racial stock.

A final important point is the probable relationship between Neolithic populations of the Baikal region and the modern Evenks (Okladnikov 1955) revealed by archaeologic research. These studies showed a direct development of material culture, including some very specific features, such as ornamentations and fine details of clothing, from a Neolithic Baikal template. Although this cannot be used as direct evidence of a common origin, it should be considered in any discussion of Europoid admixture among ancient Baikal populations. If the Baikal groups represented the undifferentiated racial stock which gave rise to all Siberian populations, then it would be more reasonable to suppose that their Neolithic material culture would not show closer affinities to that of the Evenks than to other groups. The Mongoloids of the Neolithic series, therfore, are related not to all Siberians but only to the Baikal complex. They differ from modern Evenks only in Europoid admixture and played a major role in Evenk race formation and ethnic origin.

Conclusions

Anthropometric and cranial data on Siberian peoples are numerous enough to reveal morphologically distinct, geographically delimited trait complexes among the aboriginal populations of Siberia. These have been called anthropologic types and groups of populations, but they really represent local races. They have different origins and thus demonstrate that race formation in Siberia has been a complex process. It is probable that the ancient populations of Siberia, along with all other ancient Mongoloids, differed from their modern descendants in having the Mongoloid complex of physical traits in a less developed form and resembled North American Indians more than modern Siberians. The study of this ancient population stratum in Siberian and surrounding areas demands much more scientific investigation but will be of great value in understanding the peopling of North America and the genetic relationships between ancient populations of Asia and America.

References

Aksyanova, G.A. (1975). Anthropological Study of Western Siberian Nenets. [In Russian.] Moscow: Polevie Issledovaniya Instituta Etnographii, in 1974.

Alexseev, V.P. (1960). Craniology of the Khakass in Connection with the Question of Its Origin. [In Russian.] Moscow: Trudy Kurgizskoy Archeologitscheskoy, Vol. 4.

Alexseev, V.P. (1961). Paleoanthropology of the Altay–Sayan Mountain Region in the Neolithic and Bronze Ages. [In Russian.] Moscow: Trudy IE, n. s., Vol. 71.

Alexseev, V.P. (1964a). The Craniology of the Asiatic Eskimo. *Arctic Anthropol.* 2(2):120–125.

Alexseev, V.P. (1964b). The Craniology of the Orochi. *Arctic Anthropol.* 2(2):126–132.

Alexseev, V.P. (1971). Forest Nenets (Somatological Observations). [In Russian.] *Voprosy Antropologii*, Vol. 39.

Alexseev, V.P. (1973). Recent Craniological Material on Finno–Ugrian Peoples, Their

Racial Differentiation and Ethnogenesis. *In Studies of the Anthropology of the Finno–Ugrian peoples.* Helsinki.

Alexseev, V.P., and T.S. Baluyeva. (1976). Materials on the Craniology of Arctic Eskimos (Differentiation of the Arctic Race). [In Russian.] *Sovietskaya Etnographia,* No. 1.

Alexseev, V.P., and G.F. Debets. (1964). *Craniometry.* [In Russian.] Moscow.

Alexseev, V.P., Y.D. Benevolenskaya, G.M. Davidova, I.I. Gokman, and V.K. Zhomova. (1968). Anthropological Investigations in the Lena Valley. [In Russian.] *Sovietskaya Etnographia,* No. 5.

Alexseeva, T.I. (1972). Biological Aspects of the Study of Human Adaptability. Moscow: Antropologia 70-h godov (Symposium). [In Russian.]

Alexseeva, T.I. (1975a). The Problem of Biological Human Adaptability. [In Russian.] *Priroda,* No. 6.

Alexseeva, T.I. (1975b). Biological Adaptation of Human Populations to Environmental Conditions of the Asiatic Arctic (Anthropological Aspect). [In Russian.] *Geographitscheskie Aspekty Ekologiitscheloveka.* Moscow.

Alexseeva, T.I., V.P. Volkov-Dubrovin, O.M. Pavlovskiy, N.S. Smirnova, V.A. Spitzin, and L.K. Schekochikhina. (1971). Anthropological Investigations in the Eastern Baikal Region. [In Russian.] *Voprosy Antropologii,* Vol. 37.

Alexseeva, T.I., V.P. Volkov-Dubrovin, Z.A. Golubtschikova, O.M. Pavlovskiy, N.S. Smirnova, and V.A. Spitzin. (1972). Anthropological Study of Forest Nenets. [In Russian.] *Voprosy Antropologii,* Vol. 42.

Bunak, V.V. (1924). Anthropological Type of the Cheremises. [In Russian.] *Russkiy Antropol. Zh.* Vol. 13.

Bunak, V.V. (1941). *Anthropometry.* [In Russian.] Moscow.

Bunak, V.V. (1956). Human Races and the Pathways of their Formation [in Russian]. *Sovetskaya Etnographia,* No. 1.

Cheboksarov, N.N. (1947). Main Directions of Racial Differentiation in East Asia. [In Russian.] Moscow–Leningrad: Trudy IE, n.s., Vol. 2.

Cheboksarov, N.N. (1949). Northern Chinese and Their Neighbors. [In Russian.] Moscow: Kratkie Soobtscheniya Institute Etnographii, Vol. 5.

Cheboksarov, N.N. (1951). Main Principles of Anthropological Classifications. [In Russian.] Moscow: Trudy IE, n.s., vol. 16.

Chepurkovskiy, Y.M. (1913). The Geographical Distribution of Head Form and Pigmentation in the Peasant Population, Chiefly Great Russian, in Connection with Slavic Colonization. [In Russian.] Moscow: Trudy OLEAE, Vol. 28.

Debets, G.F. (1929a). Anthropological Composition of Transbaikal Populations in the Late Neolithic. [In Russian.] *Russkiy Antropol. Zh.* Vol. 19.

Debets, G.F. (1929b). Craniological Study of the Tannu-Tuvins. [In Russian.] *Northern Asia,* nos. 5–6.

Debets, G.F. (1934). Physical Anthropology of the Soviet North. [In Russian.] *Sovetskiy Sever,* No. 6.

Debets, G.F. (1947). The Selkups. [In Russian.] Moscow–Leningrad: Trudy IE, n.s., Vol. 2.

Debets, G.F. (1948). Paleoanthropology of the USSR. [In Russian.] Moscow–Leningrad: Trudy IE, n.s., Vol. 4.

Debets, G.F. (1951). Anthropological Investigations in the Kamchatka Region. [In Russian.] Moscow: Trudy IE, n.s., Vol. 17.

Debets, G.F. (1955). An Ancient Skull from Yakutia. [In Russian.] Moscow: Kratkie Soobtscheniya Instituta Etnographii, Vol. 25.

Debets, G.F. (1961). On the Ways of Human Occupation of the Northern Parts of the Eastern European Plain and the Eastern Baltic. [In Russian.] *Sovetskaya Etnographia*, No. 6.

Debets, G.F. (1975). Paleoanthropological Materials from Ancient Whale-Hunting Culture Cemeteries at Uellen and Ekven. *In* S.A. Arutjunov and D.A. Sergeev, Eds., *Problems of Ethnic History of the Bering Sea Region* (Ekven Cemetery). [In Russian.] Moscow: Science Publishers.

Deniker, J. (1889). Essai d'une Classification des Races Humaines Baseé Uniquement sur les Caractères Physiques. *Bull. Soc. Anthropol. Paris.* Ser. 3. 12.

Deniker, J. (1900). *Les Races et les Peuples de la Terre.* Paris.

Dolgikh, B.O. (1960). Familial and Tribal Composition of Siberian Peoples in the Seventeenth Century. [In Russian.] Moscow: Trudy IE, n.s., Vol. 55.

Hrdlička, A. (1924). *Catalogue of Human Crania in the United States National Museum Collections.* Vol. 63. Washington, D.C.: Smithsonian Institute.

Hrdlička, A. (1942a). *Catalogue of Human Crania in the United States National Museum Collections.* Vol. 91. Washington, D.C.: Smithsonian Institute.

Hrdlička, A. (1942b). Crania of Siberia. *Am. J. Phys. Anthropol.* 29(4):435–481.

Hrdlička, A. (1944). *Catalogue of Human Crania in the United States National Museum Collections.* Vol. 94. Washington, D.C.: Smithsonian Institute.

Hrdlička, A. (1945). *The Aleutian and Commander Islands and Their Inhabitants.* Philadelphia: Wistar Press.

Klevtsova, N.I. (1976). Comparison of Somatic Features of Siberian Mongoloids. [In Russian.] *Voprosy Antropologii*, Vol. 52.

Klevtsova, N.I., and N.S. Smirnova. (1974). Morphological Body Features of Chukchis and Eskimos. [In Russian.] *Voprosy Antropologii*, Vol. 48.

Levin, M.G. (1936). Data on the Craniology of the Maritime Orochs. [In Russian.] *Antropol. Zh.* No. 3.

Levin, M.G. (1937). The Craniological Type of the Ulchs (Nani). [In Russian.] *Antropol. Zh.* No. 1.

Levin, M.G. (1941). The Craniological Type of the Khanty and Mansi. [In Russian.] Moscow: Kratkiye Soobshcheniya Instituta i Museya Antropologiy Moskovskogo Universitata.

Levin, M.G. (1947a). The Physical Types of the Okhotsk Coast. [In Russian.] Moscow–Leningrad: Trudy IE, n.s., Vol. 2.

Levin, M.G. (1947b). The Physical Type of the Yakuts. [In Russian.] Kratkiye Soobshcheniya IE, No. 3.

Levin, M.G. (1947c). Contributions to the Physical Anthropology of the Eskimos. [In Russian.] *Sovetskaya Etnographiya*, Nos. 6–7.

Levin, M.G. (1949). Cranial Types of the Chukchis and Eskimos. [In Russian.] Sbornik MAE, Vol. 10.

Levin, M.G. (1954). The Anthropology of Southern Siberia. [In Russian.] Kratkiye Soobshcheniya IE, No. 20.

Levin, M.G. (1956). Anthropological Materials from the Verkholensk Burial Site. [In Russian.] Moscow: Trudy IE, n.s., Vol. 33.

Levin, M.G. (1958). Ethnic Anthropology and the Problem of Ethnogenesis of the Peoples of the Far East. [In Russian.] Moscow: Trudy IE, n.s., Vol. 36. English translation H.N. Michael (Ed.). (1963). *Ethnic Origins of the Peoples of Northeastern Siberia.* Arctic Institute of North America, Anthropology of the North: Translations from Russian Sources, No. 3. Toronto: University of Toronto Press.

Levin, M.G. (1964). *The Anthropological Type of the Ancient Eskimos.* [In Russian.] Moscow: Soremannaya Antropologiya.

Mahalanobis, P. (1928). On the Need for Standardization in Measurement on the Living. *Biometrica* 20A:1–31.

Mamonova, N.N. (1973). Paleoanthropological Data on the Ancient Population of the Angara Valley. [In Russian.] *Problemy Urala i Sibiri.* Moscow.

Martin, R. (1928). *Lehrbuch der Anthropologie.* 3 Vols. Jena: Gustav Fischer.

Movsesyan, A.A., N.N. Mamonova, and Y.G. Rychkov. (1975). A Program and Method for the Study of Skull Anomalies. [In Russian.] *Voprosy Antropologii,* Vol. 51.

Okladnikov, A.P. (1955). The Neolithic and Bronze Ages of the Western Baikal Region [in Russian]. Parts I & II. Moscow: Materialy i Issledovaniya po Arkheologii SSSR, No. 43.

Petrov, G.I. (Ed.). (1933). *Materials from the Buryat-Mongolian Anthropological Expedition of 1931.* Part I. [In Russian.] Leningrad.

Roginsky, Y.Y. (1934). Materials on the Anthropology of the Evenks of the Northern Baikal Region. [In Russian.] *Antropol. Zh.* No. 3.

Roginsky, Y.Y. (1937). The Problem of the Origin of the Mongoloid Racial Type. [In Russian.] *Antropol. Zh.* No. 2.

Rosov, N.S. (1956). Materials on the Craniology of the Chulym Taters. [In Russian.] Moscow: Trudy IE, n.s., Vol. 33.

Rosov, N.S. (1961a). Anthropological Investigations of the Aboriginal Populations of Western Siberia. [In Russian.] *Voprosy Antropologii,* Vol. 6.

Rosov, N.S. (1961b). *Materials on the Anthropology of the Population of the Chulym Basin.* [In Russian.] *Voprosy istorii Sibiri i Dalnego Vostoka.* Novosibirsk.

Rudenko, S.I. (1914). Physical Anthropological Studies of the Inhabitants of Northeastern Siberia. [In Russian.] Petrograd: Zapiski Akademii Nauk, otdeleniye fiziko-matematicheskikh nauk, Vol. 3, No. 3.

Rychkov, Y.G. (1961). Anthropology of the Western Evenks. [In Russian.] Trudy IE, n.s., Vol. 71.

Rychkov, Y.G., and A.A. Movsesyan. (1972). Genetico-anthropological Analysis of the Distribution of Cranial Anomalies by Siberian Mongoloids in Connection with the Problems of Their Origins. [In Russian.] *Chelovek, Evolutsija i Vinetvividovaja Differentsiatsija.* Moscow.

Schwidetzky, I. (1914). *Das Menschenbild der Biologie.* Stuttgart: Gustav Fischer Verlag.

Sergeev, V.P. (1932). Dynamics of the Physical Development of the Buryats. [In Russian.] *Antropol. Zh.* No. 2.

Smirnova, N.S. (1973). Characteristic Morphological Features of Body Build in Different Ethno-territorial Groups: The Population Morphology of the Body. [In Russian.] Moscow: Dokladi Sovetskoy Delegatsii ua IX Ueydunarodnom Kongresse Antropologitscheskiy i Etnographitscheskiy Nauk.

Tokareva, T.Ya. (1937). Materials on the Craniology of the Aleuts. [In Russian.] *Antropol. Zh.* No. 1.

Tomtosova, R.F. (1974). A Comparative Consideration of the Craniological Type of the Yakuts. [In Russian.] *Voprosy Antropologii,* Vol. 46.

Trofimova, T.A. (1932). The Ainu Problem. [In Russian.] *Antropol. Zh.* No. 2.

Trofimova, T.A. (1947). The Tatars of Tobolsk and Barabinsk Regions. [In Russian.] Trudy IE, n.s., Vol. 1.

Yakimov, V.P. (1950). Human Skull of Bronze Period from Yakutija. *In:* A.P. Okladnikov, *Antiquities of Lena Valley.* [In Russian.] Moscow–Leningrad: Lenskie Drevnosti, Vol. 3.

Yakimov, V.P. (1956). First Steps in the Human Occupation in the Eastern Baltic. [In Russian.] Trudy IE, n.s., Vol. 32.

Yakimov, V.P. (1960). Anthropological Material from the Neolithic Cemetery on Southern Oleniy Island. [In Russian.] Sbornik MAE, Vol. 19.

Yarkho, A.I. (1929). Hair, Eye, and Skin Pigmentation of the Peoples of the Altay–Sayan Upland. [In Russian.] *Russkiy Antropol. Zh.* Vol. 17 (3–4).

Yarkho, A.I. (1932a). Unification of the Evaluation of Soft Facial Parts. [In Russian.] *Antropol. Zh.* No. 1.

Yarkho, A.I. (1932b). The Turks of Gandia. [In Russian.] *Antropol. Zh.* No. 2.

Yarkho, A.I. 1947. *Turkic Tribes of the Altay–Sayan Region: A Physical Anthropological Sketch.* [In Russian.] Abakan.

Yuzefovich, A.N. (1937). Two Types of Yakut Skulls. [In Russian.] *Antropol. Zh.* No. 2.

Zolotareva, I.M. (1960). *Ethnic Anthropology of the Buryats.* [In Russian.] Uan-Ude: Etnographitscheskiy Sbotnik.

Zolotareva, I.M. (1962). Anthropological Investigation of the Nganasans. [In Russian.] *Sovetskaya Etnographiya,* No. 6.

Zolotareva, I.M. (1965). Anthropological Investigations of the Dolgans. [In Russian.] *Sovetskaya Etnographiya,* No. 3.

Zolotareva, I.M. (1968). An Anthropological Account of the Yukagirs. [In ·Russian.] *Problemy Antropologii i Istoritscheskoy Etnographii Asii.* Moscow.

Zolotareva, I.M. (1974). Anthropological Differentiation of the Eastern Samoyedian Peoples. [In Russian.] *Antropologiya i Genogeographia.* Moscow.

Zolotareva, I.M. (1975a). Territorial Variants of the Anthropological Type of the Yakuts (in Connection with the Problem of Its Origin). [In Russian.] *Etnogenes i Etnitscheskaya Istoria Narodov Severa.* Moscow.

Zolotareva, I.M. (1975b). Anthropology of Several Peoples of Northern Siberia. [In Russian.] *Yukagiri (Istoriko-Ethnographitscheskiy Otscherk).* Novosibirsk.

Chapter 4

Aleuts and Eskimos: Survivors of the Bering Land Bridge Coast

William S. Laughlin, Jørgen B. Jørgensen, and Bruno Frøhlich

Introduction

The Aleuts and Eskimos occupy a strategic area and they occupy it in a strategically informative fashion. Their geographic distribution together with their distinctive physical configurations suggest interpretations of their genetic constitution that are relevant to their origins and to their adaptive success.

The data indicate that their ancestral population migrated from Siberia along the southern coast of the Bering Land Bridge, that their developmental homeland focused in the Umnak–Kuskokwim segment, and that the differentiation of the two major populations of Aleuts and Eskimos from the linear and continuous ancestral chain of isolates began fairly early, some 9000 or 10,000 years ago.

The division of Eskimos into two major linguistic groupings occurred sometime later. It appears increasingly likely that the Eskimos and Aleuts have maintained their old population centers in southwestern Alaska, with its high density and major linguistic diversity, while extending the northern portion of their population eastward over to Greenland. In this way they have preserved the same linear arrangement they formerly had when confined to the southern coast of the Bering Land Bridge and the eastern coast of the Bering Sea prior to the rupture of the bridge by the formation of Bering Strait (Figure 4.1). The similarities and differences between the Aleutian Aleuts, impounded at the western terminus of this continuous linear chain of isolates, and the Greenlandic Eskimos, impounded at the eastern terminus, provide the two most buffered and geographically isolated samples of the variability contained in the old Bering Sea

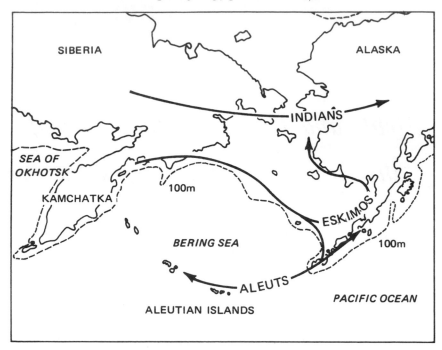

Figure 4.1. Migration routes across Beringia. The ancestors of the American Indians crossed in the interior of the Land Bridge, whereas the ancestoral Aleut–Eskimo populations followed the southern coast.

population. Fortunately, the submergence of the Bering Land Bridge did not submerge the enterprising Aleuts and Eskimos; it created more economically useful coastline and enabled them to expand.

Greenland and the Aleutians: Terminal Refugia

The Aleuts and Eskimos form a continuous, long, linear coastal distribution extending from Attu in the west to Angmagsalik in the east. We should like to call attention to the interesting fact that each end of this distribution was represented by a large population. The Aleuts numbered some 16,000 at the time of their discovery and the Greenland Eskimos numbered some 10,000. The Aleuts and the Greenland Eskimos therefore constituted approximately 30 percent of the entire stock. A consequence of the high proportion of this stock being locked into well-insulated chambers at either end of their population chain provides several analytical advantages.

These two reservoir samples enable us to employ their common elements or shared traits to detect deviations in the more exposed areas between the Aleutians and Greenland and to estimate some characteristics of the ancestral parent population from which they are derived. A well-delineated example is found in the blood groups. Blood group B is found in all the isolates of Greenland and in the Aleutian isolates. This increases the likelihood that the ancestral Aleut–Eskimos brought blood group B with them from Asia. A deviant pattern is seen in eastern Canada and the Smith Sound region (Laughlin 1951a, Mourant 1954), where the values for B are markedly depressed. This may be reliably attributed to the effects of very small and dispersed population size and of pronounced isolation in the case of the polar Eskimos. The similarity between the Aleutians and Greenland extends to other blood group antigens as well. The low frequency of blood type N is well attested and can therefore be used in estimating admixture (Boyd 1950, pp. 418–419) and simultaneously in probing affinities with the American Indian and Asiatic populations.

Greenland, like the Aleutian Islands, is a refugium with limited access. Eskimos were the first human occupants of Greenland and they remained the exclusive human occupants from about 4200 years ago to the arrival of the mediaeval Norse shortly before 1000 A.D. (Knuth 1967, Meldgaard 1977). They migrated around the coast of Greenland in two columns, one clockwise and the other counterclockwise. As a consequence the two terminal isolates of northeast Greenland and of southeast Greenland differed more from each other than any other two contiguous isolates (Laughlin and Jørgensen 1956). The genetic integrity of the Greenlandic Eskimos is beyond reasonable doubt and they provide an excellent sample of the northern division of Inyupik Eskimos, who extend from the Bering Strait to Greenland and Labrador. There have been incursions into Greenland from Canadian Eskimos and there has therefore been some replenishment of their gene pool (Gilberg 1976). All North American approaches to Greenland were inhabited by Eskimos who served as an effective filter for limited entry into Greenland.

The Aleutian Island refugium has been the exclusive domain of Aleuts for some 9000 years (Laughlin 1975). Only Aleuts occupied the chain of islands and the western portion of the Alaska Peninsula to Port Moller on the north side, including the Shumagin Islands on the south side, and no Aleuts passed beyond the common border separating Aleut from Eskimo. The Aleut occupation was also coastal, with volcanic mountains and unproductive highlands serving the same confining barrier functions as the inland ice of Greenland. The islands themselves are arrayed linearly and longitudinally. The distribution differs from Greenland in that only one east–west axis was possible. However, it is identical in that the original entry into the Aleutians took place nearly one third of the distance from the eastern end of the distribution and proceeded in two directions. The earliest occupation dated at 8700 years ago consisted of a unifacial core and blade industry with lamps, stone dishes, and other household artifacts displaying several Asiatic traits (Laughlin and Okladnikov 1976, Okladnikov and Vasilievsky 1976). This culture is continued in transitional forms to the arrival of the Russian discoverers in 1741 A.D.

There are now sufficient radiocarbon dates, as well as artifactual and skeletal evidence, to estimate the rate of expansion. The western end of the Aleutian chain

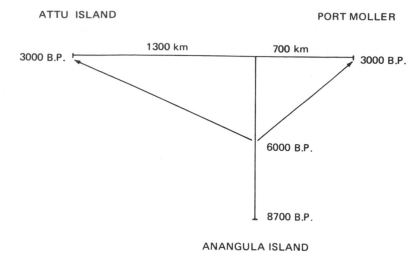

Figure 4.2. Temporal and spatial distribution of early Aleut movements in the Aleutian Islands. Both the eastern (Port Moller) and the western (Attu Island) terminals were occupied about 3000 B.P., whereas the oldest occupation dates to 8700 B.P. at Anangula, which was the terminus of the Bering Land Bridge.

appears to have been occupied no earlier than approximately 3000 years ago, and the eastern limit of the Aleuts at Port Moller appears to have been occupied about 3000 years ago (Figure 4.2) (Laughlin 1966). Assuming that the Aleuts, with a normally low rate of population growth, had generated enough population to begin expansion some 6000 years ago, it therefore appears that approximately 3000 years were expended in migrating 1300 km from Samalga Pass to Attu Island in the west and 700 km to Port Moller in the east. This formulation assumes that the population growth rate is a determining factor in population expansion and that the population migrates faster where resources are more scarce. The 6000-year estimate for beginning expansion is based on the absence of key artifacts, such as the Anangula burin made on a prismatic blade, in the later sites. The original entry into the Aleutian Islands may have taken place when Umnak Island, overlooking Samalga Pass, was part of a larger island cluster or part of the earlier Bering Land Bridge. The Aleut language evidently developed in the relative isolation of the Aleutians. The sharp boundary between Aleut and Eskimo is the necessary result of eastward expanding Aleuts encountering westward expanding Eskimos.

By the time that actual skeletal remains are recovered, some 4000 years ago, the Aleuts have developed a number of biologically interesting characteristics. These include a low cranial vault; great variation in head size, including the largest skull known; and the highest known frequency of three-rooted first lower molars (Hrdlička 1945, Turner 1967, 1971). It is significant with respect to the large head size that there is living in Nikolski, within 40 km of the Kagamil cave where the largest skull was found, an Aleut man who has a head of similar size.

The Aleuts have been in the Aleutians twice as long as the Eskimos have been in Greenland. The Eskimos entered Greenland near to or after the establishment of the modern sea level; the Aleuts entered the Aleutians when the sea level was considerably lower, some 30 m, than the modern sea level. It might be expected that the Aleuts would display more biologic distinctions. This is apparently the case but more quantification and precise study should precede further generalizations. The effects of selection in the high Arctic, population size, replenishing migrations, and degree of isolation should be considered along with the time depths.

Implications of the Aleut-Eskimo Cline for Origins (Coastal and Asiatic)

Two bodies of data assign the coastal origin of the Aleut–Eskimo stock. First is the large body of distinctions between the Aleut–Eskimo stock and the American Indians. Second is the clinal continuity of the Aleut–Eskimo stock. The clinal configuration is a result of a configuring element, which in this case is the shore of the Bering Sea.

Had the Eskimos and Aleuts originated in various groups of Indians moving out to the coast then there would not be a cline. Instead, there would be a variegated pattern of traits, such as those found along the North Pacific coast.

This cline in cephalic index is part of the evidence indicating a long, linear coastline origin, beginning most likely on the south side of the Chukchi Peninsula and migrating south along the eastern Bering Sea coast to Samalga Pass in the eastern Aleutians.

Cephalic Index: Clinal Value and Natural Partitions

The cephalic index of living Aleuts and Eskimos usefully illustrates the integrity of the Aleut–Eskimo stock and its major partitions (Table 4.1). This is an ideal clinal distribution, somewhat rare in human populations.

It is a linear progression, ranging from the lowest values in Greenland to the highest values in the Aleutian Islands. There is only one exception in a small sample of 12 Yupik females. The corresponding male series, 212 in number, fits the progression precisely. The epicenter for the highest cephalic index lies in the eastern Aleutians (84.19) and declines from there to the western Aleutians, and to Kodiak and continuously on to Greenland and Labrador.

The adjacent isolates of the three groups are closer to each other than to other isolates within their respective groups. For example, the Northwest Inyupiks are

W.S. Laughlin, J.B. Jørgensen, and B. Fröhlich

Table 4.1. Aleut–Eskimo Cephalic Index

	Male		Female	
	$\bar{x}$	$\mathcal{N}$	$\bar{x}$	$\mathcal{N}$
Inyupik				
Greenland	76.84	714	76.12	70
Labrador	76.90	30	75.63	23
Central	77.42	321	76.69	107
Northwest	78.35	284	78.38	140
Yupik				
Asiatic–St. Lawrence	80.24	123	79.67	128
Central	80.59	212	83.78	12
Kodiak	81.61	65	82.90	32
Aleut				
East	84.19	30	84.52	27
West	83.09	19	82.36	29

closer to Asiatic–St. Lawrence Yupiks than to Central or to Kodiak Yupik Eskimos.

The variation between the three groups is higher than within the three groups. For example, the range for Inyupik Eskimo is 1.51 index units from Greenland to Northwest; 1.37 units from Asiatic–St. Lawrence to Kodiak; and 1.00 index unit between the Eastern and Western Aleuts. The difference between the adjacent Northwest and the Asiatic–St. Lawrence isolate samples is 1.89, and the difference between Kodiak and Eastern Aleut samples is 2.58.

Most of these mean values were previously presented (Laughlin 1951a), but were not partitioned by linguistic division or aggregated into the means used here. An important difference lies in the mean values for the Kodiak Eskimos. Hrdlička's small and selected sample of 11 Kodiak Eskimos with a mean index of 87.5 probably reflects artificial head flattening and cannot be uncritically compared with the others (Hrdlička, 1945). The mean used here is based on three isolates of Kodiak Island: Old Harbor, Kaguyak, and Karluk, measured in the years 1960, 1961, and 1962 (Laughlin, Jørgensen, and Meier, manuscript). The value of 81.61 may possibly include six individuals in whom there is some head flattening. Their deletion would still leave the cephalic index slightly higher than that of the next contiguous group of Eskimos in this Yupik division. Meier has called attention to the problem of flattening in the Kodiak crania, especially evident in a series from Chirikoff Island. The effect of head flattening, a problem discussed by Hrdlička, has apparently distorted the values of archaeologically derived Koniag Eskimos, at least of the later Neo-Koniags.

The Aleuts series is composed of Laughlin's measurements in 1948–1949, and those of Harper in 1973 (Laughlin 1951b, Harper 1975). There is some redundancy but its maximum effect would be considerably less than one index unit. The separation between Aleutian Aleuts and Kodiak Eskimo (Koniags) is obvious.

There is one cryptic reference to earlier measurements of living Aleuts, probably made in 1910 in the eastern Aleutians, found in Jochelson (1933, p. 79): "The measurement of 138 living people (men and women) taken by Mrs. Jochelson gave a cephalic index of 84, with a standard deviation of 3.3, as individually the cephalic index ranged from 76 to 94. Thus we see that mixture with the Russians did not influence the head measurements of the Aleut." Although the difference between eastern and western Aleuts was likely greater in earlier times, it was still visible in 1948 and in 1973. The combined 1948 and 1973 values for cephalic index of 83.19 for 67 males, and 83.34 for 70 females, represents our best contemporary values.

All the living groups reported here were measured in the twentieth century, most within a 50-year period. They therefore enjoy the advantage of more contemporaneity than exists in an archaeologic series of crania. Clearly, the Aleut–Eskimo stock is justifiably known for its uniformity of which one aspect is the gradient in cephalic index.

Explanations and interpretations of this excellent cline are at hand. At the outset it should be clearly understood that variations in this form trait enjoy validity inside the Aleut–Eskimo population system. The two apparent dehiscences or discontinuities fall where expected, between major linguistic–geographic divisions, i.e., between the two language divisions (Aleut and Eskimo) and between the two major dialectal divisions of Eskimo (Yupik and Inyupik). The greater divergence between Aleuts and Eskimos than between Yupik and Inyupik Eskimos is congruent with radiocarbon dates and time depth inferred from linguistic divergence.

It is possible now to suggest a cause of the sharp interface between the Aleut language and the Eskimo language on the Alaska Peninsula. The Aleut population expanded from an old central nucleus at Samalga Pass both to the west and to the east. The distinctiveness of the Aleut language was developed in the comparative isolation of the Aleutian Islands. When the Aleuts expanded, apparently slowly judging from the archaeologic record and the radiocarbon dates, they eventually encountered Eskimos on the Alaska Peninsula moving in the opposite direction, from east to west. The present boundary may well be a fairly old boundary, but likely no older than 3000 years B.P. and it may of course have wandered.

In overall perspective, the cline in cephalic index is congruent with the early (Late Pleistocene) existence of a population of Bering Sea Mongoloids on the east coast of the Bering Sea, arrayed from north to south, and with a clinal array of head form with the most narrow headed in the north and the more broad headed in the south. The Greenlandic population is a sample derived from the northern portions of the array, and the Aleut population is a sample of the southernmost portion of the array. The relatively recent increase in head breadth in the southern region, in the areas of high population density, has not erased the gradient distinctions between the north, middle, and south portions of the Late Pleistocene array.

Trait Differences Between Aleut-Eskimos and American Indians

In no other part of the New World is there as profound a dichotomy in physical characteristics between groups as that between Aleuts and Eskimos, on the one hand, and Indians, on the other hand. These differences are profound in the sense that they are manifested in several different systems in different parts of the body (red cell antigens, haptoglobins, enzymes, albumin variants, teeth, mandible, cranial vault, spinal column) and in both size and shape. In some cases the Aleut–Eskimo stock has achieved extreme or highest values for the entire human species. Thus, they have the world's highest frequency of separate neural arches, the most narrow nasal bones, the thickest tympanic plates, the highest frequency of the mandibular torus, an absence of auditory exostoses, the highest frequency of three-rooted first lower molars and the largest mandibles. In contrast to Indians, they have no Diego, no albumin Nascapi or Mexico, and they do have blood group B which is absent in Indians. They have provided the largest modern skull on record and it is supported by others of comparable size in living Aleuts.

From the perspective of the entire human species these differences are of course neither major nor unusual in themselves. It is unusual that they should be assembled in a single stock and sharply delimited in their distribution from Indian populations also derived from Siberia. Thus, from the perspective of two adjacent groups, Eskimos and Indians, the differences are substantial. Many of them extend as far back in time as there are skeletons preserved, and they may reliably be inferred considerably further back in time.

The lower jaw of the Aleuts and Eskimos has a remarkably low and broad ascending ramus (Laughlin 1963, Laughlin *et al.* 1976). This morphologic configuration is common to Aleuts and Eskimos. It is not found in American Indians but it is found in Asia. It is clearly an objective and significant feature separating Aleuts and Eskimos from Indians and, simultaneously, it is a feature showing greater affinities with Asia than with New World populations.

This characteristic form is expressed by the ramus index, which is derived from the minimum breadth of the ascending ramus divided by the biapical height of the ascending ramus. The Paleo-Aleuts and the Neo-Aleuts display the world's highest values for this character, exceeding even that of the Eskimos (Table 4.2.). Examination of the distribution of this index reveals a gap between the Aleuts and Eskimos, on the one hand, and the Indians on the other. The American Indians have a relatively high and narrow ascending ramus, more similar to that of Europeans than to Bering Sea Mongoloids. Interestingly, when only one dimension is considered, the minimum breadth of the ascending ramus, there is one Indian group from Florida with a breadth of 39.4, essentially the same size as the Paleo-Koniags of Kodiak Island. However, when the form of the ramus is expressed in this index the Florida Indians are promptly separated by their lower index of 57.4 from the Paleo-Koniags with their higher index of 65.1. Significantly, a gap in the distribution occurs between Uelen Eskimo mandibles (61.1) and those of Alaska Indians (59.0).

Table 4.2. *Mongoloid Male Mandibular Measurements[a]*

	Minimum Breadth of Ramus			Ramus Index[b]	
	n	*x̄*		*n*	*x̄*
Paleo-Aleut	40	42.8	Paleo-Aleut	30	67.9
Neo-Aleut	74	42.1	Neo-Aleut	74	66.8
Ekven, Chukotka (Eskimo)[c]	53	41.2	Koniag, Kodiak	34	65.6
Uelen, Chukotka (Eskimo)[c]	27	40.6	Paleo-Koniag, Kodiak	55	65.1
Alaska Peninsula	18	40.3	Eskimo, in general	420	63.2
Koniag, Kodiak	34	40.1	Ekven, Chukotka (Eskimo)[c]	53	62.5
Eskimo, in general	422	39.8	Alaska Peninsula	17	62.1
Paleo-Koniag, Kodiak	92	39.7	Uelen, Chukotka (Eskimo)[c]	25	61.1
Florida Indians	100	39.4	Alaska Indians	40	59.0
Alaska Indians	40	38.9	Sioux Indians	31	57.9
Sioux Indians	36	37.8	Florida Indians	100	57.4
California Indians	100	37.1	Mongol, Eastern	33	56.4
Mongol, Eastern	35	37.1	Pueblo Indians, miscellaneous	100	55.9
Pueblo Indians, Pecos	125	36.9	California Indians	100	54.9
Old Peru, Coast Indians	175	35.3	Old Peru, Coast Indians	175	53.5
Old Peru, Mountains Indians	23	35.2	Arkansas Indians	62	53.2
Pueblo, miscellaneous	100	35.2	Chinese, Canton	58	52.0
Arkansas Indians	62	34.7	Old Peru, Mountains Indians	23	50.9
Chinese and Tibetans, miscellaneous	156	34.1			
Old Peru	26	34.0			
Chinese, Canton	58	33.7			

[a] Based on Hrdlička (1941).
[b] Ramus index = (breadth × 100)/(biapical height).
[c] Arutiunov and Sergeev (1975).

There are other aspects of the ramus which separate the Aleuts and Eskimos from Indians, such as the inclination of the ramus, which is relatively vertical in Indians. The ramus index, however, includes more information in its construction and provides a quantified expression of form.

The high frequency of the mandibular torus in Aleuts and Eskimos is informative for its variation within stock and for comparisons with Indians in whom it is relatively infrequent and in those few cases never developed as markedly as in Aleuts and Eskimos.

In Greenlandic Eskimos Laughlin and Jørgensen (1956) found 67 percent in 293 male mandibles, and 47 percent in 291 female mandibles. This frequency does not include trace or submedium grades. At the opposite end of the geographic distribution, the frequency of the mandibular torus is also high. Moorrees (1957), in his 1948 study of living Aleuts, found a frequency of 42.1 percent in males and 27.5 percent in females. Internally the frequencies varied from 61.4 percent in eastern Aleuts to 25.7 percent in western Aleuts. In archaeologically derived mandibles, the frequency for Aleut males given by Hrdlička is 65.5 percent (78 of 119) for Neo-Aleuts, and 57.1 percent (20 of 35) for Paleo-Aleuts.

Another kind of trait linking Aleuts and Eskimos, and favoring Asiatic affinities more than Indian affinities, is the three-rooted first lower molar. This trait enjoys the advantage of objective designation, it is accessible in the living by x ray, and it can also be documented for dry mandibles in the absence of the tooth itself because the socket for the distolingual cusp remains patent. The world's highest frequencies are reported for the Aleuts, next for the Eskimos, and least for the Indians, although few observations exist for the latter. Turner (1967) provides the most exhaustive study of this trait in skeletal series:

	Male frequency	n	Female frequency	n
Aleut average	0.54	39	0.14	18
Eskimo average	0.26	37	0.17	89

Pedersen (1949) was the first to call attention to this trait in a study of living Eskimos. He found a frequency of 12.5 percent (8 of 64) individual teeth in East Greenland Eskimos.

The three traits discussed here, ramus index, mandibular torus, and three-rooted first lower molar, constitute a sample of the mandible of the Aleuts and Eskimos. The indices are expressions of form, the torus is a discontinuous trait expressed in bone, and the third root is a discontinuous trait expressed in dental tissue.

Viewed as a functioning whole, the cranium is an incomplete skull. Half, or more, of the genetic information contained in the dentition is contained in the mandible. Exactly half of the teeth are in the lower jaw, but the large amount of information found in the first lower molar, considering the fissural patterns, cusp numbers, accessory cusps, deflecting wrinkle, prostylid, etc., when added to that of the other teeth suggests that the mandibular dentition is more informative than the upper. The most economical sample of the dentition is therefore the first lower molar. The theoretical aspects are supported in practice by the fact that the first lower molar is usually retained when other teeth have been lost, either pre- or postmortem.

Both genetic and evolutionary reasons further recommend the lower jaw as a strategic sampling site for intra- and interpopulation analysis. The relative independence of the lower jaw in human evolution is well illustrated in the Aleut–Eskimo stock, where a large mandible with a typical form of ascending ramus may be combined with widely different vault forms. In the more narrow headed Eskimos the large mandible is associated with a keel-shaped cranial vault and high temporal lines marking the origin of the temporalis muscle, which inserts into the coronoid process of the mandible. The world record for approximation of the temporal lines is that of a Greenlandic Eskimo from Smith Sound, in whom the two temporal lines on either side of the vault are within 7 mm of each other (Hrdlička 1910, Riesenfeld 1955). The same size mandible, or even larger, does not result in high temporal lines in the southeastern Eskimos and Aleuts because the vault is broader. This same relationship is seen internally within the Aleutian Islands in the contrast between the vault form of the earlier Paleo-Aleuts and the later Neo-Aleuts. The Paleo-Aleuts are more narrow headed and their temporal lines are closer together. The later Aleuts are broader headed and their temporal lines are much farther apart. Clearly, the vault form varies independently of the lower jaw.

Size and form of the lower jaw are closely related in the Aleuts and Eskimos. Using minimum breadth of the ascending ramus as the measure of size, the Aleuts have the largest mandibles and also the highest ramus index. Hrdlička (1941) noted a male western Eskimo from Nelson Island (north of the Kuskokwim River) with ramus breadth and index identical to the Heidelberg mandible, a breadth of 50.5 mm and an index of 72.7. This example is informative for illustrating the range of variation in Eskimos; it is not evidence of affinity with the Heidelberg specimen (Stewart 1968).

The interpretation of continuously distributed variables, such as the cephalic index, is relatively simple where the constituent isolates or populations are continuous and contiguous. There is no gap in the population distribution of Aleuts and Eskimos. They occupy the entire coastline from the Aleutians to Greenland and Labrador without interruption. The gaps in the cephalic index of living Aleuts and Eskimos, which occur between speakers of the Aleut and Eskimo language and between the two major divisions of Eskimo, are genuine in every respect. Within the three groups compared, only minor alterations can be achieved by moving isolates into different subdivisions. The gaps remain between the three major divisions. This is all the more significant because there have been significant changes in vault form in southwestern Alaska, invariably from narrow to broader, between earlier and later groups of the same population. Clearly, the vault form is a more labile aspect of the head than the mandible.

The linguistic boundaries reflect mating boundaries and provide excellent categories for the larger number of isolates within each of these three divisions. However, there is no precise way of knowing how long these linguistic boundaries have been in effect. Aleut and Eskimo are mutually unintelligible languages, whereas the two divisions of the Eskimo language can be understood, with varying degrees of difficulty, by speakers of contiguous groups.

Summary and Conclusions

The Aleut population of the Aleutians and the Eskimo population of Greenland provide excellent vantage points to view the entire Aleut–Eskimo stock. They are highly buffered or filtered and isolated reservoirs, containing some 30 percent of the entire Aleut–Eskimo stock, and they are the anchor populations at the opposite ends of a long, linear, continuous distribution. The physical traits they share provide useful commentary on the original population from which they are both derived.

The traits shared over this great distance are manifested in several different tissue systems, in different parts of the body. They are numerous and in some cases, such as the cephalic index, they form a cline.

The differences between the Aleut–Eskimo stock and the Indians have been a point of interest since the first anthropologic description of a human skull, the description by J.B. Winslow of a female Eskimo skull from Dog Island, west Greenland, published in 1722 (Winslow 1722, Herve 1910, Furst 1913, Pedersen 1949). The shared traits within the Aleut–Eskimo stock and the differences between the Aleut–Eskimo stock and Indians suggest separate routes of migration across the Bering Land Bridge into the New World from Siberia, and they may indicate differences in the times of migration as well.

It is suggested that not only did the ancestors of the Aleut–Eskimo stock follow a Bering Sea coastal route across Beringia, but also that they maintained their adaptation to the coast and expanded from the former Alaskan coast of the Bering Sea over to Greenland as the rising sea levels made more coastline available to them. Their slow rate of population growth prompted them to migrate in an extending chain of isolates. Consequently, the Greenland Eskimos are a sample of the northern end of the old Bering Sea population and the Aleuts are a sample of the southern end of the same population.

Research strategies can take advantage of the Aleut–Eskimo distribution in many ways. There is, however, one problem that profoundly effects our sample with respect to time depth. Owing to the fact that the Aleuts and Eskimos have been committed to a coastal marine adaptation, including extensions up such major rivers as the Kuskokwim and Yukon, their earliest occupations have been submerged. Contemporary sea levels were not established until some 5000 to 4500 years ago. Consequently most of our genuinely coastal sites are less than 5000-years-old, with few notable exceptions. As a rough estimate it appears that one third of Aleut prehistory and two thirds of Eskimo prehistory is underwater.

The problem of divergence is significantly qualified by the problem of submergence. In broad perspective, with high gloss for the linguistic as well as the physical divergence, three grades of divergence might eventually lead to useful estimates of the time expended in effecting evolutionary changes. First, the divergence between Aleuts and Eskimos within the Aleut–Eskimo population; second, the divergence between the Aleut–Eskimo population as a whole and American Indians, within the New World; and third, the divergence between New World and Old World populations should provide a provisionally useful index of the time dimension of human evolution in the New World.

We do not know the extent of variation in the original population chain of Bering Sea hunters and fishers who invigilated the liquidation of their expanding coastline, nor whether it was 14,000, 12,000, or only 10,000 years ago. As an intracranial exercise preparatory to more sophisticated phrasing of a research approach, it might be tentatively noted that the Aleuts and Eskimos have been diverging for some 9000 years and the Aleut–Eskimo stock has apparently been diverging from Indians for some 18,000 years. The divergence between the Aleut and Eskimo segments of the Old Bering Sea continuum probably took place on the Alaskan–Aleutian coast of the Bering Sea, but some of the differences between this stock and the Indians may have been generated in Siberia before they moved onto Beringia. The entire New World population, the First Americans, still look very much like their Asiatic relatives. In marked constrast, the Australian Aborigines, who have apparently been in Australia for some 30,000 years or more, do not look (genetically or visually) like populations on the mainland of Asia.

References

Arutiunov, S.A., and D. Sergeev. (1975). *Problems in the Ethnic History of the Bering Sea (Ekven Cemetery)*. Moscow: Science Publishers. [In Russian.]

Boyd, W.C. (1950). *Genetics and the Races of Man*. Boston: D.C. Heath.

Furst, C.M. (1913). Observations à Propos des Remarques de M. G. Herve sur un Crane de l'Ile aux Chiens Décrit par Winslow (1722). *Rev. Anthropol.* No. 12:416–418.

Gilberg, R. (1976). The Polar Eskimo Population, Thule District, North Greenland. *Meddelelser Om Grønland* 203(3):1–87.

Harper, A.B. (1975). *Secular Change and Isolate Diversity in the Aleutian Population System*. Ph.D. Thesis, University of Connecticut, Storrs, Conn.

Herve, M.G. (1910). Remarques sur un Crane de l'Ile aux Chiens. *Rev. Ecole Anthropol.* pp. 52–59.

Hrdlička, A. (1910). Contribution to the Anthropology of Central and Smith Sound Eskimo. *Anthropol. Pap. Am. Mus. Nat. Hist. N.Y.* 5:177–280.

Hrdlička, A. (1941). Lower Jaw: I. The Gonial Angle; II. The Bigonial Angle. *Am. J. Phys. Anthropol.* 23:281–467.

Hrdlička, A. (1945). *The Aleutian and Commander Islands and Their Inhabitants*. Philadelphia: The Wistar Institute of Anatomy and Biology.

Jochelson, W. (1933). *History, Ethnology and Anthropology of the Aleut*. Washington, D.C.: Carnegie Institution, Publ. No. 432.

Knuth, E. (1967). Archaeology of the Musk-Ox Way. *Contributions du Centre D'Etudes Arctiques et Finno-Scandinaves*, No. 5, 85 pp.

Laughlin, W.S. (1951a). Blood Groups, Morphology and Population Size of the Eskimos. *Cold Spring Harbor Symp. Quant. Biol.* 15:165–173.

Laughlin, W.S. (1951b). The Alaska Gateway Viewed from the Aleutian Islands. *In* W.S. Laughlin, Ed., *The Physical Anthropology of the American Indian*. New York: Viking Fund.

Laughlin, W.S. (1963). Eskimos and Aleuts: Their Origins and Evolution. *Science* 142: 633–645.

Laughlin, W.S. (1966). Genetical and Anthropological Characteristics of Arctic Populations. *In* J.S. Weiner and P.Baker, Eds., *The Biology of Human Adaptability*. Oxford: Oxford University Press, pp. 469–495.

Laughlin, W.S. (1975). Aleuts: Ecosystem, Holocene History and Siberian Origin. *Science* 189:507–515.

Laughlin, W.S., and J.B. Jørgensen. (1956). Isolate Variation in Greenlandic Eskimo Crania. *Acta Genet. Statist. Med.* 6:3–12.

Laughlin, W.S., J.B. Jørgensen, and R.J. Meier. (Manuscript). The Anthropometry of Three Kodiak Eskimo Isolates.

Laughlin, W.S., and A.P. Okladnikov. (1976). Origin of Aleuts. *Nature* (Acad. Sci. USSR) No. 1, pp. 119–131. [In Russian.]

Laughlin, W.S., A.P. Okladnikov, A.P. Derevyanko, A.B. Harper, and I.V. Atseev. (1976). Early Siberians from Lake Baikal and Alaskan Population Affinities. *Am. J. Phys. Anthropol.* 45:651–659.

Meldgaard, J. (1977). The Prehistoric Cultures in Greenland: Discontinuities in a Marginal Area. In *Continuity and Discontinuity in the Inuit Culture of Greenland*. Groningen, Netherlands: Arctic Centre, University of Groningen, pp. 19–52.

Moorrees, C.F.A. (1957). *The Aleut Dentition*. Cambridge, MA: Harvard Univ. Press.

Mourant, A.E. (1954). *The Distribution of the Human Blood Groups*. Springfield, Ill.: Charles C. Thomas.

Okladnikov, A.P., and R.S. Vasilievsky. (1976). *On Alaska and the Aleutian Islands*. Novosibirsk: Science Publishers, Siberian Section.

Pedersen, P.O. (1949). The East Greenland Eskimo Dentition: Numerical Variations and Anatomy. *Meddelelser Om Grønland*. 142:1–256.

Riesenfeld, A. (1955). The Variability of the Temporal Lines, Its Causes and Effects. *Am. J. Phys. Anthropol*. 13(4):599–620.

Stewart, T.D. (1968). The Evolution of Man in Asia as Seen in the Lower Jaw. *Proceedings VIIIth International Congress of Anthropological and Ethnological Sciences*. Tokyo: Science Council of Japan. Vol. 1, *Anthropology*, pp. 263–266.

Turner, C.G. II. (1967). *The Dentition of Arctic Peoples*. Ph.D. Thesis, University of Wisconsin, Madison, Wisc.

Turner, C.G. II. (1971). Three-Rooted Mandibular First Permanent Molars and the Question of American Indian Origins. *Am. J. Phys. Anthropol*. 34(2):229–242.

Winslow, J.B. (1722). Conformation Particulière du Crane d'un Sauvage de l'Amérique Septentrionale. *Memoires de l'Academie Royale des Sciences*, pp. 322–324.

Part 2

Affinities

Chapter 5

Blood Polymorphisms and the Origins of New World Populations

Michelle Lampl and Baruch S. Blumberg

Introduction

An objective of this volume is to discuss methods that may be used to identify contemporary Asian populations that have affinities with contemporary native American populations by virtue of their descent from the same ancestral population. If such populations exist, their similarity to American populations can be demonstrated if they have retained genes present in the ancestral populations and in their related populations in the Americas. These would be particularly useful if they included genes of polymorphic loci found only in native Americans and their Asian counterparts. In this chapter we will describe genes that may be useful in such comparisons. It is obvious that if gene frequencies have altered very differently in the two continents by virtue of drift, gene admixture, or differences in selection in the two environments, then a comparison using this technique would not be possible. Hence, this discussion is based on the assumption that it is possible to identify similarities between North American and Asian populations by comparisons using the appropriate genes.

There are two systems (serum albumins and white blood cell HL-A antigens) for which a relatively large amount of data is available and these are discussed in greatest detail. In addition, other systems that may be useful are mentioned.

Albumin Variants

The albumin polymorphisms may be one of the most useful markers for the identification of American Indian affinities. Two genes of relatively wide distribution, albumin Naskapi and albumin Mexico, have been identified in North

and Central American Indian populations. Two other genes of relatively restricted distribution, albumin Makiritare and albumin Yanomama-2, have been
found in South American populations. Although rare variants (i.e., fewer than
one per 100) have been found in European, Asian, and Oceanic populations (i.e.,
albumin B, albumin Kashmir), in no case have they reached the 1 percent figure
characteristic of genetic polymorphisms. Hence, if a polymorphism for albumin
is identified in an Asian population whose affinities with American Indians is
suspected, and in particular if the polymorphic variant has the same electrophoretic mobility and other characteristics as Naskapi, Mexico, Makiritare, or
Yanomama, then this would provide a strong argument in favor of the association
of that population with American Indians (Schell and Blumberg 1977).

Human serum albumin phenotypes are determined by autosomal codominant
alleles which segregate at a single locus. Albumin variants are distinguished by
their electrophoretic mobilities (often using several techniques for the separations)
and approximately 25 such variants have so far been identified (Table 5.1). Most
of these variants occur rarely (one in several thousand) and presumably are
maintained in the population primarily as a consequence of recurrent mutation.

Table 5.1. *Relative Mobilities of 25 Albumin Variants in Three
Starch Gel Electrophoretic Systems[a]*

	Buffer system		
Variant	pH 5.0	pH 6.0	pH 6.9
1. RS-I	1	2	6
2. Pollibauer	1	2	4
3. Belem I	2	1	4
4. B	3	3	1
5. Roma	3	1	3
6. Gainesville	3	3	4
7. Paris (Gombak)	4	3	5
8. Kashmir (Afghanistan)	5	2	1
9. Otsu	6	4	2
10. Santa Ana	7	4	3
11. SO/BS	7	5	3
12. Cartago	7	3	5
13. Xavante	8	6	7
14. Pushtoon	8	4	6
15. Cayemite	9	5	6
16. Mexico	9	4	5
17. Uinba	9	6	7
18. Yanomama-2	10	6	6
Normal albumin	10	6	7
19. Medan	11	6	8
20. Maku	12	7	10
21. Reading (New Guinea)	13	9	9
22. Makiritare-3	13	8	9
23. Naskapi	14	10	10
24. Gent	15	11	10
25. Kyoto	16	11	10

[a] From Weitkamp (1973).

The present chapter considers the four variants that reach polymorphic frequencies: Two of these have been observed in widely separated geographic regions (Naskapi and Mexico), whereas the other two have been primarily confined to the populations in which they were first discovered (Makiritare and Yanomama-2).

The first albumin variant to be found at polymorphic frequency was reported by Melartin and Blumberg (1966). It has a mobility faster than the common albumin A. It was found originally in the relatively isolated Naskapi and Montagnais Indians of Quebec Province, Canada and was termed albumin Naskapi. This study also contained the first report of individuals homozygous for an albumin other than the common albumin A. In 1967, a slow moving variant (albumin Mexico) was found in polymorphic frequency among Indians of the American Southwest and Mexico (Melartin *et al.* 1967). These variants (Naskapi and Mexico) have not been reported in populations other than American Indians, even though many tens of thousands of sera have been examined by the appropriate electrophoretic methods.

Albumin Naskapi is found predominantly among northern populations, whereas albumin Mexico is found among southeastern United States and middle American populations. Populations in the southwest United States have both Naskapi and Mexico (Table 5.2, Figure 5.1). Table 5.2 is adopted from Schell and Blumberg (1977) and Figure 5.1 is taken from Schell *et al.* (1978).

The highest frequency of the albumin Naskapi gene (13 percent) is found in the eastern regions of North America, among the Algonkian-speaking Naskapi of Schefferville, Quebec. High gene frequencies (8 percent) are also found among the Montagnais Indians from the same region. Naskapi living at Northwest River, however, have a lower frequency of Al^{Na} (3 percent). Albumin Naskapi was not found in samples of Mohawk and Micmac Indians from the St. Lawrence River area, and this river may represent the southern limit of the distribution of this variant in the east (Schell *et al.* 1978).

In the Ungava region of northern Quebec a few cases of albumin Naskapi have been found among Eskimos living close to Indian groups with a high frequency of the Naskapi variant. These appear to result from admixture, as no variant has been found among the more remote Eskimo populations from Alaska, Canada, and Greenland (McAlpine *et al.* 1974, Melartin 1967, Melartin and Blumberg 1966, Persson *et al.* 1971). Albumin Naskapi is also apparently absent from the Haida and Tlingit Indians of the Canada Pacific coast (Melartin 1967, Melartin and Blumberg 1966).

Albumin Naskapi, however, is found at high gene frequencies among Athabascan Indians of Alaska (2.8–5.1 percent) (Melartin and Blumberg 1966, Scott and Wright 1969) and Athabascan-speaking Slave and Beaver Indians of Canada (9.8 percent) (Bowen *et al.* 1971). Lower frequencies (less than 3.4 percent) are found among a number of other central and western Indian groups; Sioux (Melartin and Blumberg 1966), Assiniboine (Bowen *et al.* 1971), Ojibwa (Polesky and Rokala 1967, Szathmary *et al.* 1974, Tanis *et al.* 1974), Cree (Bowen *et al.* 1971), Chippewa (Tanis *et al.* 1974, Weitkamp 1973, Weitkamp *et al.* 1968), and

Table 5.2. *The Distributions of Albumin Mexico and Albumin Naskapi in North and Central America*

Linguistic affiliation, Tribal name (location)	Number tested	Number Naskapi[b]	Number Mexico[b]	Naskapi gene frequency	Mexico gene frequency	Reference
Algonkian						
Ojibwa (No. Ontario)	117	6	0	0.026	0.0	Szathmary et al. (1974)
Ojibwa (So. Ontario)	94	1	0	0.005	0.0	Szathmary et al. (1974)
Ojibwa (?)	120	8	0	0.033	0.0	Tanis et al. (1974)
Ojibwa/Anglo (No. Minnesota)	250	11 + (1)	0	0.026	0.0	Polesky and Robala (1967)
Chippewa/Cree (Saskatchewan)	610	36 + (1)	0	0.031	0.0	Weitkamp (1973)
Chippewa/Cree (Saskatchewan)	102	2	0	0.009	0.0	Weitkamp et al. (1968)
Chippewa (?)	87	2	0	0.011	0.0	Tanis et al. (1974)
Cree, Plains (Alberta)	605	25 + (1)	0	0.022 (0.023)[c]	0.0	Bowen et al. (1971)
Cree, Northern (Alberta)	187	10	0	0.026 (0.034)[c]	0.0	Bowen et al. (1971)
Blackfoot (Montana)	97	4	0	0.020	0.0	Polesky and Rokala (1967)
Naskapi (Schefferville, Quebec)	151	37 + (1)	0	0.130	0.0	Melartin and Blumberg (1966)
Montagnais (Schefferville, Quebec)	112	14 + (2)	0	0.080	0.0	Melartin and Blumberg (1966)
Naskapi (N.W. River, Labrador)	66	2	0	0.03	0.0	Schell et al. (1978)
Micmac (Restigouche, Quebec)	102	0	0	0.0	0.0	Schell et al. (1978)
Iroquois						
Mohawk (Caugh Nawaga, Quebec)	112	0	0	0.0	0.0	Schell et al. (1978)
Siouian						
Sioux (?)	160	2	0	0.007	0.0	Melartin and Blumberg (1966)
Assinboine (Alberta)	100	1	0	0.005 (0.007)[c]	0.0	Bowen et al. (1971)
Omaha (Macy Reservation, Nebraska)	96	0	0	0.0	0.0	Schell et al. (1978)
Eskimoan						
Eskimo (Igloolik, Canada)	356	0	0	0.0	0.0	McAlpine et al. (1974)
Eskimo (E. Greenland)	78	0	0	0.0	0.0	Persson et al. (1971)
Eskimo (Thule, Greenland)	297	0	0	0.0	0.0	Persson et al. (1971)
Eskimo (W. Greenland)	116	0	0	0.0	0.0	Persson et al. (1971)
Eskimo (Wainwright, Alaska)	111	0	0	0.0	0.0	Melartin and Blumberg (1966)
Eskimo (Anaktuvuk)	55	0	0	0.0	0.0	Melartin and Blumberg (1966)
Eskimo (Ungava Bay, Canada)	124	3	0	0.012	0.0	Melartin et al. (1968)
Eskimo (Barrow, Alaska)	88	0	0	0.0	0.0	Melartin et al. (1968)
Eskimo (Beaver, Alaska)	14	0[d]	0	0.0	0.0	Melartin et al. (1968)

Population	N					Reference
Eskimo (St. Lawrence Is., Alaska)	82	0	0.0	0	0.0	Melartin et al. (1968)
Eskimo (Nat'l. Guard, Alaska)	107	0	0.0	0	0.0	Melartin et al. (1968)
Aleut (Sitka and Nat'l. Guard, Alaska)	15	0[d]	—	0	—	Melartin et al. (1968)
Eskimo (Frobisher Bay Baffin Island N.W. Terr.)	101	0	0.0	0	0.0	Melartin et al. (1968)
Na-Dené (non-Athabascan branch)						
Tlingit (Sitka and Mt. Edgecomb, Alaska)	91	0[d]	0.0[d]	0	0.0[d]	Melartin (1967)
Haida (Queen Charlotte Is., British Columbia)	365	0	0.0	0	0.0	Melartin and Blumberg (1966)
Na-Dené (Athabascan branch)						
Alaskan Athabascans (Alaska)	230	11 + (1)	0.028	0	0.0	Melartin and Blumberg (1966)
Alaskan Athabascans (Alaska)	137	14	0.051	0	0.0	Scott and Wright (1969)
Slave and Beaver (Alberta)	143	22 + (2)	0.090 (0.098)[c]	0	0.0	Bowen et al. (1971)
Navajo (Arizona and New Mexico)	95	12	0.063	6	0.005	Melartin et al. (1968)
Navajo (Sherman School, Ca.)	468	30	0.032	4	0.006	Johnston et al. (1969)
Navajo (?)	192	11	0.03	7	0.010	Tanis et al. (1974)
Apache (Sherman School, Ca.)	103	3	0.015		0.034	Johnston et al. (1969)
W. Apache (San Carlos, Arizona)	538	9	0.016	19	0.036	Schell et al. (1978)
Yuman River						
Cocopah (Arizona)	164	0	0.0	13	0.039	Polesky et al. (1968)
Maricopa (Arizona)	101	0	0.0	9 + (1)	0.054	Polesky et al. (1968)
Maricopa (Sherman School, Ca.)	9	1	—	0	—	Johnston et al. (1969)
Mohave (Sherman School, Ca.)	19	1	—	1	—	Johnston et al. (1969)
Yuman Upland						
Havasupi (Sherman School, Ca.)	6	0	—	0	—	Johnston et al. (1969)
Walapai (Sherman School, Ca.)	8	0	—	0	—	Johnston et al. (1969)
Yavapai (Sherman School, Ca.)	3	0	—	0	—	Johnston et al. (1969)
Uto–Aztecan						
Pima (Arizona)	1528	0	0.0	100 + (7)	0.037	Polesky et al. (1968)
Pima (Sherman School, Ca.)	113	0	0.0	11	0.049	Johnston et al. (1969)
Papago (Sherman School, Ca.)	115	0	0.0	3	0.013	Johnston et al. (1969)
Papago (Sherman School, Ca.)	179	0	0.0	2	0.006	Melartin (1967)
Papago (Arizona)	546	0	0.0	5	0.005	Brown and Johnson (1970)
Hopi (Sherman School, Ca.)	82	1	0.006	0	0.0	Johnston et al. (1969)
Ute (Sherman School, Ca.)	17	0	—	1	—	Johnston et al. (1969)
Paiute (Sherman School, Ca.)	4	0	—	0	—	Johnston et al. (1969)
Chemehueve (Sherman School, Ca.)	2	0	—	0	—	Johnston et al. (1969)
Mission (Sherman School, Ca.)	1	0	—	0	—	Johnston et al. (1969)

Table- Continued

Table 5.2.—*Continued*

Linguistic affiliation, Tribal name (location)	Number tested	Number Naskapi[b]	Number Mexico[b]	Naskapi gene frequency	Mexico gene frequency	Reference
Washoan						
Washoe (Sherman School, Ca.)	1	0	0		—	Johnston et al. (1969)
Zunian						
Zuni (New Mexico)	655	0	16	0.0	0.012	Melartin (1967)
Eastern Muskogean (tribal affiliation indeterminate)						
Dania (Fla.)	150	0	0	0.0	0.0	Pollitzer et al. (1970)
Big Cypress (Fla.)	135	0	0	0.0	0.0	Pollitzer et al. (1970)
Brighton (Fla.)	128	0	0	0.0	0.0	Pollitzer et al. (1970)
Macro-Nahua						
Nahua (Puebla, Hidalgo, Veracruz, Mexico)	440	0	6	0.0	0.006	Lisker et al. (1971)
Otomi (Hidalgo, Mexico)	100	0	5	0.0	0.025	Lisker et al. (1971)
Cora (Nayrati, Mexico)	83	0	2	0.0	0.012	Lisker et al. (1971)
Macro-Maya						
Maya (Yucatan, Mexico)	263	0	0	0.0	0.0	Melartin et al. (1967)
Huasteco (Veracruz, Tamulipas, Mexico)	188	0	4	0.0	0.01	Lisker et al. (1971)
Tzeltal–Tzotzil (Chiapas, Mexico)	144	0	0	0.0	0.0	Lisker et al. (1971)
Chol (Chiapas, Mexico)	138	0	0	0.0	0.0	Lisker et al. (1971)
Dialect not reported						
(Guatemala City and San Marcos, Guatemala)	204	0	1	0.0	0.002	Johnston et al. (1973)
(San Antonio Ilotenango, Guatemala)	186	0	3	0.0	0.008	Johnston et al. (1973)

Macro–Mixteco						
Zapoteco (Oaxaca, Mexico)	255	0	3	0.0	0.005	Lisker et al. (1971)
Zapotec (Guelatao, Mexico)	123	0	1	0.0	0.004	Melartin et al. (1967)
Mixteco (Oaxaca, Mexico)	48	0	0	—	—	Lisker et al. (1971)
Mazateco (Oaxaca, Mexico)	22	0	0	—	—	Lisker et al. (1971)
Mazatec (Huautla de Jimenez, Mexico)	20	0	0	—	—	Melartin et al. (1967)
Mixe (Oaxaca, Mexico)	21	0	0	—	—	Lisker et al. (1971)
Tarasco						
Tarasco (Michoacan, Mexico)	167	0	5	0.0	0.014	Lisker et al. (1971)
Mixed Linguistic Groups						
Mestizo–Zapotec (Pochutla, Mexico)	36	0	1	—	—	Melartin et al. (1967)
Mestizo–Mixtec						
(San Pedro Mixtec, Mexico)	20	0	0	—	—	Melartin et al. (1967)
(Ometepec, Mexico)	20	0	0	—	—	Melartin et al. (1967)
(Cauajiniicuilapa, Mexico)	20	0	0	—	—	Melartin et al. (1967)
Mestizo (Tlaxcala, Hidalgo, and Mexico, D. F., Mexico)	185	0	10	0.0	0.027	Melartin et al. (1967)
Mestizo (Tlaxcala–Hidalgo, Mexico)	257	0	12	0.0	0.023	Lisker et al. (1971)
Mestizo (Veracruz, Mexico)	109	0	1	0.0	0.004	Lisker et al. (1971)
Mestizo (Campeche, Mexico)	109	0	0	0.0	0.0	Lisker et al. (1971)
Mestizo (West Coast, Mexico)	99	0	1	0.0	0.005	Lisker et al. (1971)
Mestizo (Mexico City, Mexico)	1313	0	27	0.0	0.010	Lisker et al. (1971)

[a] Adapted from Schell and Blumberg (1977).
[b] Homozygotes in parentheses.
[c] Published gene frequency, calculated with correction for related individuals.
[d] A very low frequency was reported originally; this value has since been corrected to the value shown here.

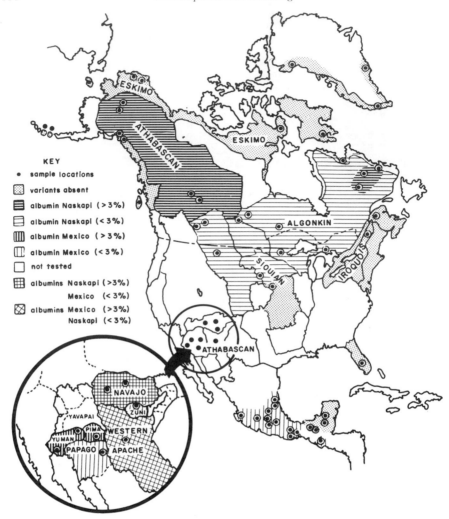

Figure 5.1. Distribution of albumins Naskapi and Mexico in native American populations. (From Schell *et al.* 1978.)

Blackfoot (Polesky and Rokala 1967), although the variant is absent in others, i.e., Omaha (Schell *et al.* 1978). It also occurs in relatively high frequency in the Athabascan-speaking Navajo and Apache (Schell *et al.* 1978).

This wide geographic dispersion is consistent with some linguistic and cultural–historical data. From archaeologic evidence it is postulated that the movement of the Navajo and Apache into the Southwest from the North took place about 500–1000 years ago (Aikens 1970). The Navajo and Apache, furthermore, are linguistically related to Canadian and Alaskan Athabascan-speaking Indians, and glottochronologic analyses support these dates (Hoijer 1956, Hymes 1957).

In the southern regions of North America and in Central America the slow albumin variant, albumin Mexico, has been found in a distribution extending from the southwest United States to Guatemala. Albumin Mexico has not (at this time) been found in any other native northern North American population. Polymorphic frequencies of albumin Mexico have been found among the Uto–Aztecan-speaking Pima (Johnston *et al.* 1969, Polesky *et al.* 1968) and Papago (Brown and Johnson 1970, Johnston *et al.* 1969, and Melartin 1967), the Zuni (Brown and Johnson 1970), and the Yuman River-speaking Maricopa and Cocopah (Weitkamp 1973).

The highest frequencies of albumin Mexico in middle America are reported for two Mestizo samples (Lisker *et al.* 1971, Melartin *et al.* 1967) and the variant is present at lower frequencies in populations from Tamulipas, Mexico on the Gulf coast (Lisker *et al.* 1971) to Guatemala City, Guatemala (Johnston *et al.* 1973).

The distribution of albumin Mexico parallels archaeologic data for past cultural contacts (Weaver 1972, and Wormington 1970). The evidence suggests that a culture centered near the Valley of Mexico (area of modern Mexico City) had contacts with the peoples of the Guatemalan highlands, to the north with the Hohokam culture in southwestern United States and with people from the area of Tamulipas, Mexico. The Pima and Papago are thought to be descendants of the Hohokam (Spencer and Jennings 1965).

Among several southwestern United States groups (i.e., Navajo, Apache, Mohave, and Maricopa), both albumins Mexico and Naskapi are found concurrently at varying frequencies (Johnston *et al.* 1969).

It is interesting that albumin Mexico is found among the Navajo and, to a greater extent, among the Apache, but albumin Naskapi is only rarely found among the Uto–Aztecan tribes. This is consistent with ethnohistorical evidence that, as a result of warfare and kidnapping, Pima and Papago and others were assimilated into the Navajo and Apache groups but the reverse happened less often.

In South America the two geographically restricted variants (albumins Yanomama-2 and Makiritare) are found in polymorphic frequencies in the ethnic groups after which they are named and in some nearby groups. Yanomama-2 has been found at a frequency of 7.6 percent among 64 Yanomama villages in southern Venezuela and Northern Brazil. The highest frequency of Yanomama-2 (40 percent) has been found in a northern village (Tanis *et al.* 1974).

Albumin Makiritare (which is identical with albumin Warao) has been found among the Makiritare, the Warao of Venezuela, and the Trio and Wajana tribes of Surinam (Arends *et al.* 1969, 1970, Geerdink *et al.* 1974). The variant reaches polymorphic frequencies among several tribal groups, but its distribution is confined to the northeastern area of South America. Although the average frequency for the albumin in the Makiritare is about 1 percent it reaches a frequency of 4 percent in several individual villages.

Hence, polymorphic variants of albumin have been found only in American Indians (but not Eskimos). The identification of similar variants in Asian populations could contribute to our understanding of affinities with Native Americans.

The HL-A System (Human Major Histocompatibility Complex)

The HL-A system is thought to be the major human histocompatibility system (apart from ABO) in humans and is present as an antigenic polymorphism on many human tissues. The HL-A antigens were first found as isoagglutinogens on platelets and white blood cells, and routine determinations utilize these cells. Subsequent research has revealed their medical importance in determining whether and to what extent the tissues of a potential transplant donor are compatible with the putative recipient. Histocompatibility antigens show a high degree of polymorphism, and the great variation of antigen frequencies in different populations makes the HL-A system a potentially powerful genetic marker for population studies. Research into the histocompatibility system is complicated by the large number of antigenic specificities and the difficulties in standardizing reagents. Despite this, these systems may prove to be among the most important genetic systems for studying relations between human populations. We use it here as an example of how such data may be applied to the study of Asian populations that may be related to Americans. A revised system of nomenclature for the HL-A system has recently been introduced (Amos et al. 1976). To maintain consistency with the published population data the earlier nomenclature is used here.

It is now generally assumed that the HL-A system consists of four autosomal linked loci, now termed A, B, C, and D, each with a complex series of allelic genes giving rise to distinct antigens. (The actual sequence on the chromosome is D, B, C, A.) Of these, the antigens defined by the A, B, and C loci are serologically determined and the D locus antigens are responsible for the mixed lymphocyte response (MLR locus). An HL-A haplotype generally refers to a section of a chromosome containing the A (or LA) locus and the B (second or fourth locus) (i.e., a pair of antigens). The reagents for the C locus were not generally available until recently. The use of haplotype designations, as we shall see later, can lead to very specific identifications of populations. There are some population-related antigens at each of the loci. HL-A1 is characteristic of Caucasians, whereas HL-A3 and HL-A8 characterize non-Caucasian populations. HL-A9 is high among Orientals, New Guineans, New Zealand Maoris, Eskimos, and American Indians. W10 is high among populations of Asians and North and South American Indians; and W15 is commonly found in Asia and South America (Bodmer and Bodmer 1974, Colombani and Degas 1973, Dausset and Colombani 1973).

There is a striking amount of homogeneity with respect to the HL-A system in American Indians; only two or three alleles of each allele series account for most of the variation. Summarizing from numerous sources (Bodmer 1973, Bowen et al. 1971, Cann et al. 1973a, b, Colombani and Degas 1973, Corley et al. 1973, Daveau et al. 1975, Dossetor et al. 1973, Kissmeyer-Nielson et al. 1973, Kostyu et al. 1975, Layrisse et al. 1973a, b, Perkins et al. 1973, Rubenstein et al. 1967, Spees et al. 1973, Tittor et al. 1973, Troup et al. 1973, Van der Does et al. 1973, Zaretskaya and Fedrunova 1973) it can be said that at the first locus HL-A2 is the most frequent gene, found at frequencies higher among American Indian populations than any other populations yet studied. HL-A9, W28, and W31 are the other common genes. W10, W16, and one of the antigens of the cross-reacting set of HL-A5, W5, and W15, account for the majority of the second series variability.

The HL-A antigens vary geographically among the New World populations. W31 of the first series and W15 of the second series are found at higher frequencies among South American populations, whereas W28, HL-A9, and W5 are found at highest frequencies among North and Central American Indian populations. There appears to be a north–south cline for HL-A9; the highest frequency for any population has been observed in Eskimos. Conversely, HL-A2 is observed at high frequencies in central and southern populations tested but is found at much lower frequencies among Eskimos.

Bodmer and Bodmer (1974) suggest that HL-A1, -3, -10, -11, W29, and 19.6 are absent in American Indians and their presence in a population designated as Indian represents admixture. Studies on Mexican Indians, Chilean Mapuchi, and Maya support this (Amos *et al.* 1970, Rubenstein *et al.* 1967).

The data available on HL-A antigens in Mongoloid populations other than American Indians indicate that the alleles HL-A2, -9, -10, -11, and W31 of the

Table 5.3. *Population Distribution of HL-A Haplotypes*

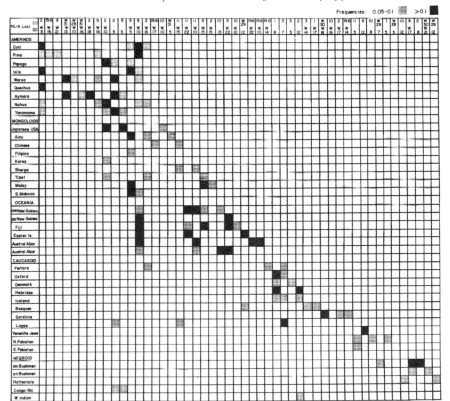

first locus and HL-A5, -12, W5, W10, W15, and W22 of the second locus characterize this group (Dausset and Colombani 1973, Albert *et al.* 1970, Neel *et al.* 1972).

At the International Workshop and Conference on Histocompatibility Testing held at Evian, France in 1972, papers were presented in which a large number of populations had been tested using the same or appropriately compared antisera. Frequencies of haplotypes for each of these populations were estimated and the most common haplotypes shown in tabular form (Bodmer and Bodmer 1974, Dausset and Colombani 1973). We have rearranged these data in Table 5.3 to show how the native American populations tested could be readily distinguished from most other populations, including the closely related Mongoloid and Oceanic groups. In this table high (>0.1) and intermediate (0.05–0.01) frequencies are indicated. From this, it can be seen that there are about 10 haplotypes that occur at these levels only in native Americans or at most in one other population. Furthermore, there are many haplotypes that occur commonly in one or more other population(s) but not at all (or very rarely) in native Americans. From this, the discriminating power of the HL-A haplotypes is apparent; it would be a very useful tool for comparing eastern Asian populations with native Americans.

The Gm System

The Gm system resembles HL-A in genetic complexity and immunologic importance. Gm factors contribute to the inherited variations of the heavy chains of IgG. The genetics of the Gm system is complex and similar to that of the Rh red blood cell antigens. The frequencies of some of the Gm factors vary strikingly among populations, and some of the Gm haplotypes (also called allotypes) are distinct to specific populations and characteristically either present or absent in certain populations. This makes the Gm system a most useful population marker. There is presently, however, a number of problems that make it difficult to use these factors to their full potential. As more antisera were developed, more factors were determined. In earlier studies only one or two factors may have been identified, and in subsequent studies four or five factors may have been tested. This makes interpopulation comparisons difficult. Furthermore, Szathmary *et al.* (1974) have pointed out the difficulties of comparing data because of the use of different varieties of antisera ("Ragg" and "Snag") which may have caused differences of classifications.

Early in the investigations the population associations of single factors were described (Grubb 1970). Gm(1) is found at nearly 100 percent frequency in non-European populations; Gm(6) is rare except in Negroes and Gm(16) is rare except in Japanese (Van Loghem 1970).

Haplotype combinations have also been used to identify populations. Mongoloid markers reported in the literature are Gm(1,17,21), Gm(1,2,17,21), Gm(1, 13,17), Gm(1,11,16), Gm(1,3,5,13,14), Gm(1,3,5,11), Gm(1,11,13,15,16,17,1,5),

Gm(1,13), Gm(1,13,15,16,17), Gm(1,10,11,17,25), Gm(1,4,5,10,11,14,23,25), and Gm(1,4,22). Caucasoid markers include Gm(1,17,21), Gm(1,2,21), Gm(3,5, 13,14), Gm(4,5,13,14), and Gm(4,5,8,10,11) (Daveau *et al.* 1975, Giblett 1969, Grubb 1970, Steinberg 1969, Steinberg *et al.* 1974, Szathmary *et al.* 1974).

There are certain Gm factors thought to be characteristic of American Indian populations. Factors 1; 1,2; 1,2,21; and 1,21 are the most frequently and ubiquitously found. The Mongoloid marker 1,11,16, however, has been identified among populations of Ojibwa, and Szathmary *et al.* (1974) reports that Schanfield identified Gm(1,11,16) in an Indian group in the Valley of Mexico and in the Papago of the American Southwest.

The Gm system has promising potential for distinguishing native American populations from others, if comparison and "control" populations are studied with the same reagents in the same laboratory.

Red Blood Cell Antigens

The red blood cell antigens are the most extensively and intensively tested of the human polymorphisms. Specific gene frequency distributions for American Indians have been known since the earliest investigations of the red blood cell antigens. These have been thoroughly reviewed recently by Mourant *et al.* (1976) and will be presented only briefly here. In the ABO system, O occurs in extremely high frequencies. In fact, it is assumed that a great majority of prehistoric Indians were of the O blood type and that the presence of A and B usually indicates genetic admixture with nonnative Americans. Native American populations are characterized by a high frequency of M of the MNSs system and, in particular, by the alleles *MS* and *Ms*. The frequency of the R^2 gene of the Rh system is higher than anywhere else in the world and the frequency of $R^{\circ}r$ is extremely low or absent. The Kell antigen was probably absent in precontact Indians and occurs in extremely low frequency or is absent in most contemporary American Indian populations. There is an extremely high frequency of Fy^a of the Duffy system and the Diego gene (Di^a) occurs only in native American, Asian, and some Oceanic populations and, in general, is absent from other populations. There is in general a cline of increasing frequency of the Di^a gene from north to south in the Americas. The Lewis A and Lutheran A are thought to occur in American Indian populations only as results of genetic admixture.

Serum Protein Polymorphisms

A large number of serum protein polymorphisms is now known and in some cases alleles found only (or rarely) in Mongoloid populations and in native Americans have been identified (Mourant *et al.* 1976). The transferrin variant "D Chinese" is found in people of east Asia and also in many American Indian populations but rarely in Europeans or other populations. Transferrin B_{0-1} has been reported only in people of Asiatic origin (including Finns, Lapps, and Hungarians) and in native Americans.

The rare Gc variant, Gc Chippewa, has been found so far only in Chippewa and Ojibwa Indians (Szathmary *et al.* 1974).

The serum lipoprotein systems (Ag, Lp) have been used only occasionally in native American population studies, and no generalization can be made about their distribution at present. However, the Ag system is determined by alleles segregating at four or more closely linked locuses and complex combinations of "haplotypes" are known. If populations can be tested with the same antisera using the same techniques, then it is likely that interesting and useful distinctions between populations can be made.

Conclusions

On the basis of the currently available data, there are several polymorphic systems and/or alleles that appear to be unique to native American populations. These include the polymorphisms at the albumin locus, several of the haplotypes of the HL-A system, probably some of the haplotypes of the Gm system, and certain fairly rare alleles (such as Gc Chippewa) of other systems. There is also a large number of new polymorphic traits the distributions of which are not known but which may in due course be shown to have characteristics that make them useful for distinguishing native Americans from other populations.

Using these traits, and others with distinctive frequencies in native Americans, a study could be designed to identify contemporary Asian populations that might have ancient affinities with native Americans.

Acknowledgments

This work was supported by USPHS grants CA-06551, RR-05539, and CA-06927 from the National Institutes of Health, and by an appropriation from the Commonwealth of Pennsylvania.

References

Aikens, C.M. (1970). Hogup Cave. Anthropological Papers, No. 93. Dept. of Anthropol., University of Utah.

Albert, E.D., M.R. Mickey, A.C. McNichols, and P. Terasaki. (1970). Seven New HL-A Specificities and Their Distribution in Three Races. P. Terasaki, Ed., *Histocompatibility Testing 1970*. Copenhagen: Munksgaard, pp. 221–230.

Amos, D., G. Cabresa, W.B. Bias, J.M. McQueen, S.L. Lancaster, J.G. Southworth, and R.E. Ward. (1970). The Inheritance of Human Leukocyte Antigens. III. The Organization of Specificities. *In* P. Terasaki, Ed., *Histocompatibility Testing 1970*. Copenhagen: Munksgaard, pp. 259–275.

Amos, D.B., R. Batchelor, W.F. Bodmer, R. Ceppellini, and J. Dausset. (1976). WHO–IUS Terminology Committee. Nomenclature for Factors of the HL-A System. *Transplantation* 21:353–358.

Arends, T., M.L. Gallango, M. Layrisse, J. Wilberg, and H.D. Heinen. (1969). Albumin Warao: New Type of Human Alloalbuminemia. *Blood* 33:414–420.

Arends, T., L. Weitkamp, M. Gallango, J. Neel, and J. Schultz. (1970). Gene Frequencies and Microdifferentiation among the Makiritare Indians. II. Seven Serum Protein Systems. *Hum. Genet.* 22:526–532.

Bodmer, W.F. (1973). Population Genetics of the HL-A System: Retrospect and Prospect. *In* J. Dausset and J. Colombani, Eds., *Histocompatibility Testing 1972*. Copenhagen: Munksgaard, pp. 611–620.

Bodmer, J., and W.F. Bodmer. (1974). Population Genetics of the HL-A System. *In*. B. Ramot, Ed., *Genetic Polymorphisms and Disease in Man*. New York: Academic Press.

Bowen, P., F. O'Callaghan, and C.S.N. Lee. (1971). Serum Protein Polymorphism in Indians of Western Canada. *Hum. Hered.* 21:242–253.

Brown, K.S., and R.S. Johnson. (1970). Population Studies on Southwestern Indian Tribes. III. Serum Protein Variation of Zuni and Papago Indians. *Hum. Hered.* 20:281–286.

Cann, H.M., J.G. Bodmer, and W.F. Bodmer. (1973a). The HL-A Polymorphism in Mayan Indians of San Juan La Laguna, Guatemala. *In* J. Dausset and J. Colombani, Eds., *Histocompatibility Testing 1972*. Copenhagen: Munksgaard, pp. 367–376.

Cann, H.M., K.M. Kidd, R. Lisker, R. Rodvany, and R. Payne. (1973b). Genetic Structure of the HL-A System in a Nahua Indian Population in Mexico. *Tiss. Antigens* 3:364–372.

Colombani, G., and L. Degas. (1973). Variations of HL-A Antigens in Populations. *Rev. Eur. Etud. Clin. Biol.* 17:551–563.

Corley, R.B., E.K. Specs, G. Cabrera M., J.L., Swanson, and D.B. Amos. (1973). HL-A Antigens of the Guatemalan Ixils. *In* J. Dausset and J. Colombani, Eds., *Histocompatibility Testing 1972*. Copenhagen: Munksgaard, pp. 351–358.

Dausett, J., and J. Colombani (Eds.) (1973). *Histocompatibility Testing 1972*. Copenhagen: Munksgaard.

Daveau, M., L. Rivat, A. Langaney, N. Afifi, E. Bois, and C. Ropartz. (1975). Gm and Inv. Allotypes in French Guiana Indians. *Hum. Hered.* 35:88–92.

Dossetor, J.B., W.T. Howson, J. Schlaut, P.R. McConnachie, J.D.M. Alton, B. Lockwood, and L. Olson. (1973). Study of the HL-A System in Two Canadian Eskimo Populations. *In* J. Dausset and J. Colombani, Eds., *Histocompatibility Testing 1972*. Copenhagen: Munksgaard, pp. 325–332.

Geerdink, R.A., H.A. Bartstra, and J.M. Schillhorn Van Veen. (1974). Serum Proteins and Red Cell Enzymes in Trio and Wajana Indians from Surinam. *Am. J. Hum. Genet.* 26:581–587.

Giblett, E. (1969). *Genetic Markers in Human Blood*. Philadelphia: F.A. Davis.

Grubb, R. (1970). *The Genetic Markers of Human Immunoglobulins*. Berlin, New York: Springer-Verlag.

Hoijer, H. (1956). The Chronology of the Athabascan Languages. *Intl. J. Am. Ling.* 22:219–232.

Hymes, D. (1957). A Note on Athabascan Glottochronology. *Intl. J. Am. Ling.* 23:290–297.

Johnston, F.E., O. Alareon, F. Benedict, M. Dary, M. Galbraith, and P.S. Gindhart. (1973). Albumin Mexico (Al^{Me}) in the Guatemalan Highlands. *Am. J. Phys. Anthropol.* 38:27–30.

Johnston, F.E., B.S. Blumberg, S.S. Agarwal. L. Melartin, and T.A. Burch. (1969). Alloalbuminemia in Southwestern U.S. Indians: Polymorphism of Albumin Naskapi and Albumin Mexico. *Hum. Biol.* 41:263–270.

Kissmeyer-Nielson, F., K.E. Kjerbye, L.U. Lamm, J. Jorgensen, G. Bruun Petersen, and H. Gurtler. (1973). Study of the HL-A System in Eskimos. *In* J. Dausset and J. Colombani, Eds., *Histocompatibility Testing 1972.* Copenhagen: Munksgaard, pp. 317–324.

Kostyu, D.D., D.B. Amos, and S. Hinostroza. (1975). An Analysis of the 4c Complex of HL-A Based on Indian Populations. *Tiss. Antigens* 5:420–430.

Lisker, R., L. Cobo, and G. Mora. (1971). Distribution of Albumin Variants in Indians and non-Indians of Mexico. *Am. J. Phys. Anthropol.* 35:119–124.

Layrisse, Z., M. Layrisse, I. Malave, P. Teraski, R.H. Ward, and J.V. Neel. (1973a). Histocompatibility Antigens in a Genetically-Isolated American Indian Tribe. *Am. J. Hum. Genet.* 25:493–509.

Layrisse, Z., P. Terasaki, J. Wilbert, H.D. Heinen, B. Rodriquez, A. Soyano, K. Mittal, and M. Layrisse. (1973b). Study of the HL-A System in the Warao Population. *In* J. Dausset and J. Colombani, Eds., *Histocompatibility Testing 1972.* Copenhagen: Munksgaard, pp. 377–386.

McAlpine, P.J., S.H. Chen, D.W. Cox, J.B. Dossetor, E. Giblett, A.G. Steinberg, and N.E. Simpson. (1974). Genetic Markers in Blood in a Canadian Eskimo Population with a Comparison of Allele Frequencies in Circumpolar Populations. *Hum. Hered.* 24:114–142.

Melartin, L. (1967). Albumin Polymorphism in Man. *Acta Pathol. Microbiol. Scand.*, Suppl. 191.

Melartin, L., and B.S. Blumberg. (1966). Albumin Naskapi: A New Variant of Serum Albumin. *Science* 153:1664–1666.

Melartin, L., B.S. Blumberg, and R. Lisker. (1967). Albumin Mexico, a New Variant of Serum Albumin. *Nature (London),* 215:1288–1289.

Melartin, K., B.S. Blumberg, and J.R. Martin. (1968). Albumin Polymorphism (Albumin Naskapi) in Eskimos and Navajos. *Nature (London)* 218:787–789.

Mourant, A.E., A.C. Kopec, and K. Domaniewska-Sobczak. (1976). *The Distribution of the Human Blood Groups and Other Polymorphisms.* London: Oxford Univ. Press.

Neel, J.V., T. Arends, C. Brewer, N. Chagnon, H. Gershowitz, M. Layrisse, Z. Layrisse, J. MacClever, E. Migliazza, W. Oliver, F. Salzano, R. Spielman, R. Ward, and L. Weitkamp. (1972). Studies on the Yanomama Indians. *Proc. IV Int. Congr., Hum. Genet. Paris. Excerpta Medica* 1972:96–111.

Perkins, H.A., R.O. Payne, K.K. Kidd, and D.W. Huestis. (1973). HL-A and Gm Typing of Papago Indians. *In* J. Dausset and J. Colombani, Eds., *Histocompatibility Testing 1972.* Copenhagen: Munksgaard, pp. 317–324.

Persson, I., L. Melartin, and A. Gilberg. (1971). Alloalbuminemia: A Search for Variants in Greenland Eskimos. *Hum. Hered,* 21:57–59.

Polesky, H.F., and D.A. Rokala. (1967). Serum Albumin Polymorphism in North American Indians. *Nature (London)* 216:184–185.

Polesky, H.F., D.A. Rokala, and R.A. Burch. (1968). Serum Albumin Polymorphism in Indians of Southwestern United States. *Nature (London)* 220:175–176.

Pollitzer, W.S., D. Rucknagel, R. Tankian, D. Shreffler, W. Leyshon, K. Namboodiri, and R. Elston. (1970). The Seminole Indians of Florida: Morphology and Serology. *Am. J. Phys. Anthropol.* 32:65–81.

Rubenstein, P., R. Costa, A. van Leeuwen, and J.J. van Rood. (1967). The Leukocyte Antigens of Mapuchi Indians. *In* E.S. Curtoni, P.L. Mattiuz, and R.M. Tosi, Eds., *Histocompatibility Testing 1967.* Baltimore: Williams and Wilkins, pp. 251–255.

Schell, L.M., and B.S. Blumberg. (1977). The Genetics of Human Serum Albumin. *In* V.M. Rosenoer, M.A. Rothschild, and M. Oratz, Eds., *Albumin—Structure, Function and Uses in Man*. London, New York: Pergamon Press, pp. 113–141.

Schell, L.M., S.S. Agarwal, B.S. Blumberg, H. Levy, P.H. Bennett, W.S. Laughlin, and J.P. Martin. (1978). Distribution of Albumin Variants Naskapi and Mexico among Aleuts, Frobisher Bay Eskimos, and Micmac, Naskapi, Mohawk, Omaha and Apache Indians. *Am. J. Phys. Anthrop.* 49:111–118.

Scott, E.M., and R.C. Wright. (1969). The Absence of Close Linkage of Methemoglobinemia at Other Loci. *Am. J. Hum. Genet.* 21:194–195.

Spees, E.K., D.D. Kostyu, R.C. Elston, and D.B. Amos. (1973). HL-A Profiles of the Pima Indians of Arizona. *In* J. Dausset and J. Colombani, Eds., *Histocompatibility Testing 1972*. Copenhagen: Munksgaard, pp. 345–350.

Spencer, R.F., and J.D. Jennings. (1965). *The Native American*. New York: Harper and Row.

Steinberg, A.S. (1969). Globulin Polymorphisms in Man. *Ann. Rev. Genet.* 3:25–52.

Steinberg, A.G., A. Tiilikainen, M.R. Eskola, and A.W. Eriksson. (1974). Gammaglobulin Allotypes in Finnish Lapps, Finns, Aland Islanders, Maris (Cheremis) and Greenland Eskimos. *Am. J. Hum. Genet.* 26:223–243.

Szathmary, E.J.E., D.W. Cox, H. Gershowitz, D.L. Ruchnagel, and M.S. Schanfield. (1974). The Northern and Southeastern Ojibwa: Serum Proteins and Red Cell Enzyme Systems. *Am. J. Phys. Anthropol.* 40:49–66.

Tanis, R., R.E. Ferrel, J.V. Neel, and M. Morrow. (1974). Albumin Yanomama-2 a "Private" Polymorphism of Serum Albumin. *Am. J. Hum. Genet.* 38:179–190.

Tittor, W., J. Sobenes, G.S. Smith, P. Sturgeon, E. Zeller, and R.L. Walford. (1973). Distribution of HL-A Antigens, Blood Group Antigens, and Serum Protein Groups in Quechua Indians of Peru. *In* J. Dausset and J. Colombani, Eds., *Histocompatibility Testing 1972*. Copenhagen: Munksgaard, pp. 387–390.

Troup, B.M., R.L. Harvey, R.L. Walford, G.S. Smith, and P. Sturgeon. (1973). Analysis of the HL-A, Erythrocyte and Gammaglobulin Antigen Systems in the Zuni Indians of New Mexico. *In* J. Dausset and J. Colombani, Eds., *Histocompatibility Testing 1972*. Copenhagen: Munksgaard, pp. 339–344.

Van der Does, J.A., J. D'Amaro, A. Van Leeuwen, P. Meera Khan, L.F. Bernibi, E. Van Loghem, L. Nijenhuis, G. Van der Steen, J.J. Van Rood, and P. Rubenstein. (1973). HL-A Typing in Chilean Aymara Indians. *In* J. Dausset and J. Colombani, Eds., *Histocompatibility Testing 1972*. Copenhagen: Munksgaard, pp. 391–396.

Van Loghem, E. (1970). Stability of Gm Polymorphism. *In* R. Grubb and G. Samuelson, Eds., *Human Anti-Human Gammaglobulins*. Oxford: Pergamon Press, pp. 29–37.

Weaver, M.P. (1972). *The Aztecs, Maya and Their Predecessors*. New York: Seminar Press.

Weitkamp, L.R. (1973). The Contribution of Variations in Serum Albumin to the Characterization of Human Populations. *Israel J. Med. Sci.* 9:1238–1248.

Weitkamp, L.R., E.B. Robson, D.C. Shreffler, and G. Corney. (1968). An Unusual Serum Albumin Variant: Further Data on Genetic Linkage Between Loci for Human Serum Albumin and Group Specific Component (Gc). *Am. J. Hum. Genet.* 20:392–397.

Wormington, H.M. (1970). *Prehistoric Indians of the Southwest*. Denver, Colo.: Denver Museum of Natural History.

Zaretskaya, Yu., and V. Fedrunova. (1973). Some General Characteristic of the HL-A System in the Russian Population. *Tiss. Antigens* 3:218–221.

Chapter 6

Dental Traits in Ainu, Australian Aborigines, and New World Populations

Kazuro Hanihara

Introduction

I have studied the dentition of several populations that belong to the Mongoloid, Caucasoid, Negroid, and Australoid racial stocks. In my earlier studies, the investigations focused on nonmetric crown characters for which the frequencies varied from population to population (Hanihara 1970).

In the course of these investigations, I found some crown characters the frequencies of which were higher in Mongoloids than in other populations and I referred to this group of characters as the "Mongoloid dental complex." The Mongoloid dental complex includes the shovel shaping in the upper central incisors, cusp 6, cusp 7, deflecting wrinkle, and protostylid of the lower first molars (Hanihara et al. 1975). In contrast, the Caucasoid populations are, as is well known, characterized by higher frequencies of the Carabelli's cusp in the upper first molars.

Combining frequencies of these six crown characters, I compared populations on the basis of distance coefficients calculated by C.A.B. Smith's method. As a result, it becomes quite evident that the Mongoloid populations, such as the Japanese, Pima Indians, and Eskimos, are grouped into one cluster. Ainu and Australian Aborigines seem to be included in this cluster. Caucasians and American Negroes are, however, located far from each other and from the Mongoloid cluster.

The present study compares the tooth crown measurements of six populations, Japanese, Ainu, Pima Indians, Australian Aborigines, American Caucasians, and American Negroes. The investigations are based on the distance coefficients and principal component analysis and are carried out to analyze affinities between populations.

Materials and Methods

The materials used in this investigation were plaster casts of male permanent dentition in which all the teeth except for the third molars were in good condition. In each population 20 samples were selected at random and the mesiodistal crown diameters were measured by myself. Table 6.1 gives means and standard deviations of each tooth in each population investigated.

Table 6.1.   *Mesiodistal Crown Diameters in the Male Permanent Dentition[a]*

Tooth	Japanese M	SD	Ainu M	SD	Pima M	SD	Australian Aborigine M	SD	Caucasian M	SD	American Negro M	SD
I^1	8.65	0.5257	8.39	0.4246	9.14	0.3268	9.21	0.4025	8.68	0.5700	8.96	0.4905
I^2	7.03	0.5964	7.19	0.3660	7.65	0.3457	7.47	0.5102	6.50	0.6143	7.02	0.7829
C	8.27	0.3469	7.91	0.3227	8.68	0.4599	8.36	0.6460	7.74	0.5102	8.18	0.4217
P^1	7.52	0.5074	6.99	0.3838	7.85	0.5405	7.59	0.5937	7.01	0.4032	7.60	0.4779
P^2	7.00	0.3973	6.48	0.3275	7.50	0.4861	7.14	0.4987	6.68	0.4866	7.21	0.5210
M^1	10.86	0.5366	10.41	0.4537	11.22	0.6003	11.28	0.5484	10.67	0.6816	11.00	0.6199
M^2	9.56	0.5356	9.09	0.5566	10.63	0.6121	10.95	0.5800	10.30	0.6962	10.63	0.5665
I_1	5.59	0.3281	5.38	0.2607	5.90	0.4174	5.74	0.3775	5.40	0.4752	5.53	0.3389
I_2	6.20	0.3663	6.01	0.3355	6.72	0.4683	6.43	0.3420	5.92	0.4913	6.10	0.4111
C	7.18	0.3837	7.10	0.3220	7.68	0.2863	7.40	0.3699	6.84	0.4626	7.21	0.4861
P_1	7.35	0.4135	6.96	0.3069	7.52	0.7562	7.40	0.6341	7.04	0.4198	7.76	0.5083
P_2	7.21	0.4622	6.70	0.3284	7.68	0.5240	7.62	0.5547	7.20	0.4834	7.74	0.5678
M_1	11.59	0.4000	11.23	0.5183	11.89	0.4723	11.97	0.4954	11.23	0.7384	12.02	0.6556
M_2	10.63	0.6458	10.45	0.7015	11.52	0.5435	11.53	0.6507	10.94	0.7119	11.43	0.7526

[a] $N = 20$ in each population. M, mean; SD, standard deviation.

In regard to the distance statistics, Boyce (1969), Sneath and Sokal (1973), and Corruccini (1973) recommend Penrose's shape distance and Q-mode correlation coefficients as the best methods for classification problems. In the present study, therefore, these two statistics were employed for analyzing affinities in dental crown diameters. In addition to this, some other statistical procedures were also used to make affinities among populations more clear. One of the methods was principal component analysis, or PCA, and the other was quantification theory model IV, devised by Hayashi (1952, 1954). Both methods are very similar to each other and are used in order to reduce multiple dimensions to a simple dimension with a minimum loss of information.

Distance Coefficients

First, Penrose's (1954) size and shape distances were calculated. The results are shown in Tables 6.2 and 6.3.

In regard to the size component, larger distances are found between Ainu and Pimas, and between Caucasians and Pimas. In contrast, smaller distances are found between Ainu and Caucasians and between Australian Aborigines and American Negroes. On the basis of size distances, affinities between the six populations can be illustrated in one-dimensional space. The uppermost part of Figure 6.1 was made by using coordinates resulting from analyses based on the quantification theory model IV. In this figure, the populations are simply arranged in the order of tooth size; overall size of the teeth is smallest in Caucasians and largest in Pimas. It is quite interesting that Australian Aborigines do not have the largest teeth among these modern populations; and the Ainu, occasionally said to be similar to Australian Aborigines, have smaller teeth and are rather close to Caucasians where tooth size is concerned. These findings are quite comparable with those obtained from large samples (Hanihara 1976).

Penrose's shape distance gives a quite different picture from that of the size distance. It is likely that Japanese, Ainu, and Pimas represent one cluster, and Australian Aborigines, Caucasians, and American Negroes represent another cluster. This is seen in the middle position of Figure 6.1.

In addition to Penrose's distance, Q-mode correlation coefficients were calculated among the six populations using the mean values of the mesiodistal crown diameters in each tooth.

Table 6.2. *Penrose's Size Distances*

	Japanese	Ainu	Pima	Australian Aborigine	Caucasian	American Negro
Japanese	—					
Ainu	0.4032	—				
Pima	0.9974	2.6689	—			
Australian Aborigine	0.5434	1.8829	0.0684	—		
Caucasian	0.4107	0.0543	2.0088	1.3359	—	
American Negro	0.2285	1.2388	0.3144	0.0672	0.8080	—

Table 6.3. *Penrose's Shape Distances*

	Japanese	Ainu	Pima	Australian Aborigine	Caucasian	American Negro
Japanese	—					
Ainu	0.1044	—				
Pima	0.1357	0.1919	—			
Australian Aborigine	0.3308	0.3878	0.1598	—		
Caucasian	0.4107	0.6320	0.2863	0.0756	—	
American Negro	0.3478	0.5932	0.3615	0.1872	0.1168	—

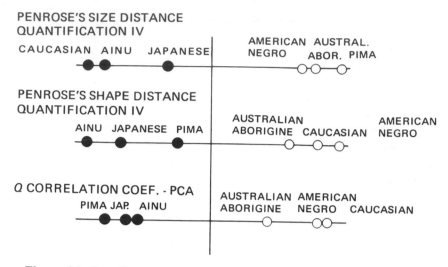

Figure 6.1. One-dimensional expressions based on three distance coefficients.

As is well known, the usual correlation coefficients, or *R*-mode correlation coefficients, give relationships between a pair of measurements or items of observation. In contrast, *Q*-mode correlation coefficients are calculated between a pair of populations, so that they represent similarity or dissimilarity in the shape component between populations. Thus the distances between populations as represented by *Q*-mode correlation coefficients deviate from +1.0 to −1.0, the former representing complete concordance in shape, and the latter complete discordance. Corruccini (1973) and other authors recommend applying a principal component analysis to express the group constellations in a space of reduced dimensions. In this case, eigenvectors of a *Q*-mode correlation matrix can be used as coordinates of one-, two-, or three-dimensional scattergrams.

In the present study, the *Q*-mode correlation coefficients between pairs of the six populations are shown in Table 6.4, and the first and second eigenvectors in Table 6.5. The bottom part of Figure 6.1 has been drawn by using the first eigenvector. In general, *Q*-mode correlation coefficients give a picture very similar to

Table 6.4. *Q-Mode Correlation Coefficients Between Populations*

	Japanese	Ainu	Pima	Australian Aborigine	Caucasian	American Negro
Japanese	—					
Ainu	0.17	—				
Pima	0.45	0.58	—			
Australian Aborigine	−0.50	−0.18	−0.61	—		
Caucasian	−0.61	−0.75	−0.59	0.30	—	
American Negro	−0.81	−0.62	−0.71	0.05	0.30	—

Table 6.5. *Eigenvectors in PCA*

	First Vector	Second Vector
Japanese	−0.35	−0.50
Ainu	−0.43	0.41
Pima	−0.50	0.02
Australian Aborigine	0.32	0.56
Caucasian	0.46	0.03
American Negro	0.37	−0.51

that of Penrose's shape distances, although the arrangements of populations are somewhat different from each other.

As with the Penrose distance statistics, Japanese, Ainu, and Pimas seem to represent one cluster here, and Caucasians, Aborigines, and American Negroes another cluster. In addition, using the first and second eigenvectors, a two-dimensional scattergram can be drawn (Figure 6.2). In this case, 75.6 percent of the total variance is provided by the first two eigenvectors.

It is quite evident, as in the one-dimensional expression, that Japanese, Pimas, and Ainu represent one cluster. However, the Australian Aborigines are located somewhat far from the Caucasians and American Negroes, so that the three clusters are likely to be separated on the basis of the two-dimensional expression of the group constellation, namely, the Japanese–Pima–Ainu cluster, the Caucasian–American Negro cluster, and the Australian Aborigine cluster. A close

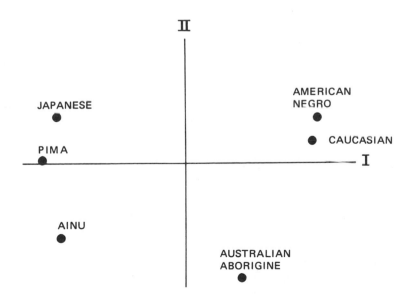

Figure 6.2. Two-dimensional expression based on *Q*-mode correlation coefficients.

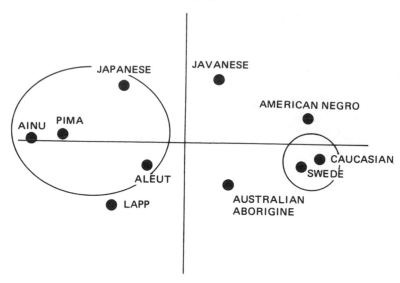

Figure 6.3. Two-dimensional expression based on Q-mode correlation coefficients.

affinity between Caucasians and American Negroes might be partly caused by admixture between the two populations.

The distance analysis also was carried out on some populations additional to the six populations discussed above. The data for the additional populations were taken from Moorrees (1957), and the Q-mode correlation coefficients and PCA were applied.

As is evident from Figure 6.3, Japanese, Ainu, Pimas, and Aleuts seem to represent one large cluster, which may be referred to as the Mongoloid cluster. It is of interest that Lapps are very close to this cluster. On the other side of the figure, American Caucasians and Swedes are located very close to each other so that they seem to be included in another cluster, which might be called the Caucasoid cluster. As already seen, American Negroes are located close to this cluster. Two other populations, Javanese and Australian Aborigines, appear to belong to neither the Mongoloid cluster nor the Caucasoid cluster.

Tooth Crown Diameters as Expressed by Principal Component Scores

As the second step of the investigation, characteristics in the mesiodistal crown diameters in each population were analyzed by applying a PCA.

First, a correlation matrix was obtained from the pooled data for the six populations, and a PCA was applied to this matrix. As a result of this procedure a matrix of rotated factor loadings was obtained (Table 6.6). In this case, all the permanent teeth are explained by the first five components which represent 78.2 percent of the total variance.

Table 6.6. *Rotated Factor Loadings for the First Five Components*

Component Tooth	1	2	3	4	5
I^1	0.2401	−0.3380	0.1494	0.1442	0.6955
I^2	0.1125	−0.8499	0.0932	0.2358	0.3050
$\underline{C}$	0.2508	−0.2207	0.1735	0.7809	0.2277
P^1	0.7861	−0.1815	0.2153	0.2997	0.1695
P^2	0.7990	0.0129	0.2714	0.0566	0.2725
M^1	0.4197	−0.0612	0.6764	0.1262	0.2646
M^2	0.3460	−0.3049	0.7050	−0.1773	0.1663
I_1	0.1905	−0.1186	0.2552	0.1295	0.8303
I_2	0.1906	−0.0836	0.1600	0.3623	0.7864
$\overline{C}$	0.3272	−0.0998	0.1700	0.6983	0.3999
P_1	0.7903	−0.1745	0.2520	0.2313	0.1168
P_2	0.7333	0.0051	0.3585	0.1949	0.2059
M_1	0.2288	0.0956	0.7371	0.2955	0.2118
M_2	0.2014	−0.0812	0.8128	0.1939	0.1032

The first component shows a higher correlation with the premolar size, the second with the upper lateral incisor size, the third with the molar size, the fourth with the canine size, and the fifth with the incisor size except for the upper lateral incisor.

Second, simple equations were set up using larger values of the rotated factor loadings as shown in Table 6.7. The symbols P^1, P^2, etc., in the equations are standardized mesiodistal crown diameters for the corresponding teeth. The five functions, therefore, can be calculated, each of which represents a principal component score. Table 6.8 shows mean values of the scores in each population.

Table 6.7. *Functions for Calculating Principal Component Scores[a]*

Function 1	$Y_1 = 0.7861P^1 + 0.7990P^2 + 0.7903P_1 + 0.7333P_2$
Function 2	$Y_2 = 0.8499I^2$
Function 3	$Y_3 = 0.6764M^1 + 0.7050M^2 + 0.7310M_1 + 0.8128M_2$
Function 4	$Y_4 = 0.7009\underline{C} + 0.6983\overline{C}$
Function 5	$Y_5 = 0.6955I^1 + 0.8303I_1 + 0.7864I_2$

[a] Symbols P, I, M, and C represent standardized mesiodistal crown diameters of corresponding teeth.

Table 6.8. *Principal Component Scores in Each Population*

	1	2	3	4	5
Japanese	−0.0298	−0.1472	−1.0981	0.0292	−0.2966
Ainu	−2.6476	0.0592	−2.5447	−0.6001	−1.3549
Pima	1.9083	0.6591	1.3975	1.3664	1.8098
Australian Aborigine	0.8218	0.4205	1.8281	0.4961	1.1214
Caucasian	−1.6172	−0.8310	−0.8025	−1.2386	−1.0994
American Negro	1.5666	−0.1601	1.2215	−0.0541	−0.1784

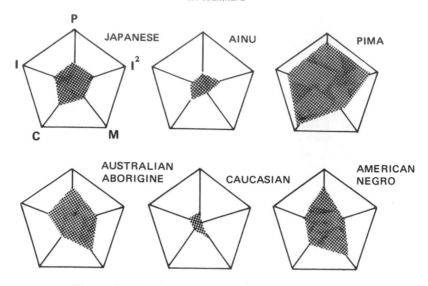

Figure 6.4. Pentagons based on mean PC scores.

Using the five scores, characteristics in the shape component of dentition can be expressed by pentagons as shown in Figure 6.4. The axes of the pentagons connecting the center point and each vertical angle represent each component, in which the innermost point represents -1 standard deviation of the scores, and the outermost point $+1$ standard deviation, so that the middle point of the axes corresponds to the grand mean of the principal component scores. Thus, the mean values of the scores in each population are expressed in standard deviation units, and the dotted areas represent characteristics of populations as expressed by the principal component scores.

Based on the pentagons drawn by such a procedure, the following results are obtained: The dentition of Japanese is of medium overall size and the canines are relatively large; the Ainu dentition is generally small but the upper lateral incisor is relatively large and the molars are particularly small; Pimas have the largest teeth among the six populations, above all the upper lateral incisors and canines are particularly large but the molars are relatively small; Australian Aborigines have an unexpectedly small dentition but the molars are quite large; Caucasians are smallest in overall size, the upper lateral incisors and canines being small in particular but the molars relatively large; American Negroes have relatively large teeth, among which the premolars and molars are particularly large, and the upper lateral incisors and canines are relatively small.

As a whole, it is of interest to note that the shapes of the dotted areas are similar each other among Japanese, Ainu, and Pimas; this is also true of the shapes for Australian Aborigines, Caucasians, and American Negroes. In the former group the upper lateral incisors and canines are larger and the molars smaller in a relative sense, and in the latter group the premolars and molars are larger but the upper lateral incisors are smaller. Such differences between the two groups of populations might be reflected in the scattergrams as shown in Figures 6.1 and 6.2.

Conclusions

The following points can be emphasized in the basis of the distance and principal component analyses of the mesiodistal crown diameters in the permanent dentition:

1. Japanese, Ainu, and Pimas are classified into one cluster that represents Mongoloid characteristics. This conclusion is quite parallel to that from the nonmetric traits of the dentition.

2. Australian Aborigines, Caucasians, and American Negroes may be classified into another cluster, although the size component is different among them. According to a detailed analysis of the distance coefficients, however, Australian Aborigines are somewhat different from Caucasians and American Negroes, so that the former may represent a third cluster.

3. The cluster tentatively referred to as the Mongoloid cluster is characterized by larger front teeth and smaller molar teeth, whereas, the Australian Aborigine–Caucasian–American Negro cluster is characterized by smaller front teeth and larger molar teeth.

4. Although this investigation is mainly concerned with the shape component, I should like to refer to the overall size of the teeth. The tooth size in Australian Aborigines has previously been said to be largest among modern human populations. However, as shown by Penrose's size distances and the pentagons in Figure 6.4, this is not necessarily true. Overall size of the teeth is largest in Pimas and next largest in Australian Aborigines. In Australia, I have obtained the same sort of data from the aboriginal skulls from Queensland. They showed no difference from the dentition of aborigines in Central Australia (Hanihara 1976). This seems to show that the data presented here are quite reasonable. However, the size component of the teeth should be analyzed in more detail in future investigations.

5. The results obtained from analyses of the shape components generally agree with those from the nonmetric crown traits, including the Mongoloid dental complex. This shows that both metric and nonmetric traits in the dentition reveal almost the same trends in comparative studies of different populations. However, there is an apparent discrepancy between the results from the two kinds of dental traits in Australian Aborigines; namely, this population shows a close affinity to Caucasians and American Negroes in the metric traits, and some affinity to Mongoloids in the nonmetric traits. Such a discrepancy may also support the view that Australian Aborigines represent a third cluster as mentioned above.

6. Many more populations need to be analyzed. It is necessary to investigate dentition in the people of Siberia, in the Eskimos, in the Aleuts, and in several tribes of American Indians. In addition to this, the dental traits in the people of southeastern Asia are also quite important in relation to the theme of this volume.

Acknowledgments

I should like to express my sincere gratitude to A.A. Dahlberg of the University of Chicago, A.L. Altemus of Howard University, and the late M.J. Barret of the University of Adelaide for their kind permission to study material from Pima Indians, Caucasians, American Negroes, and Australian Aborigines.

References

Boyce, A. J. (1969). Mapping Diversity: A Comparative Study of Some Numerical Methods. *In* A.J. Cole, Ed., *Numerical Taxonomy.* London: Academic Press. pp. 1–30.

Corruccini, R.S. (1973). Size and Shape in Similarity Coefficients Based on Metric Characters. *Am. J. Phys. Anthropol.* 38:743–753.

Hanihara, K. (1970). Mongoloid Dental Complex in the Deciduous Dentition, with Special Reference to the Dentition of the Ainu. *J. Anthropol. Soc. Nippon,* 78:3–17.

Hanihara, K. (1976). Statistical and Comparative Studies of the Australian Aboriginal Dentition. Bulletin No. 11, The University Museum, the University of Tokyo.

Hanihara, K., T. Masuda, T. Tanaka, and M. Tamada. (1975). Comparative Studies of Dentition. *In Anthropological and Genetic Studies on the Japanese,* Part III: *Anthropological and Genetic Studies of the Ainu.* JIBP Synthesis Vol. 2, pp. 256–262. Tokyo: Univ. of Tokyo Press.

Hayashi, C. (1952). On the Prediction of Phenomena from Qualitative Data and the Quantification of Qualitative Data from the Mathematico-statistical Point of View. *Ann. Inst. Stat. Math.* 3:69–98.

Hayashi, C. (1954). Multidimensional Quantification with the Applications to Analysis of Social Phenomena. *Ann. Inst. Stat. Math.* 5:121–143.

Moorrees, C.F.A. (1957). *The Aleut Dentition.* Cambridge, MA: Harvard Univ. Press.

Penrose, L.S. (1954). Distance, Size and Shape. *Ann. Eugen.* 18:337–343.

Sneath, P.H.A., and R.R. Sokal. (1973). *Numerical Taxonomy, The Principles and Practice of Numerical Classification.* San Francisco: W.H. Freeman.

Chapter 7

Genetic Distances, Trees, and Maps of North American Indians

James N. Spuhler

Introduction

The purpose of this chapter is to analyze the variation in blood group gene frequencies among samples of local, North American Indian populations in order to explore their historical biologic affinities.

Man may have more than 100,000 structural gene loci in his chromosomes (Bodmer and Cavalli-Sforza 1976, p. 132). Many of these are polymorphic in most human breeding populations; that is, the second most frequent allele has a relative frequency greater than 1 percent (Bodmer and Cavalli-Sforza 1976, p. 308). Gene frequencies in local populations change by mutation, gene flow, natural selection, and genetic drift. Continued operation of the microevolutionary processes causes local breeding populations to differ in gene frequencies (Cavalli-Sforza and Bodmer 1971, Crawford and Workman 1973). The frequencies of 9 to 25 of these polymorphic genes are accurately known for the Indian groups considered here.

Similarity of gene frequencies between local breeding populations may result from recent common ancestry, gene exchange, or convergence through natural selection in similar environments. Mutation is the ultimate source of all gene variation, but given inherited variation, differential mutation is not important in explanations of genetic differences between local human populations. The geography of gene frequencies among North American Indians depends largely on gene flow between local populations as one major factor, on genetic drift as another, and possibly on differing local selection intensities as a third force. If there is much gene flow, local races do not develop and a regional or continental population may remain genetically uniform; if there is less gene flow, clines in gene frequencies may form; if still less gene flow, local races may differentiate through genetic drift and selection (Wright 1968, 1969, 1977). Clinal distribution of single-locus gene frequencies may result from gene flow, differences in selection intensity, or (in genetically structured populations) from genetic drift. If clines result historically from gene flow and neither from selection nor drift, the gene frequencies at several loci should show related clinal distribution if the parental populations differ in the frequency of the genes being exchanged (Bodmer and Cavalli-Sforza 1971, Spuhler 1973).

Gene flow may rapidly change gene frequencies in a mixed population. If historical information is available for the parental populations, rates of gene flow can be accurately measured (Balakrishnan 1974, Chakraborty 1975, Elston 1971, Szathmary and Reed 1972, Thompson 1973). For most gene frequencies of anthropologic utility in establishing historical affinities, the rates of change resulting from selection are too small to be measured over a few generations— human population geneticists make natural selection a primary evolutionary operator for theoretical rather than human empirical grounds. Theoretical models for genetic drift are well established and these models, along with known or suspected gene flow, are made to account for much of the observed gene geography in North American Indian populations (Cavalli-Sforza 1973, Spuhler 1951, 1972, 1973).

Linguistic, Geographic, and Genetic Data

The names of the 62 North American Indian groups whose gene frequencies, linguistic classification, cultural area, and geographic relationships are considered in this chapter are listed below, ordered according to the linguistic classification of Voegelin and Voegelin (1966). Language phyla are designated by Roman numerals, families by Roman numbers plus lower case letters, and languages by Arabic numerals (numbers and names of families lacking gene frequency are omitted).

Names and identification numbers of the groups are in italics. The numbers for 12 samples are followed by *a* or *b*, (for example *14a*); 11 of these represent samples—usually smaller—from a closely related group belonging to the same tribe; the sample from Penobscot (*28b*) is included as an example of a group experiencing considerable European admixture over several generations. None of these 12 samples is included in the "overall" continental analyses by language family and culture area, but each is considered in discussions of individual families and areas. Additional information is keyed as follows:

a. The number of individuals in the blood group sample, N; the number of genes or gene complexes identified, G; the number of living individuals belonging to the group or tribe, P (with the date of the census). The population numbers are taken from the original reference or from Swanton (1952), Landar (1973), or the 1970 United States Census.

b. Geographic coordinates of the group. Where a place name is given the coordinates refer to that place, usually where all or a major part of the blood samples representing the group were collected. Otherwise the coordinates refer to the estimated geographic or population center of the group, usually based on maps in Kroeber (1939), Swanton (1952), Jenness (1955), and Driver (1969).

c. Culture area of the group.

d. Reference to the original source giving gene frequency estimates for the group.

e. For each of the 53 groups included in the "overall" calculations (see pp. 145–147), the mean genetic distance and the mean correlations (defined below, p. 147) of the group with all the other 52 groups are given. The overall mean between-group genetic distance is 0.7162 ± 0.1438 with skewness 1.1298 and kurtosis 3.7347 and the range is from 0.5246 to 1.1217. The overall mean of the mean correlation is $+0.2838 \pm 0.1438$, with skewness -1.1297 and kurtosis 3.7346 and the range is from -0.1217 to $+0.4754$.

f. Other comments.

I. AMERICAN ARCTIC–PALEO-SIBERIAN PHYLUM

Ia. Eskimo–Aleut Family

1. Central-Greenlandic Eskimo

   1. North Alaska Coast Eskimo. (a) N = 241, G = 15, P = 34,000 (1973). (b) 68°20′N, 159°00′W, Brooks Range, Alaska. (c) Arctic. (d) Allen (1959). (e) 0.813, +0.187. (f) The samples were obtained from Wainwright, Barrow, near Summit Lake (Anaktuvuk Pass, Brooks Range), and Beaver.

   2. Copper Eskimo. (a) N = 320, G = 15, P = ca. 1000 (1958). (b) 69°15′ N, 105°00′W, Cambridge Bay, Canada. (c) Arctic. (d) Chown and Lewis (1959). (e) 0.732, +0.268. (f) Samples collected at Cambridge Bay, Holman Insand, Reid Island, and Coppermine.

   3. Southampton Eskimo (Iglulik). (a) N = 130, G = 15, P = 215 (1959). (b) 65°N, 85°W, Southampton Island. (c) Arctic. (d) Chown and Lewis (1960). (e) 0.777, +0.223. (f) Only "purebloods" were sampled. Settled partly from Repulse and Wager Bay areas around 1908 and Coats Island and southern Baffin Island about 1924.

   4. East Arctic Eskimo (Ungava District). N = 514, G = 15, P = ca. 1000 (1960). (b) 60°N, 75°W. (c) Arctic. (d) Chown and Lewis (1956). (e) 0.692, +0.308.

   5. James Bay Eskimo (Inuit). (a) N = 64, G = 15, P = ? (b) 59°12′N, 77°51′W, Moose Factory, east coast of Hudson Bay. (c) Arctic. (d) Chown and Lewis (1956). (e) 0.714, +0.286. (f) Samples obtained at Old Factory, Richmond Gulf, Great Whale, Povungnetuk, Belcher Island, Fort Harrison, and Fort George. Little mixture with whites, practically none with Indians.

   6. West Greenland Eskimo. (a) N = 186, G = 15, P = 17,000 (1973). (b) 64°10′N, 51°32′W, Southwest Greenland. (c) Arctic. (d) Ahrengot and Eldon (1952). (e) 0.703, +0.297.

   7. Thule Eskimo (Ita). (a) N = 152, G = 15, P = 700 (1970). (b) 78°20′ N, 74°42′W, Etah, Greenland. (c) Arctic. (d) Gürtler (1971). (e) 0.892, +0.108.

8. East Greenland Eskimo. (a) N = 246, G = 15, P = 3000 (1970).
(b) 70°30'N, 22°W, Scoresbysund, Greenland. (c) Arctic. (d) Fernet *et al.* (1977). (e) 0.613, +0.387.
9. Siberian Eskimo (Asiatic). (a) N = 124, G = 15, P = 1000 (1973).
(b) 64°25'N, 172°15'W, Chaplino, Chukotskiy Peninsula. (c) Arctic.
(d) Rychkov and Sheremetyeva (1972). (e) 0.790, +0.210. (f) The sample includes 23 from Naukan, 46 from Chaplino, and 55 from Sireniki. Gene frequencies adjusted slightly by new maximum likelihood estimates.
10. Commander Island Aleut. (a) N = 59, G = 15, P = 400 (1970).
(b) 50°40'N, 167°13'W. (c) Arctic. (d) Rychkov and Sheremetyeva (1972a). (e) 0.666, +0.334. (f) Sample includes 39 from Bering and 20 from Medny Island. Gene frequencies adjusted slightly by new maximum likelihood estimates.
11. Western Aleut. (a) N = 142, G = 15, P = 100 (1970). (b) 52°40'N, 177°30'W. (c) Arctic. (d) Laughlin (1951, 1957), Mourant *et al.* (1976).
(e) 0.627, +0.373. (f) The sample includes 41 from Atka, Attu; 47 from "Inter Island"; and 54 from Nikolski, Unalaska. The rhesus results reported by Mourant *et al.* (1976; cited as personal communication from Laughlin 1957) include the same individuals whose ABO and MN groups were reported by Laughlin (1951).

Ib. Chukchi–Kamchatkan Family
12. Coastal Chukchi. (a) N = 96, G = 15, P = 12,000 (1936). (b) 65°49'N, 170°32'W, Nunyama, Chukotskiy Peninsula. (c) Arctic (Siberia).
(d) Rychkov and Sheremetyeva (1972b). (e) 0.794, +0.206. (f) The sample includes 48 from Nunyamo and 48 from Sireniki. Gene frequencies adjusted slightly by new maximum likelihood estimates.

II. NA-DENÉ PHYLUM

IIa. Athapaskan Family

2. Slave
13. Slave. (a) N = 156, G = 25, P = 12,000 (1962). (b) 61°N, 120°W.
(c) Subarctic. (d) Alfred *et al.* (1972a). (3) 0.647, +0.353. (f) Data collected at Upper Liard Reserve, near Watson Lake, Yukon. Gene frequencies by maximum likelihood estimates.
3. Kutchin
14. Kutchin (Old Crow). (a) N = 92, G = 25, P = ca. 900 (1970).
(b) 67°15'N, 139°58'W, Old Crow, Yukon, Alaska. (c) Subarctic.
(d) Lewis *et al.* (1961). (d) 1.122, −0.122; maximum and minimum, respectively, of the overall series.
14a. Kutchin (Brooks Range). (a) N = 78, G = 25, P = 258 (1958). (b) 68° 20'N, 158°W, Brooks Range. (c) Subarctic. (d) Allen (1959). (e) —, —.
(f) The majority were regarded as full blood.
14b. Athapaskan (Northern). (a) N = 204, G = 25, P = ? (b) 55°N,

(b) 68°20′N, 158°W, Brooks Range. (c) Subarctic. (d) Allen (1959). (e) —, —. (f) The majority were regarded as full blooded.

14b. Athapaskan (Northern). (a) N = 204, G = 25, P = ?. (b) 55°N, 125°W, British Columbia. (c) Subarctic. (d) Lewis and Chown (1952). (e) —, —.

4. Tuchone

15. Tuchone. (a) N = 92, G = 25, P = 500 (1970). (b) 62°N, 140°W. (c) Subarctic. (d) Alfred *et al.* (1972a). (e) 0.542, +0.458. (f) Samples collected in the Ross River reserve, Alaska. Gene frequencies recalculated by maximum likelihood.

5. Beaver–Sarsee

16. Beaver. (a) N = 53, G = 25, P = 300 (1962). (b) 57°N, 115°W. (c) Subarctic. (d) Alfred *et al.* (1972b). (e) 0.741, +0.259. (f) Data collected from groups near Doig and Prophet Rivers, British Columbia. Gene frequencies newly estimated by maximum likelihood.

17. Sarsee. (a) N = 95, G = 25, P = 1000–2000 (1962). (b) 51°10′N, 114°51′W, near Morley, Alberta, Canada. (c) Plains (late), Subarctic (early). (d) Chown and Lewis (1955a, b). (e) 0.731, +0.269. (f) Included in Plains culture area in "overall" computations.

6. Carrier–Chilcotin

18. Carrier. (a) N = 104, G = 25, P = 1000–3000 (1962). (b) 54°N, 127°W. (c) Subarctic. (d) Chown and Lewis (1952). (e) 0.643, +0.357. *19. Chilcotin.* (a) N = 186, G = 25, P = 500–1000 (1962). (b) 53°N, 123°W. (c) Subarctic. (d) Alfred *et al.* (1972b). (e) 0.690, +0.310. (f) Also named Anaham.

11. Hupa

20. Hupa. (a) N = 353, G = 25, P = 998 (1970). (b) 41°18′N, 123°35′W, Hupa Reservation, California. (c) California. (d) Hulse (1960). (e) 0.672, +0.328.

15. Navaho

21 Central Navajo. (a) N = 237, G = 16, P = 35,700 (1970) [one-half total on reservation in Arizona]. (b) 36°06′N, 110°14′W, Piñon, Arizona. (c) Southwest. (d) Corcoran *et al.* (1962). (e) 0.579, +0.421.

21a. Western Navajo. (a) N = 106, G = 25, P = 35,700 (1970) [one-half total on reservation in Arizona]. (b) 36°09′N, 111°09′W, Tuba City, Arizona. (c) Southwest. (d) Brown *et al.* (1958). (e) —, —. (f) The rhesus frequencies reported by Brown *et al.* (1958) differ significantly from those reported by others on relatively unmixed samples from Southwestern groups, perhaps because of technical error or inclusion of mixed individuals (see Gershowitz 1959, p. 199, fn. 2).

21b. Eastern Navajo. (a) N = 110, G = 16, P = 87,438 (1969) [total on reservation in New Mexico]. (b) 35°41′N, 108°09′W. Crown Point, New Mexico. (c) Southwest. (d) Yeung and Spuhler (1975). (e) —, —.

16. San Carlos Apache

22a. San Carlos Apache (Western Apache). (a) N = 179, G = 16, P = 4709

(1970). (b) 33°22′N, 110°29′W, San Carlos, Arizona. (c) Southwest. (d) Kraus and White (1956). (e) —, —. (f) MN frequencies estimated.
17. Chiricahua–Mescalero Apache
    22b. Chiricahua Apache. (a) N = 23, G = 16, P = >1000 (1962). (b) 31° 50′N, 109°17′W, Chiricahua Peak, Arizona. (c) Southwest. (d) Gershowitz (1959). (e) —, —. (f) Sampled on Mescalero Reservation, New Mexico.
    22. Mescalero Apache. (a) N = 47, G = 16, P = 1317 (1969). (b) 33°07′N, 105°44′W, Mescalero, New Mexico. (c) Southwest. (d) Gershowitz (1959). (e) 0.618, +0.382.

IIb. Tlingit Language Isolate
    23. Tlingit. (a) N = 198, G = 21, P = 4462 (1930). (b) 57°08′N, 135° 18′W. (c) Northwest Coast. (d) Corcoran *et al.* (1959). (e) 0.671, +0.329.

IIc. Haida Language Isolate
    24a. Haida (Skidegate). (a) N = 153, G = 21, P = 588 (total Queen Charlotte Islands, 1930). (b) 53°13′N, 132°02′W, Skidegate, Queen Charlotte Islands. (c) Northwest Coast. (d) Thomas *et al.* (1964). (e) —, —.
    24. Haida (Old Masset). (a) N = 284, G = 21, P = 588 (total Queen Charlotte Islands, 1930). (b) 54°02′N, 132°09′W, Old Masset, Queen Charlotte Islands. (c) Northwest Coast. (d) Thomas *et al.* (1964). (e) 0.633, +0.367.

III. MACRO–ALGONQUIAN PHYLUM

IIIa. Algonquian Family
  1. Cree–Montagnais–Naskapi
    25. Cree (James Bay). (a) N = 166, G = 24, P = 4270 (1955). (b) 53°40′ N, 73°34′W, Moose Factory, James Bay, Canada. (c) Subarctic. (d) Chown and Lewis (1956). (e) 0.524, +0.475; minimum and maximum, respectively, of the overall series.
    25a. Cree (Eastern). (a) N = 50, G = 24, P = ? (b) 53°N, 91°W. (c) Subarctic. (d) Allegro *et al.* (1972). (e) —, —.
    26. Montagnais. (a) N = 84, G = 24, P = 5000 (1962). (b) 59°52′N, 67°01′W, Schefferville, Quebec. (c) Subarctic. (d) Blumberg *et al.* (1964). (e) 0.689, +0.311.
    27. Naskapi. (a) N = 152, G = 24, P = 2500 (1858). (b) 54°52′N, 67°01′W. (c) Subarctic. (d) Blumberg *et al.* (1964). (e) 0.696, +0.304. (f) Sampled at Schefferville, Quebec.
  6. Ojibwa–Ottawa–Algonquian–Salteaux
    28. Ojibwa (Southwestern Chippewa). (a) N = 491, G = 24, P = 2378 (1960). (b) 47°39′N, 99°27′W, Great Leach Lake Indian Reservation, Minnesota. (c) Northeast. (d) Rokala (1971). (e) 0.575, +0.425.
    28a. Southwestern Chippewa. (a) N = 161, G = 24, P = 2741 (1970).

(b) 48°09′N, 95°55′W, Red Lake, Minnesota. (c) Northeast. (d) Matson *et al.* (1954). (e) —, —.

8. Penobscot–Abenaki

28b. Penobscot (Eastern Abenaki). (a) N = 229, G = 24, P = 301 (1930). (b) 44°55′N, 68°42′W, Old Town, Maine. (c) Northeast. (d) Allen and Corcoran (1960). (e) —, —.

11. Blackfoot–Piegan–Blood

29. Blackfoot. (a) N = 39, G = 24, P = 1250 (1952). (b) 53°N, 112°W. (c) Plains. (d) Chown and Lewis (1953). (e) 0.765, +0.235. (f) Samples from Blackfoot Reservation, Alberta, Canada.

30. Blood (Blackfoot). (a) N = 241, G = 24, P = 2002 (1951). (b) 52°N, 114°W. (c) Plains. (d) Chown and Lewis (1953). (e) 0.702, +0.298. (f) Samples from Blood Reservation, Alberta, Canada.

IV MACRO-SIOUAN PHYLUM

IVa. Siouan Family

7. Dakota

31. Sioux. (a) N = 408, G = 19, P = 33,625 (1937). (b) 43°33′N, 102°13′ W, Pine Ridge Indian Reservation, South Dakota. (c) Plains. (d) Workman (personal communication, 1975). (e) 0.559, +0.441. (f) The samples include 265 from Pine Ridge and 143 from Rosebud Reservations, South Dakota.

32. Assiniboine (Stoney). (a) N = 145, G = 19, P = 1000–2000 (1962). (b) 52°N, 105°W. (c) Plains, earlier in Subarctic. (d) Chown and Lewis (1955a). (e) 0.584, +0.416. (f) Some modern linguists assign Assiniboine to a separate language family.

IVc. Iroquoian Family

33. Cherokee. (a) N = 78 "full blood," G = 13, P = 4494 (1960). (b) 35°33′N, 83°12′W, Cherokee Indian Reservation, North Carolina. (c) Southeast. (d) Pollitzer *et al.* (1962). (e) 0.737, +0.263. (f) "Full bloods" selected on basis of tribal records covering seven generations and removal of individuals with ABO alleles A_2 and B and rhesus chromosomes with $D-$.

IVd. Caddoan Family

1. Caddo

34. Caddo. (a) N = 47 (18 "full blood"), G = 19, P = 250 (1960). (b) 34°N, 95°W. (c) Plains. (d) Gray and Laughlin (1960). (e) 0.573, +0.427. (f) Individuals classed as "full blood" had all four grandparents listed as Caddo tribal members.

2. Wichita

35. Wichita. (a) N = 49, G = 19, P = 485 (1970). (b) 38°N, 98°W. (c) Plains. (d) Gray and Laughlin (1960). (e) 0.581, +0.419.

3. Pawnee–Arikara
36. Pawnee. (a) N = 80, G = 19, P = 1928 (1970). (b) 42°N, 99°W.
(c) Plains. (d) Gray and Laughlin (1960). (e) 0.560, +0.440.

V. HOKAN PHYLUM
Va. Yuman Family
2. Up-River Yuman
37. Maricopa. (a) N = 124, G = 14, P = 500 (1962). (b) 33°04'N,
111°44'W, Gila Crossing, Arizona. (c) Southwest. (d) Matson *et al.* (1968).
(e) 1.055, −0.055.
38. Yuma (Quechan). (a) N = 182, G = 14, P = 1000 (1962). (b) 32°40'
N, 114°40'W, Fort Yuma, California. (c) Southwest. (d) Brown *et al.*
(1958). (e) 0.627, +0.373. (f) See note (f) under *21a* above.
4. Southern and Baja Californian Yuman
39. Diegueño (Ipai). (a) N = 58, G = 14, P = 322 (1930). (b) 33°05'N,
116°46'W, Santa Ysabel Reservation, California. (c) California. (d) Pantin and Kallsen (1953). (e) 0.794, +0.206.

VI. PENUTIUM PHYLUM
VIf. Sahaptin–Nez Perce Family
40. Yakima. (a) N = 95, G = 14, P = 1000–2000 in Washington State
(1962). (b) 46°35'N, 120°30'W, White Swan, Yakima Indian Reservation. (c) Plateau. (d) Hulse (1957). (e) 0.811, +0.189.
VIo. Zuni Language Isolate
41. Zuni. (a) N = 662, G = 25, P = 7306 (1970). (b) 35°05'N, 108°52'
W, Zuni, New Mexico. (c) Southwest. (d) Workman *et al.* (1974). (e) 0.750,
+0.250.

VII. AZTEC–TANOAN PHYLUM

VIIa. Kiowa–Tanoan Family
1. Tiwa
42. Isleta. (a) N = 24, G = 16, P = 1974 (1969). (b) 34°55'N, 106°45'W,
Isleta Pueblo, New Mexico. (c) Southwest. (d) Yeung and Spuhler (1975).
(e) 0.608, +0.392.
2. Tewa
43. Hopi–Tewa, Hano. (a) N = 123, G = 16, P = 500 (1966). (b) 35°53'
N, 110°39'W, Oraibi, Arizona. (c) Southwest. (d) Brown *et al.* (1958).
(e) 0.874, +0.126. (f) See note (f) under *21a* above.
3. Towa
44. Jemez. (a) N = 26, G = 16, P = 1076 (1969). (b) 35°37'N, 106°44'
W, Jemez Pueblo, New Mexico. (c) Southwest. (d) Yeung and Spuhler
(1975). (e) 0.559, +0.441.

VIIb. Uto–Aztecan Family
5. Hopi
43. Hopi–Tewa, Hano. See above under Tewa.

11. Pima–Papago

45. Pima. (a) N = 904, G = 16, P = 8240 (1970). (b) 32°00′N, 111°00′ W, Pima Reservation, Sacaton, Arizona. (c) Southwest. (d) Matson *et al.* (1968). (e) 0.583, +0.471.

46. Papago. (a) N = 709, G = 16, P = 5644 (1970). (b) 32°33′N, 112° 12′W, Sells, Arizona. (c) Southwest. (d) Niswander *et al.* (1970). (e) 0.598, +0.402.

14. Tarahumara

47. Tarahumara (Taracahitian). (a) N = 97, G = 16, P = 25,000 (1963). (b) 26°48′N, 107°03′W, Guachochi, Sierra Tarahumara, Chihauhua. (c) Southwest. (d) Rodriguez *et al.* (1963). (e) 0.620, +0.380. (f) Sample includes 47 from Guachochi and 50 "nomades" from the Sierra. The authors estimate 19.77 percent European admixture. The text follows Swanton's (1952, p. 636) spelling, "Tarahumare."

15. Cora

48. Cora. (a) N = 96, G = 16, P = 7000 (1960). (b) 21°50′N, 104°42′W. (c) Southwest. (d) Cordova *et al.* (1967). (e) 0.667, +0.333.

16. Huichol

49. Huichol. (a) N = 71, G = 16, P = 7043 (1959). (b) 22°02′N, 104°07′ W. (c) Southwest. (d) Cordova *et al.* (1967). (e) 0.996, +0.004.

VIII. UNDETERMINED PHYLUM AFFILIATIONS

VIIIa. Keres Language Isolate

50. Laguna. (a) N = 34, G = 16, P = 2956 (1969). (b) 35°03′N, 107°24′ W, Laguna Pueblo, New Mexico. (c) Southwest. (d) Yeung and Spuhler (1975). (e) 1.062, −0.062.

VIIIg. Salish Family

4. Okanagon–Sanpoil–Coville–Lake

51. Okanagon. (a) N = 72, G = 14, P = 1000–2000 in Washington State (1962). (b) 48°55′N, 119°25′W, Oroville, Washington (c) Plateau. (d) Hulse (1957). (e) 0.814, +0.186. (f) "Indian" sample selected to exclude non-Indian groups.

11. Snoqualmi–Duamish–Nisqualli

52. Swinomish (Skagit). (a) N = 102, G = 14, P = 200–300 (1962). (b) 48°25′N, 122°27′W, Swinomish Indian Reservation, Washington. (c) Northwest Coast. (d) Hulse (1957). (e) 0.829, +0.171. (f) "Indian" sample selected to exclude non-Indian groups.

VIIIh. Wakashan Family

1. Nootka

53. Nootka. (a) N = 198, G = 21, P = 669 (1968). (b) 49°37′N, 126°05′ W, Ahousat, Flores Island, off west coast of Vancouver Island, Canada. (c) Northwest Coast. (d) Alfred *et al.* (1969). (e) 1.064, −0.064.

These 53 North American Indian groups belong to eight language phyla (all phyla recognized by Voegelin and Voegelin, 1966, for the area in North America covered by this volume). They represent 12 of the 36 language families and 4 of the 31 language isolates in the area. Therefore, 67 percent of North American language families and 87 percent of language isolates are not represented in this study because gene frequency data are lacking. However, the 16 language families and isolates included a 65 percent sample (population data modified from Mooney by Kroeber, 1939, pp. 174–175).

Gene Frequency Data

The raw data analyzed in this chapter are population frequencies of genes or gene complexes controlling red blood cellular antigens. Most of the frequencies are tabulated in detail by Mourant *et al.* (1976); the sources of recently published and unpublished frequencies are given in the list of tribes. Theoretical and practical aspects of blood group genetics are clearly presented by Cavalli-Sforza and Bodmer (1971), Giblett (1969), Mourant (1954), Mourant *et al.* (1976), and Race and Sanger (1975).

Limits of space and available data restrict the sample for overall analyses to 53 populations representing 50 North American Indian or Eskimo–Aleut tribes and three Siberian groups. The groups included are selected from 83 tribes resident in the continent (79 of these are listed in Post *et al.* 1968; data for Isleta, Jemez, Laguna, Sioux, and Zuni are from later sources) based on about 170 separate investigations. Most of the omitted tribes are not usable because they are not observed for the minimum set of 13 gene frequencies or because of considerable mixture with Europeans or Africans (see section on mixture below).

Six sets of gene or haplotype frequencies are used in this study. The minimum set including 13 frequencies (A, B, and O of the ABO series; M and N of the MNSs series; $R^1 + R'$, $R^2 + R''$, $R^0 + r$, and $R^z + r^y$ of the rhesus haplotypes) are used in the 53-group overall analyses, in the Eskimo language family, as well as in the Arctic, California, Eastern, and Southwest culture areas.

Frequencies of two Diego genes, Di^a and Di^b, are added to the minimum set in computations of the Aztec–Tanoan language phylum, making a total of 15 gene frequencies

A set of 21 gene frequencies is used for the Na-Dené, Penutian, Keresan, and Wakashan language phyla, and for the Subarctic and Northwest Coast culture areas by substituting MS, Ms, NS, and Ns for M and N and by adding two Duffy [Fy^a, $Fy^b + Fy$], two Kell [K, k], and two Diego frequencies to the minimum set.

The maximum set, including 24 gene frequencies [as in the Na-Dené set except substitution of A^1 and A^2 for A and addition of two Lutheran frequencies, Lu^a and Lu^b], is used to study the genetic affinities of the Algonquian-speaking tribes.

The 19 gene frequencies employed to study the Siouan language phylum and the Plains culture area are identical with the Na-Dené set excluding the two Diego genes.

The 17 gene frequencies used in the Hokan language phylum analysis are identical with the Na-Dené set except that the Diego and Kell frequencies are absent.

In part, therefore, differences in the results from studies on the six culture areas and the nine language stocks, on the one hand, and from the pooled 53 member groups, on the other hand, result from differences in the kinds and numbers of genes in the several gene frequency sets. Given sample sizes of at least 100, improved results are obtained by increasing the number of gene frequencies (loci) tested on each individual rather than by increasing the number of individuals tested.

Gene Frequency Statistical Analyses

Numerical methods used to infer biologic history and affinity from the study of gene frequencies are well developed (Sneath and Sokal 1973) and have been applied to several tribes (Ward and Neel 1970, Harpending and Jenkins 1974, Morton 1973, Workman and Niswander 1970, Workman *et al.* 1974), regions (Harpending and Jenkins 1973, Morton *et al.* 1973), continents (Spuhler 1972), the major races of man (Lewontin 1972, Nei and Roychoudhury 1972), and the world (Cavalli-Sforza and Edwards 1967).

The first step in these studies is to convert gene frequencies into genetic distances. A genetic distance reduces to a single number gene frequency data on as many as 96 chromosomal loci (e.g., in Nei and Roychoudhury 1972) observed on thousands of individuals from population pairs. There are many measures of genetic distance; all are mathematically related, and most are highly correlated with the others (Gower 1972, Goodman 1974). I use two measures of genetic distance: D_{ij} (a numerical example of D_{ij} is given by Harpending and Jenkins 1973) and a measure based on the logit transformation, d_{jk}:

$$D_{ij} = \text{VAR } (X_i) + \text{VAR } (X_j) - 2 \text{ COV } (X_i, X_j) \qquad (7.1)$$

for populations X_i and X_j.

The *total genetic distance* is computed by taking the arithmetic mean of all pairwise distances.

The *within-group distance* is computed by taking the means of all pairwise distances within each defined group and averaging these group means to form a grand mean.

The *between-group distance* is computed by taking the mean gene frequency for each group and computing distances between these means.

D_{ij} is the distance measure used as input for the maximum and minimum cluster analyses.

The logit centroid tree uses gene frequencies transformed by the logit function:

$$f(p) = \log_e [p/(1 - p)] \tag{7.2}$$

as a measure of genetic distance; that is, the distance d_{jk} between populations j and k is:

$$d_{jk} = [f(p_j) - f(p_k)] = \log_e [p_j/(1 - p_j)] - \log_e [p_k/(1 - p_k)] \tag{7.3}$$

There are many computer programs that convert genetic distances into cladograms, dendrograms, or genetic trees (Cavalli-Sforza and Bodmer 1971, Sneath and Sokal 1973). All are restricted to evolutionary divergence by bifurcation and all neglect gene flow; all reduce and somewhat distort the raw data given in tables of genetic distances in order to locate population points in two (or at most three) dimensions. A cophenetic correlation coefficient may be used to estimate the distortion between the tree and the unreduced matrix of distances (Sneath and Sokal 1973)—this correlation is above +0.9 for most of the trees discussed in this chapter, the lowest cophenetic correlation included here being +0.68 in the Numerical Taxonomy System (NT-SYS) phenogram for 53 groups based on nine gene frequencies.

Agglomerative trees (see below) are most economical of computer time; the nonagglomerative routines (for instance, that of Cavalli-Sforza and Edwards 1967) are statistically more elegant but prohibitively expensive in computing costs for more than a dozen or so populations.

In this study I use four different agglomerative programs: the logit centroid (Spuhler *et al.* 1972), the maximum linkage, the minimum linkage (Sneath and Sokal 1973), and the NT-SYS phenogram program (Rohlf *et al.* 1972). The trees graphed by the four techniques are usually similar in broad topology but differ in detail. Harpending and Jenkins (1973) give an easy to follow illustration of how to make maximum and minimum linkage trees. The logit centroid tree employs an agglomerative method; all pairwise contemporary populations are searched to find the two daughter populations with minimum age of separation (genetic distance); this age is assigned to the first interior node of the tree with defined sample size, gene frequencies, and standard deviation. The process is repeated on the remaining $n - 2$ populations iteratively until the root of the tree is reached. Availability of estimated Ur-frequencies of the stem population for each centroid tree makes possible the combination of several trees into a composite tree. Details on the graphing of logit centroid trees are given in Spuhler *et al.* (1972).

A genetic map is a two-dimensional interpretation of a genetic distance matrix in which observed gene frequencies are transformed to new frequencies that are imaginary, scaled deviations from the sample mean such that the imaginary gene frequencies are uncorrelated and the populations may be plotted on axes representing the two or three most variable imaginary genes. In several respects, a genetic map gives a more satisfactory picture of biologic affinities among several populations than does a genetic tree (Morton *et al.* 1974). Computational details used in genetic mapping are presented in Dixon (1973), Nie *et al.* (1970), Sneath and Sokal (1973), and Tatsuoka (1971).

Two mapping methods are used in this study: principal components analysis (PCA), and NT-SYS cluster analysis (Rohlf *et al.* 1972). Most of the phyla and areas were also mapped with nonmetric multidimensional scaling (Kruskal 1964), giving distributions highly correlated with the PCA maps. The distance measure used as input for the NT-SYS cluster analysis of p variables in populations i and j is:

$$D_{ij} = \sqrt{\left[\sum_{k=1}^{p} \frac{(X_{ik} - X_{jk})^2}{p} \right]} \qquad (7.4)$$

The cluster algorithm applied to the D_{ij} distances is a weighted pair-group method, using arithmetic means for linkage (WPGMA).

A stepwise discriminant analysis (BMDO7M, Dixon 1973) is used to estimate within and between genetic distance for language phyla and culture areas as well as the posterior probabilities that a given group belongs to each of the several phyla or areas.

In canonical correlations (BMDO7M, Dixon 1973) observed variates x_i are transformed linearly into theoretical variates λ_i so that the members of each group of variates are independent among themselves, each member of one group is independent of all but one member of the other, and the correlations between members of different groups are maximized. The two sets of transformed variates are called canonical variates. Discriminant analysis finds linear combinations of a set of variables that best differentiate among several groups; canonical analysis makes a linear combination of each of two sets of variables that correlate maximally with each other (Tatsuoka 1971).

Indian and Non-Indian Gene Flow

Table 7.1 and Figure 7.1 give a simple, hypothetical, numerical example of some consequences of gene flow from other Indian and from European or African sources. Frequencies of the ABO, MN, and rhesus D system are given in Table 7.1.

Table 7.1. *Hypothetical Examples of Indian and Non-Indian Gene Flow*

Group	Gene Frequencies							
	A_1	A_2	B	O	M	N	$D+$	$D-$
X	0.00	0.00	0.00	1.00	0.90	0.10	1.00	0.00
Y	0.00	0.00	0.00	1.00	0.85	0.15	1.00	0.00
Z	0.38	0.00	0.00	0.62	0.80	0.20	1.00	0.00
E	0.19	0.06	0.07	0.68	0.55	0.45	0.60	0.40
$\frac{1}{2}$X + $\frac{1}{2}$Z	0.19	0.00	0.00	0.81	0.85	0.15	1.00	0.00
$\frac{1}{2}$X + $\frac{1}{2}$E	0.095	0.030	0.035	0.840	0.725	0.275	0.80	0.20
$\frac{1}{2}$Y + $\frac{1}{2}$Z	0.19	0.00	0.00	0.81	0.825	0.175	1.00	0.00
$\frac{1}{2}$Y + $\frac{1}{2}$E	0.095	0.030	0.035	0.840	0.70	0.30	0.80	0.20
$\frac{1}{2}$Z + $\frac{1}{2}$E	0.285	0.030	0.035	0.650	0.675	0.325	0.80	0.20

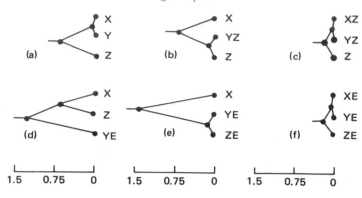

Figure 7.1. Hypothetical gene flow.

The Indian frequencies are hypothetical but closely approximate those observed in some American Indian groups without European or African mixture where A_2, B, and D- are absent, A_1 present in some but absent in other groups, M high in frequency with variation between groups. The European data were collected by Reed (1968) in the San Francisco Bay area, California and were employed by Szathmary and Reed (1972) in their study of Caucasian-Ojibwa admixture in Ontario; the main topographical results emphasized here would not be changed by use of European gene frequencies from English, French, or Spanish samples. The trees throughout this chapter are graphed by the logit centroid method, as defined on p. 146.

In the tree (Figure 7.1a) representing unmixed Indians, groups X and Y join at the proximal node (with lesser genetic distance 0.058 ± 0.035) and Z joins at the distal node (with greater genetic distance 0.702 ± 0.342). An Indian–Indian gene flow (Figure 7.1b) from group Z into Y changes the topography of the tree with YZ and Z joining at the proximal node (0.141 ± 0.070) the distance between unmixed X and Z which now meet in the distal node remaining unchanged from the value shown for these two groups in Figure 7.1a. An Indian–Indian gene flow (Figure 7.1c) from Z into both X and Y keeps the topography of Figure 7.1a but shortens both proximal (0.023 ± 0.014) and distal (0.164 ± 0.066) distances to their respective nodes.

A European mixture into Y (Figure 7.1d) changes the topography of the tree (compare 7.1a) with X and Z joining proximally at the same distance as in 7.1a and YE joining distally at an increased distance (1.311 ± 0.378). A European gene flow into both Y and Z joins them in the proximal node (0.163 ± 0.089 and results in greater genetic distance to the node joining unmixed X (1.474 ± 0.216). A European mixture into all three groups in equal amounts (Figure 7.1f) keeps the form of the tree given in 7.1a but greatly shortens the distances to the proximal (0.015 ± 0.009) and to the distal (0.178 ± 0.085) nodes.

From the topography of the tree (that is to say, from the multivariate genetic distances alone), the Indian–Indian versus European–Indian gene flow (7.1c vs.

7.1f) cannot be distinguished, and (7.1b) differs from (7.1d) in distances but not in topography. Of course, in this hypothetical example, XE, YE, ZE can be distinguished on a population basis from X, Y, Z, XZ, and YZ because representative samples of the European–Indian mixed groups contain A_2, B, and D-chromosomes. The point is that localized Indian–Indian gene flow may change the topography of genetic distance trees and maps as much as or more than some degrees of European–Indian mixture. The reader easily can extend the analysis to the desired complexity.

Genetic Distance and Language Phyla

A composite logit tree of the 53 groups graphed in three steps is shown in Figure 7.2. Each group in this tree was made to graph with its language phylum despite the fact that a given group might show greater similarity in gene frequencies to a member of a different phylum, for example, Southampton Eskimo with Algonquian Blackfoot (Figure 7.3, node 73). Step 1 graphed the six Southwestern Athapaskans (Apacheans) separately (see Figure 7.13). In step 2 the eight multimembered phyla were each graphed independently. The centroid method for constructing trees provides estimates of the gene frequencies for each node including the stem for each tree (Spuhler *et al.* 1972). In step 3 the Ur-frequencies of the eight several-member phyla, the observed frequencies of the two single-member families, and the pooled observed frequencies of three major races (as estimated by Lewontin 1972) were graphed together. Results of the three steps are then united—with the stem nodes of the Apacheans and Aztec–Tanoans as reference points—to form the composite tree (Figure 7.2).

Perfect agreement between tribe and language phylum in the composite logit tree (Figure 7.2) was imposed by the method of analysis (but, of course, connections within phyla reflect the observed gene frequencies). The restriction by language holds to a much lesser extent for the canonical variables map (Figure 7.4) of the 53 groups where the X axis represents the first and the Y axis the second canonical variable. Clusters are shown by joining each group to the point representing its phylum mean coordinates. The method analyzes the 53 groups by language but allows overlapping distributions between phyla. Note that no linguistic restriction is imposed in the canonical variables map by culture areas (Figure 7.6) nor in the logit tree of the 53 groups shown in Figure 7.3. In that tree the 53 groups join to 33 proximal nodes, and of these only 16 join two groups belonging to the same language phylum (for instance, North Alaska and Copper Eskimo at node 65), 17 of the 33 cases form first linkages across language boundaries (Southampton Eskimo–Algonquian, Thule Eskimo–Salishan, Yakimo–Salishan/Arctic, Hupa–Arctic, Blood–Na-Dené/Siouan, James Bay Cree–Na-Dené Tuchone–Algonquian/Siouan, Montagnais–Na-Dené, Naskapi–Na-Dené, Chilcotin–Siouan/Na-Dené, Nootka–Aztec–Tanoan, Caddo–Aztec–Tanoan, Cherokee–Hokan, Zuni–Hokan/Siouan, Laguna–Aztec–Tanoan, Maricopa–Aztec–Tanoan/Keresan, and Kutchin–Hokan/Aztec–Tanoan/Keresan).

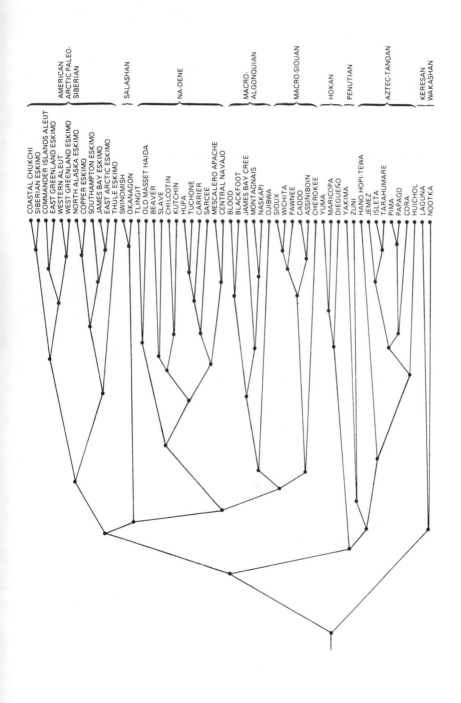

Figure 7.2. Composite tree of 53 groups.

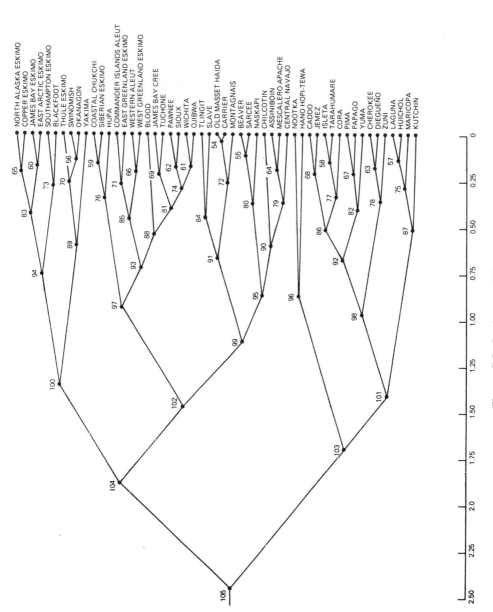

Figure 7.3. Logit tree of 53 groups.

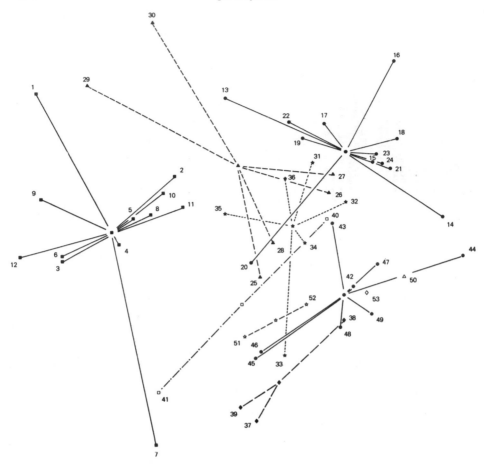

Figure 7.4. Map of 53 groups by language phyla. ■, Arctic–Siberian; ●, Na-Dené; ▲, Macro-Algonquian; ★, Macro-Siouan; ◆, Hokan; □, Penutian; ⊕, Aztec–Tanoan; △, Kenesan; ☆, Salispan; ◇, Wabashan.

Still, noteworthy clustering by language phyla is observed in the 53-group tree (Figure 7.3) constructed without any linguistic restrictions: All Arctic–Paleo-Siberian groups are in the upper two branches joining at nodes 97 and 100; the two Salishans meet in a proximal node, all Na-Dené (except Kutchin) and all Macro-Algonquian are in the upper main branch (node 104); all Hokan are in the branch joining node 101; and all Aztec–Tanoans in the 103 main branch. However, the Macro-Siouans are scattered, the Laguna Keresans join the Aztec–Tanoan Huichol at a close node, Nootka and Tanoan Hano meet at an intermediate node, and the distantly Penutian Zuni connect with geographically extreme west and east coast non-Penutians at a fairly close node.

In some cases known mixture accounts for the observed genetic affiliations across language taxa, for instance, Apache and Navajo with Jemez (Hester 1962,

Schroeder 1968), or Yuma with Pima and Papago (Kroeber 1939). In most cases the operation of gene flow versus genetic drift and/or selection cannot be disentangled on available evidence (Cavalli-Sforza 1973, Cavalli-Sforza and Bodmer 1971, Hiorns and Harrison 1977, Spuhler 1973).

Figure 7.5 plots the points for the mean canonical variables for each of the 10 language phyla along with circles with centers at the point for each phylum and with radius equal to the mean distance of the individual points in each phylum set to the center of that set. These mean distances decrease in the order from greater to lesser canonical variate map spread (one measure of genetic diversity within phyla) as follows with the largest mean set to unity:

1.00 Macro-Algonquian
0.99 Penutian
0.67 Arctic–Paleo-Siberian
0.57 Na-Dené
0.56 Aztec–Tanoan
0.53 Macro-Siouan
0.51 Hokan
0.29 Salishan
0.00 Wakashan
0.00 Keresan

It should be noted that all the samples in this analysis are small and those of size 2 or 3 (e.g., Penutian, with two widely divergent members, and Salishan, with two rather similar members) may greatly misrepresent the actual genetic diversity by language phylum.

Clearly a method of analysis and genetic mapping that estimates probabilities of multiple genetic affinities—methods that may reflect both branching descent and admixture (or other genetic forces leading to convergence in gene frequencies) —is highly useful in genetic studies of North American Indians and most other human populations.

Stepwise discrimination function analyses provide one such method. Tables 7.2 through 7.18 give the posterior probabilities for genetic relationship of the 53 tribes to their own and other language phyla. Of necessity, the method classifies one-member groups to their own language family (here, Keresan and Wakashan). No member of a two or more member set is necessarily classified in the language phylum represented by the set.

Eight of the nine Eskimo groups (Table 7.2) classify as American–Arctic–Paleo-Siberian with probabilities ranging from 0.38 to 0.98, four with probabilities greater than 0.91. The Commander Islands Aleuts show similar probabilities for both Arctic and Algonquian relationship. The Thule Eskimo classify as Hokan with a probability 22 times greater than that as Arctic. The affinities of these groups are considered in more detail in the discussion of culture areas given below.

Eight of the 12 Na-Dené tribes (Table 7.3) show highest genetic affinities, with probabilities ranging from 0.36 to 0.86 to Na-Dené, but the Slave and Chilcotin have higher probability for Algonquian connections and the Hupa

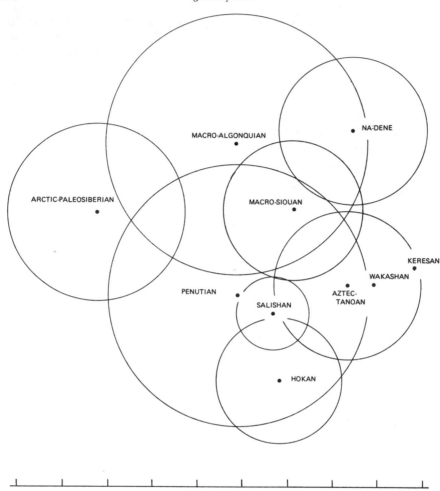

Figure 7.5. Centers of language phyla.

show equal or greater probable kinship with the other phyla in the six cases, with a conspicuously low probability (0.05) for alliance with other Athapaskans. Gifford (1926) points out that the medium-statured, high-faced, broad-headed, narrow-nosed Hupa differ in body size and shape from other nearby California Athapaskans that have the Yukian physical type (short stature, low face, narrow head, and broad nose) but resemble their non-Athapaskan neighbors of the California physical type. Although a detailed study is needed to settle the issue, my guess is that the Hupa represent a genetically non-Athapaskan people who took over the language without taking in many genes from the donor population.

It is of interest to note that the non-Athapaskan Tlingit (0.75) and Haida (0.67) show relatively high (ranking third and fourth highest of the set of 12) Na-Dené affinity.

Table 7.2. *Posterior Probability of Affinity of American Arctic–Paleo-Siberian Groups with Language Phyla*

Group	Arctic-Siberian	Na-Dené	Macro-Algonquian	Macro-Siouan	Hokan	Penutian	Aztec–Tanoan	Undetermined
North Alaska Eskimo	**0.978**	0.001	0.016	0.001	0.000	0.004	0.000	0.000
Copper Eskimo	**0.381**	0.039	0.304	0.132	0.002	0.128	0.004	0.009
Southampton Eskimo	**0.906**	0.000	0.080	0.004	0.003	0.006	0.000	0.000
East Arctic Eskimo	**0.622**	0.003	0.221	0.056	0.013	0.072	0.005	0.007
James Bay Eskimo	**0.560**	0.003	0.345	0.023	0.002	0.053	0.002	0.011
West Greenland Eskimo	**0.967**	0.000	0.011	0.006	0.002	0.013	0.000	0.000
Thule Eskimo	0.037	0.000	0.001	0.005	**0.815**	0.110	0.014	0.017
East Greenland Eskimo	**0.512**	0.006	0.322	0.128	0.011	0.010	0.010	0.001
Siberian Eskimo	**0.977**	0.000	0.001	0.001	0.000	0.021	0.000	0.000
Commander Islands Aleut	0.423	0.015	**0.428**	0.090	0.004	0.028	0.008	0.004
Western Aleut	**0.355**	0.029	0.227	0.300	0.015	0.051	0.021	0.003
Coastal Chukchi	**0.993**	0.000	0.001	0.000	0.000	0.005	0.000	0.000

The six Macro-Algonquian samples (Table 7.4) are rather diverse genetically: Only three classify as Algonquian, with probabilities from 0.28 to 0.89; two go with Na-Dené, and one with Siouan. The high A_1 gene frequencies at the *ABO* locus in the Blackfoot and Blood dominate the genetic distances within this phylum.

Table 7.3. *Posterior Probability of Affinity of Na-Dené Groups with Language Phyla*

Group	Arctic–Siberian	Na-Dené	Macro-Algonquian	Macro-Siouan	Hokan	Penutian	Aztec-Tanoan	Undetermined
Slave	0.057	0.237	**0.537**	0.118	0.000	0.007	0.005	0.002
Kutchin	0.000	0.251	0.004	0.031	0.001	0.036	0.041	**0.636**
Tuchone	0.000	**0.615**	0.067	0.209	0.001	0.007	0.077	0.024
Beaver	0.000	**0.866**	0.046	0.080	0.000	0.000	0.007	0.000
Sarcee	0.001	**0.476**	0.174	0.318	0.001	0.003	0.025	0.003
Carrier	0.000	**0.791**	0.033	0.117	0.000	0.003	0.040	0.016
Chilcotin	0.003	0.230	**0.646**	0.059	0.000	0.003	0.027	0.033
Hupa	0.058	0.045	0.103	**0.315**	0.113	0.044	0.308	0.014
Navajo	0.000	**0.624**	0.047	0.195	0.002	0.007	0.088	0.036
Mescalero Apache	0.005	**0.363**	0.359	0.246	0.001	0.006	0.014	0.006
Tlingit	0.000	**0.749**	0.017	0.098	0.000	0.041	0.019	0.075
Haida	0.000	**0.666**	0.039	0.191	0.001	0.007	0.072	0.024

Of the six Macro-Siouan tribes (Table 7.5), only Siouan Assiniboin, with Caddoan Caddo and Pawnee, show highest affinities to the macrophylum, with Iroquoian Cherokee joining to Hokan and Caddoan Wichita to Macro-Algonquian, while the Sioux show Na-Dené resemblance. All of the maximum probabilities for the six groups in the phylum are less than one half.

The Yuma, one of three Hokan groups (Table 7.6), identifies as Aztec–Tanoan, most likely as a result of gene flow from Pima and Papago. Both Maricopa and Diegueño, although clearly Hokan in gene frequencies, also show moderate Penutian similarities.

Yakima and Zuni represent a sample of only two of the 16 or so languages some classify under Penutian (Table 7.7). In this analysis Zuni goes a little more strongly with Hokan than with Penutian, and Yakima shows strong affinity with its two Salishan neighbors represented in our sample. This is an example of a two-membered set in which no member is classified by the method in their designated language phylum.

Table 7.4. *Posterior Probability of Affinity of Macro-Algonquian Groups with Language Phyla*

Group	Arctic–Siberian	Na-Dené	Macro-Algonquian	Macro-Siouan	Hokan	Penutian	Aztec–Tanoan	Undetermined
Cree	0.030	0.036	0.094	**0.298**	0.133	0.140	0.199	0.071
Montagnais	0.003	**0.364**	0.161	0.144	0.004	0.039	0.113	0.173
Naskapi	0.002	**0.440**	0.090	0.333	0.003	0.042	0.047	0.043
Ojibwa	0.023	0.071	**0.279**	0.251	0.047	0.031	0.247	0.051
Blackfoot	0.299	0.001	**0.696**	0.004	0.000	0.000	0.000	0.000
Blood	0.069	0.022	**0.889**	0.020	0.000	0.000	0.000	0.000

Table 7.5. *Posterior Probability of Affinity of Macro-Siouan Groups with Language Phyla*

Group	Arctic–Siberian	Na-Dené	Macro-Algonquian	Macro-Siouan	Hokan	Pinutian	Aztec–Tanoan	Undetermined
Sioux	0.004	**0.383**	0.180	0.332	0.003	0.007	0.083	0.007
Assiniboine	0.001	0.292	0.064	**0.447**	0.011	0.014	0.153	0.018
Cherokee	0.004	0.004	0.004	0.101	**0.447**	0.197	0.169	0.074
Caddo	0.006	0.103	0.097	**0.431**	0.055	0.040	0.230	0.038
Wichita	0.120	0.062	**0.439**	0.194	0.014	0.078	0.052	0.041
Pawnee	0.012	0.264	0.231	**0.375**	0.006	0.035	0.056	0.021

Table 7.6. *Posterior Probability of Affinity of Hokan Groups with Language Phyla*

Group	Arctic–Siberian	Na-Dené	Macro-Algonquian	Macro-Siouan	Hokan	Penutian	Aztec–Tanoan	Undetermined
Maricopa	0.003	0.000	0.001	0.013	**0.678**	0.138	0.074	0.093
Yuma	0.000	0.013	0.006	0.146	0.235	0.009	**0.577**	0.013
Diegueño	0.006	0.000	0.001	0.023	**0.703**	0.176	0.056	0.035

Table 7.7. *Posterior Probability of Affinity of Penutian Groups with Language Phyla*

Group	Arctic–Siberian	Na-Dené	Macro-Algonquian	Macro-Siouan	Hokan	Penutian	Aztec–Tanoan	Undetermined
Yakima	0.001	0.084	0.045	0.037	0.002	0.125	0.020	**0.686**
Zuni	0.118	0.000	0.001	0.022	**0.441**	0.396	0.016	0.006

Five of the eight Aztec–Tanoan groups (Table 7.8) classify most strongly in that phylum, although only Hano with a probability over 0.90. Here Pima and Papago go into Hokan, with little doubt reflecting gene exchange with Maricopa and Yuma (Kroeber 1939, Schroeder 1968). Jemez biologically is as unusual a Pueblo population as it is culturally (Eggan 1950, Parsons 1925, 1939). Their maximum probable affinity for any of the eight phyla is only 0.26 with the "undetermined phyletic set," as a result, perhaps, of the fact that Jemez is highly mixed biologically, and the "undetermined" set is a heterogeneous lot. The known Apache plus Navajo and Tanoan ancestry is reflected in substantial probabilities (0.21 and 0.14, respectively); there is some ethnohistoric evidence to suggest that the affinity with Macro-Siouan of the same order (0.20) may reflect long-term contacts with the southern Plains, some of it via Pecos (Lange 1953).

Laguna, the two Salishan representatives, and the Nootka constitute the residual set (Table 7.9) in the analysis by language phyla. In the present data

Table 7.8. *Posterior Probability of Affinity of Aztec–Tanoan Groups with Language Phyla*

Group	Arctic–Siberian	Na-Dené	Macro-Algonquian	Macro-Siouan	Hokan	Penutian	Aztec–Tanoan	Undetermined
Isleta	0.001	0.049	0.019	0.192	0.106	0.020	**0.550**	0.062
Jemez	0.002	0.206	0.112	0.195	0.012	0.069	0.144	**0.259**
Hopi–Tewa	0.000	0.033	0.002	0.034	0.009	0.000	**0.921**	0.001
Pima	0.010	0.002	0.011	0.067	**0.434**	0.212	0.134	0.129
Papago	0.009	0.003	0.009	0.098	**0.442**	0.227	0.130	0.083
Tarahumara	0.000	0.137	0.016	0.257	0.048	0.033	**0.408**	0.100
Cora	0.001	0.019	0.007	0.137	0.222	0.095	**0.362**	0.156
Huichol	0.000	0.034	0.006	0.142	0.143	0.044	**0.502**	0.129

Table 7.9. *Posterior Probability of Affinity of Undetermined Phyletic Groups with Language Phyla*

Group	Arctic–Siberian	Na-Dené	Macro-Algonquian	Macro-Siouan	Hokan	Penutian	Aztec–Tanoan	Undetermined
Laguna (Keresan)	0.000	0.117	0.006	0.216	0.055	0.027	**0.484**	0.094
Okanagon (Salishan)	0.019	0.004	0.021	0.081	0.238	**0.367**	0.084	0.185
Swinomish (Salishan)	0.002	0.018	0.014	0.078	0.061	0.301	0.089	**0.438**
Nootka (Wakashan)	0.000	0.002	0.001	0.000	0.000	0.017	0.003	**0.976**

there is no other Keresan or Wakashan group for comparison with Laguna and Nootka. Founded in 1698 or 1699 by an unknown group of Keres speakers, probably from the Middle Rio Grande Valley, Laguna classifies first with Aztec–Tanoan simply because the analysis includes no data on other Keresans and because they are now basically like (0.48) some other long-resident Southwestern groups in gene frequencies. Given the linguistic classificatory residue, Swinomish joins their Nootka Wakashan neighbors and Okanagon joins their Yakima Penutian neighbors, perhaps in conformity with the history of gene flow in that area of Washington State and British Columbia (Hulse 1957).

Overall, excluding Keresan and Wakashan, the two one-member classes, 33 of the 51 groups (64.7 percent) have highest probable genetic relationship to their own language phylum or family. In the one-member language family samples, Keresan Laguna is less dissimilar genetically from the other 53 groups than is Wakashan Nootka.

A summary of the probabilistic classification by stepwise discriminant analysis of the 53 tribes by language phylum is given in Table 7.10.

Table 7.10. *Number of Groups Classified by Gene Frequencies into Each Language Phylum*

Phylum	Percent[a]	Arctic Siberian	Na-Dené	Macro-Algonquian	Macro-Siouan	Hokan	Penutian	Aztec–Tanoan	Undetermined
Arctic–Siberian	83	**10**	0	1	0	1	0	0	0
Na-Dené	67	0	**8**	2	1	0	0	0	1
Macro–Algonquian	33	0	2	**3**	1	0	0	0	0
Macro-Siouan	50	0	1	1	**3**	1	0	0	0
Hokan	67	0	0	0	0	**2**	0	1	0
Penutian	50	0	0	0	0	1	**0**	0	1
Aztec–Tanoan	62	0	0	0	0	2	0	**5**	1
Undetermined	50	0	0	0	0	0	1	1	**2**

[a] Percentage of groups belonging to language phylum classified into own phylum.

A one-way analysis of variance demonstrates that A, O, R^1, and R' gene frequencies discriminate the 53 groups by language phyla at the 0.01 level of statistical significance; B, R^2, and R'' at the 0.05 level; and that M, N, R^0, r, R^z, and r^y gene frequencies are not discriminative at the 0.05 level in this respect.

Comparison of within- and between-group genetic distances for language phyla in the 51 groups belonging to multimember sets discloses that within-phylum mean genetic distances are larger than the mean between-phylum distance only in Na-Dené and Macro-Algonquian, two of the nine cases.

The biologic evidence does not suggest a close genetic relationship between Hokan and Siouan peoples, which were included by Sapir (1929) in his extensive and diverse Hokan–Siouan stock. However, the evidence presented by Haas

(1964) that Siouan is related to Na-Dené (specifically Athapaskan and Tlingit) is of interest in that San Carlos Apache and Athapaskan (British Columbia) map closest to the three Macro-Siouan points, although the gene frequencies place Tlingit at some distance but still well within the Athapaskan bundle that entirely embraces the three Siouan tribes. Also, there is no genetic support for an appreciable Keresan element at Zuni proposed by some archaeologists (Ford *et al.* 1972) and linguists (Harrington 1945), or for Sapir's (1929) intuitive inclusion of Keresan in Hokan–Siouan.

Genetic Distances and Culture Areas

In this section the genetic affinities of the 53 tribes are examined by culture areas. The assignment of groups and tribes to culture areas follows the forthcoming *Handbook of North American Indians* to be published by the Smithsonian Institution, Washington, D.C. Their classification is approximately that of Driver (1969) and Driver and Massey (1957), who based their classification largely on Kroeber (1939). For some of the groups included here, the tribal maps in Jenness (1955), Murdock and O'Leary (1975), Sherzer (1973), and Swanton (1952) provided supplementary information.

Ten additional groups not considered in the overall analysis are included in the trees for their culture areas (three classified in the Subarctic, one in the Northwest Coast, two in the Northeast, and four in the Southwest culture areas).

Logit trees, the canonical variates map by culture areas (Figures 7.6 and 7.7), and stepwise discriminate analyses are discussed under each area.

American Arctic

The Eskimo were land and sea mammal hunters adapted to the ice, water, and off-shore lands of the arctic coasts. At the time of European contact they numbered some 89,700 persons, not including some 16,000 Aleuts, in an area of 22,288,000 km^2, with a population density of 4.02 per 100 km^2, more than four times the density of the subarctic peoples to the south (Kroeber 1939). The Eskimo were subdivided into a very large number of local groups with much movement between groups. Swanton (1952) lists the names of some 40 major, and nearly 500 minor, Eskimo groups. No tribal organization existed among the Eskimo. The Thule culture that evolved in the Western Maritime Provinces about 1000 A.D. is fully Eskimo, is ancestral to the modern Eskimos, and spread rapidly to Greenland replacing the earlier Dorset culture on the way. Willey (1966) concludes that the Eskimo physical type is associated with the Arctic small-tool tradition dating 4500–4000 B.P., an American cultural tradition clearly of Asian origin, perhaps as old as 8700 B.P. in the western arctic (Laughlin 1963a, b), ancestral to the Dorset culture that in turn gave rise to Eskimo. The linguistically

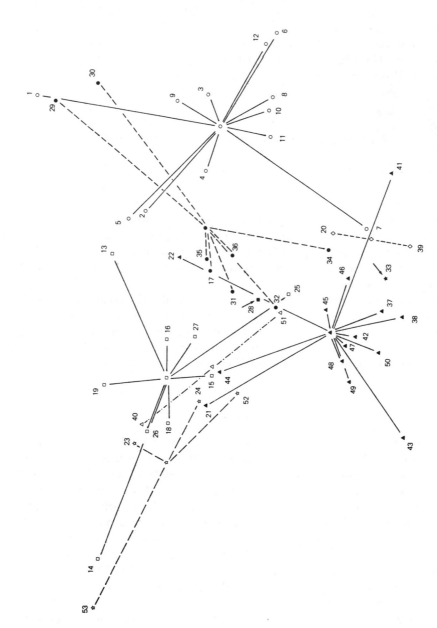

Figure 7.6. Map of 53 groups by culture area. ○, Arctic; □, Subarctic; ☆, Northwest; △, Plateau; ◇, California; ●, Plains; ■ Northeast; ★, Southeast; ▲, Southwest.

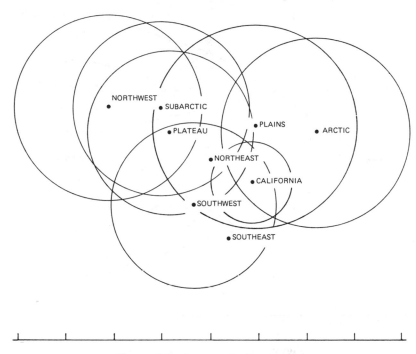

Figure 7.7. Centers of culture areas.

related Pacific Aleut subtradition developed from the Aleutian core and blade, not the Arctic small-tool tradition.

In North America the Arctic culture area has identical tribal membership and geographic distribution with the Arctic–Paleo-Siberian language phylum discussed above (p. 153 and Table 7.2). The arctic tree (Figure 7.8) shows two distinct major branches, with the North Alaskan, Central, and Thule Eskimo in the upper, and the Siberian Eskimo, Aleuts, and East and West Greenland Eskimo in the lower branch. Laughlin (1950) and others have long noted the close relationship of the Aleuts to the Greenland Eskimo. Of the 12 arctic groups the Siberian Eskimo and Coastal Chukchi show the closest accord in genetic distance, the Thule Eskimo the greatest divergence.

Discriminant analysis places only two of the 12 groups in other culture areas (Table 7.11), the James Bay Eskimo tieing in with the Plains and the Thule Eskimo with the Cherokee. The Plains likeness of the James Bay Eskimo probably is caused by Algonquian mixture (Chown and Lewis 1956) but the distinctive gene frequencies of the Thule Eskimo that parallel those of the geographically distant Cherokee and Diegueño apparently are the result of random drift in a long-isolated small gene pool. The Coastal Chukchi, along with the Siberian, North Alaska, and West Greenland Eskimo, classify in the Arctic culture area with probabilities greater than 0.90.

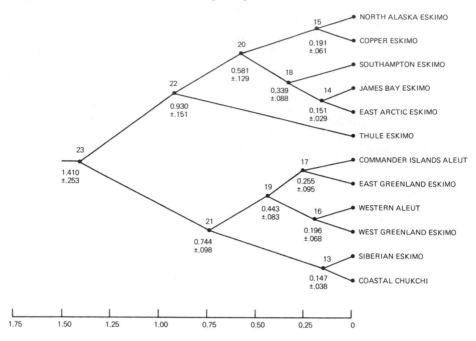

Figure 7.8. Arctic culture area tree.

Table 7.11. *Gene Frequency Classification of Arctic Groups by Culture Areas*

Group	Arctic	Sub-arctic	North-west Coast	Plateau	Cali-fornia	Plains	South-west	North-east	South-east
Coastal Chukchi	**0.939**	0.000	0.000	0.000	0.053	0.004	0.001	0.002	0.002
Siberian Eskimo	**0.980**	0.001	0.000	0.000	0.015	0.002	0.000	0.001	0.001
Commander Islands Aleut	**0.583**	0.001	0.000	0.001	0.062	0.287	0.012	0.044	0.011
Western Aleut	**0.547**	0.003	0.000	0.002	0.130	0.208	0.026	0.055	0.029
North Alaska Eskimo	**0.982**	0.001	0.000	0.001	0.000	0.015	0.000	0.000	0.000
Copper Eskimo	**0.469**	0 062	0.007	0.118	0.012	0.270	0.016	0.037	0.009
Southampton Eskimo	**0.540**	0.001	0.000	0.002	0.008	0.428	0.003	0.014	0.004
James Bay Eskimo	0.235	0.080	0.005	0.098	0.005	**0.511**	0.013	0.052	0.004
East Arctic Eskimo	**0.447**	0.012	0.000	0.030	0.025	0.405	0.020	0.041	0.020
Thule Eskimo	0.025	0.001	0.000	0.021	0.228	0.017	0.109	0.012	**0.586**
West Greenland Eskimo	**0.936**	0.000	0.000	0.000	0.028	0.026	0.001	0.001	0.007
East Greenland Eskimo	**0.461**	0.001	0.000	0.000	0.065	0.396	0.011	0.057	0.009

Based largely on the morphology of the head and face, the Eskimo are often given taxonomic recognition separate from the North American Indians (Garn 1971, Laughlin 1963b). Gene frequency analyses place the Eskimo and Aleut as a distinctive cluster at one extreme of the range occupied by our sample of 53 tribes (Figure 7.6 and 7.7). Although resident in the American area for at least 8700 years (Laughlin 1975), some Eskimo groups show close relationship with Siberian peoples—the Siberian Eskimo connect more closely to the Coastal Chukchi than the North Alaska Eskimo relate to the Copper Eskimo (Figure 7.8).

Cluster compactness, that is, the amount of genetic heterogeneity within the culture area, can be measured by finding the mean distance from each group in the area to their culture area mean on the canonical variate map (Figure 7.6). None of the present areas represented by more than two groups is highly compact; the Arctic area ranks after the Plains in lack of compactness or extent of diversity.

If distinctness is measured by finding the mean distance between the center (mean) of each culture area and the other eight areas, the Eskimo–Aleut of the arctic rank high in distinctness but no more so than the Northwest Coast at the opposite side of the canonical variate map.

Subarctic

The Subarctic was the most sparsely populated of the nine culture areas in pre-Columbian times. Kroeber (1939) estimated that the western subarea had a population numbering 33,930 living on 38,944,000 km², with a population density of 0.87 per 100 km², and that the eastern subarea contained 23,000 individuals occupying 25,677,000 km², with a density of 1.11 per 100 km². Athapaskan is the only language family represented in the Yukon and Mackenzie Subarctic. In general the Athapaskan speakers are caribou hunters living in the coniferous forest along lakes and rivers. They are organized into bands with matrilineal descent and fluid membership. Individuals and families change residence frequently. The distance between birthplaces of spouses is often large. Definitely, Athapaskan peoples are first recognized in the Western Denetasiro tradition dating 300 A.D., a tradition that probably developed from the North-west microblade tradition beginning in Alaska as early as 6500 B.C. (Willey 1966, pp. 415 and 466). Considerable prehistoric trade and intermarriage between Northwest Coast groups and the Athapaskans of the interior is evident, a contact that greatly increased after establishment of the European fur trade.

The Algonquian Eastern Subarctic culture area shows an even stronger linguistic uniformity, with only two languages, Cree and Ojibwa, spoken over the entire subarea and a similar loose band organization with considerable movement and intermarriage. Hunters with a similar basic culture pattern lived in the Eastern Subarctic back to at least 5000 B.C. European cultural influence was probably stronger in the eastern than in the western subarea.

Seven samples of the Athapaskan groups considered here live in the western subarea and four representatives of Algonquian groups occupy the eastern subarea. These subarctic peoples show the lowest correlations in North America

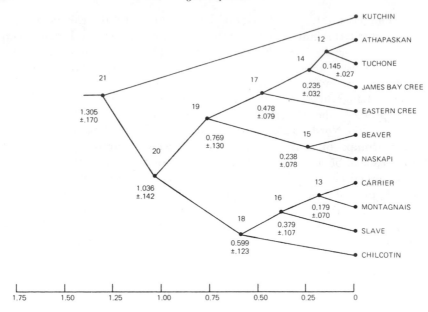

Figure 7.9. Subarctic culture area tree.

between geographic and genetic distance—the Brooks Range Kutchin (*14a*) are closer genetically to Tlingit (*23*) from Mount Edgecombe and Sitka in southern Alaska than to the Old Crow Kutchin (*14*) in the Yukon (Figure 7.6). The Kutchin are the most divergent group in the subarctic tree (Figure 7.9), where only Tuchone and Athapaskan (*14b*) of the seven northern Athapaskan groups meet at a proximal node, Beaver joining first with Algonquian Montagnais and with Slave and then Chilcotin joining more remotely to the Carrier–Montagnais branch. Eastern and James Bay Cree join adjacent nodes on the same branch. Kutchin (*14*), Slave (*18*), and James Bay Cree (*25*) are outliers on the canonical variate map (Figure 7.6).

One or more subarctic groups appear in all four major branches of the 53-group tree (Figure 7.3). As expected from previous discussion of the two linguistic stocks (pp. 153–155), the Algonquians are more heterogeneous genetically than the northern Athapaskans.

Only four of the nine groups are classified with highest probability by stepwise discriminant analysis in the Subarctic area: Tuchone, Carrier, Chilcotin in the west, and Montagnais in the east (Table 7.12). Kutchin joins the Northwest Coast with the highest probability in the table (0.92), Beaver classifies with the Ojibwa of the Northeast, Naskapi with the Plateau, Slave with the Plains, and James Bay Cree with the Southwest, each at successively lower probabilities.

The Subarctic ranks fifth from the largest in mean distance of its area mean from the means of the other eight areas and fourth from the least in compactness (Figure 7.6).

Table 7.12. *Gene Frequency Classification of Subarctic Groups by Culture Areas*

Group	Arctic	Sub-arctic	North-west Coast	Plateau	Cali-fornia	Plains	South-west	North-east	South-east
Western									
Kutchin	0.000	0.025	**0.923**	0.051	0.000	0.000	0.000	0.000	0.000
Tuchone	0.002	**0.438**	0.074	0.078	0.008	0.064	0.112	0.219	0.005
Carrier	0.000	**0.575**	0.222	0.063	0.001	0.016	0.037	0.085	0.001
Chilcotin	0.000	**0.715**	0.042	0.015	0.000	0.075	0.009	0.143	0.000
Slave	0.079	0.315	0.008	0.015	0.004	**0.364**	0.013	0.201	0.000
Beaver	0.002	0.366	0.008	0.011	0.002	0.190	0.035	**0.385**	0.000
Eastern									
Cree	0.037	0.046	0.004	0.053	0.174	0.120	**0.268**	0.175	0.124
Montagnais	0.000	**0.554**	0.325	0.053	0.000	0.008	0.016	0.043	0.000
Naskapi	0.010	0.245	0.095	**0.379**	0.010	0.084	0.085	0.076	0.017

Northwest Coast

The Northwest Coast culture area is unusual in being occupied by a population rather distinctive in physical type (Drucker 1955), in being uniform in culture but diverse in language, and in maintaining without agriculture the second highest population density in native America north of Mexico. The population included 129,000 Indians living on a long, narrow coastal belt with an area of 4,560,000 km^2 and a density of 28.30 per 100 km^2. Seven language families, each with several distinct languages, are represented in that relatively small space. The culture, equated with or derived from the Southern Arctic tradition, marked a climax of 3000 years of west coast subsistence based mostly on regular harvests of fish.

Before European contact none of their political units included more than 1000 individuals located in villages at the mouth of streams; most villages included single clans and demes; some had several clans and demes. Trade, ceremonies, and intermarriage were common across local communities and languages within the coastal area and were not rare between that area and the adjacent Subarctic and Plateau areas (Driver 1969, Drucker 1955).

The Northwest Coast tree (Figure 7.10) places the two Haida samples in separate branches, Skidegate going with Tlingit and Old Masset with the Salishan Swinomish. Nootka connects remotely to the stem of the tree.

In the 53-group tree (Figure 7.3), both Tlingit and Haida associate first with the interior Athapaskan Slave and Carrier, Swinomish connects to its co-Salishan neighbor Okanagon at a close node, and Nootka joins Hopi at a node with some depth, a relationship resulting more from random convergence than from moderately recent descent from a common gene pool. The canonical variate map (Figure 7.6) puts the four Northwest Coast groups on the periphery, with Nootka outlying at an extreme.

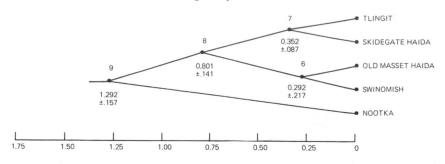

Figure 7.10. Northwest Coast culture area tree.

The posterior probabilities obtained by discriminant analysis classify Tlingit and Nootka in their own culture area but place Haida with the Subarctic and Swinomish with the Salishan Okanagon in the Plateau culture area (Table 7.13). The Northwest area ranks fourth from the top in mean distance from the centers of the other eight areas and third in degree of compactness.

Plateau

The Columbia and Fraser River drainages are the locale of the Plateau area and, as the great salmon streams of the continent south of Alaska, the two rivers provide the main subsistence basis in the culture area. At the time of contact, the native population totaled 47,650 living on 6,660,000 km², with a density of 7.15 per 100 km², a population concentration intermediate between that of the northwest coast to the west and the plains to the east.

The breeding population included many small groups speaking diverse languages representing five language families. Villages were combined into tribelets; full tribal organization was lacking before introduction from the plains in historic times. Sanpoil villages averaged only 75 persons. Movement of groups within the area was common, relations between groups amicable, and exchange of wives between communities frequent, wealthy interior tradesmen often obtaining wives from coastal villages. The Plateau culture area was a "hinterland" of

Table 7.13.   *Gene Frequency Classification of Northwest Coast Groups by Culture Areas*

Group	Arctic	Sub-arctic	North-west Coast	Plateau	Cali-fornia	Plains	South-west	North-east	South-east
Tlingit	0.000	0.141	**0.733**	0.118	0.000	0.001	0.004	0.003	0.000
Haida	0.001	**0.498**	0.173	0.090	0.004	0.026	0.079	0.126	0.003
Nootka	0.000	0.012	**0.972**	0.016	0.000	0.000	0.000	0.000	0.000
Swinomish	0.001	0.076	0.200	**0.598**	0.006	0.006	0.070	0.011	0.032

the North Pacific coast with a lively interchange between the two areas continuing from prehistory (Driver 1969, Kroeber 1939, Spencer and Jennings 1977).

Our sample of two represents only about 0.07 percent of the major tribelets belonging to the area. In the 53-group tree (Figure 7.3) Yakima joins the branch with Okanagon at the third node back, and on the canonical variate map (Figure 7.6) the two Plateau groups are positioned intermediately between the Northwest Coast and the Subarctic, with Okanagon (*51*) close to Assiniboine (*32*) of the northern Plains and closer to the Southwest than to the Plateau area center. The posterior probabilities in Table 7.14 arrange Okanagon first in the Plateau (but

Table 7.14. *Gene Frequency Classification of Plateau Groups by Culture Areas*

Group	Arctic	Sub-arctic	North-west Coast	Plateau	Cali-fornia	Plains	South-west	North-east	South-east
Okanagon	0.016	0.037	0.020	**0.362**	0.066	0.039	0.194	0.037	0.231
Yakima	0.000	0.126	**0.465**	0.395	0.000	0.003	0.007	0.004	0.001

with considerable affinity both to the Southeast and the Southwest) and place Yakima with the Northwest Coast ($P = 0.46$) and with the Plateau ($P = 0.40$). The Plateau area center ranks eighth (next smaller of the set after the Northeast, group *28*) in mean distance from the other seven area centers (Figures 7.6 and 7.7).

California

Indian California is outstanding as a culture area in its high population density without food production and its great linguistic diversity within a relatively small geographic space. The population, living on 1,941,000 km^2 and concentrated along the coastal belt and around the mouths of streams, numbered 84,000 persons, with a density of 43.30 per 100 km^2. Families were organized into small bands or villages; the Yokuts were a "tribelet," politically unified tribes occurred only among the peripheral Mohave and Yuma. Trade, bilingualism, and inter-marriage were characteristic of the area. Linguistic areas cut across the distribution of language families (Sherzer 1973). Relations with nearby residents, whether kinsmen or not, were quite close, while relations with groups 50 or more miles away were practically nonexistent in most of the area (Kroeber 1939), although trade was regular between Yumans and Uto–Aztecans in southern California (Spencer and Jennings 1977).

Our sample of two groups includes Athapaskan Hupa and Yuman Diegueño. The 53-group tree (Figure 7.3) connects Diegueño to Cherokee at a proximal node and Hupa to Coastal Chukchi–Siberian Eskimo at a penultimate node. The California area cluster on the canonical variate map (Figure 7.6) is the most compact of the multi-membered areas, and it centers closest to the Southeast (that

Table 7.15.   *Gene Frequency Classification of California Groups by Culture Areas*

Group	Arctic	Sub-arctic	North-west Coast	Plateau	Cali-fornia	Plains	South-west	North-east	South-east
Hupa	0.031	0.011	0.000	0.001	**0.408**	0.062	0.162	0.290	0.034
Diegueño	0.006	0.000	0.000	0.008	0.241	0.003	0.084	0.004	**0.653**

is, to Cherokee, group *33*) and next closer to the Southwest area mean. Discriminant analysis (Table 7.15) classifies the Hupa in the California area and Diegueño first in the Southeast culture area ($P = 0.65$) and then in California ($P = 0.24$). California ranks third in the greater distance of its area mean from the means of the other eight culture areas.

These two samples provide little evidence on the biologic relationships of the precontact California cultures as a whole.

Plains

The classical bison-hunting Plains culture area merged together in the late eighteenth and nineteenth centuries as a consequence of Indian migration to the west in response to European pressure in the east and the introduction of horses and firearms. Occupation of the Plains region is old; aside from the Pueblos, the Mandan have the longest known archaeologic tradition of any American Indian tribe, with documentation back to 1100–1200 A.D. and roots arising from the Chamberlain Aspect of ca. 700 A.D. in the Middle Missouri River basin (Spencer and Jennings 1977). None of the four language families present in the Plains is restricted to that area, and all have had earlier centers of dispersal outside of the area. In general, the diverse Plains tribes were not in close social contact; each occupied a fairly distinct territory which, although sometimes shared with congenial neighbors, was defended against most outside tribes. Oliver (1962) shows that the Plains tribes share, and are unified by, culture traits concerned with ecologic adaptation, but that their social organization differs in well-integrated patterns developed before, and retained after, consolidation of the classical period. The culture area extended over 18,541,000 km², including a sparsely scattered population totaling 76,400, with a mean density of only 4.12 per 100 km².

The gene frequency tree (Figure 7.11) of the eight Plains tribes departs markedly from a tree based on genetic linguistic affinity (Figure 7.2). The three most proximate nodes join Sioux with Wichita, Assiniboine with Caddo, and Sarcee with Blackfoot in that order; Pawnee joins the Sioux–Wichita branch more closely than does Blood.

On the canonical variate map (Figure 7.6), Blackfoot (*29*) and Blood (*30*) are extreme outliers, overlapping with North Alaskan Eskimo (*1*), so placed because of their high *ABO* locus A^1 gene frequencies. Assiniboine (*32*) and Caddo (*34*) are somewhat less peripheral, with Sarcee (*17*), Sioux (*31*), Wichita (*35*), and Pawnee (*36*) clustering together near the area center.

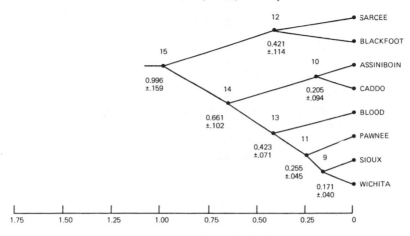

Figure 7.11. Plains culture area tree.

Discriminant analysis of the gene frequencies classifies the Blackfoot, Blood, Sarcee, Wichita, and Pawnee in the Plains area with probabilities decreasing in that order from 0.77 to 0.31 (Table 7.16). The probabilities place two of the remaining three Plains tribes in the Northeast culture area (here represented only by Ojibwa) with values less than one-half (Sioux = 0.46, Assiniboine = 0.28), whereas Caddo shows its closest affinity to the Southwest, with a value of only 0.25.

When plotted by first and second canonical variates, the Plains form the most disperse cluster; that is, the mean distance of the eight points representing Plains tribes from the Plains area center is greater than that for any of the other culture areas (Figures 7.6 and 7.7), yet the distance of the Plains area center itself is fairly close to the centroid of all 53 groups, the mean distance from the Plains area center to the other area centers being greater than those of only the Southwest, Plateau, and Southeast areas.

Table 7.16. *Gene Frequency Classification of Plains Groups by Culture Areas*

Group	Arctic	Sub-arctic	North-west Coast	Plateau	Cali-fornia	Plains	South-west	North-east	South-east
Sarcee	0.028	0.116	0.003	0.025	0.019	**0.428**	0.071	0.303	0.007
Blood	0.253	0.002	0.000	0.000	0.000	**0.729**	0.000	0.016	0.000
Blackfoot	0.215	0.003	0.000	0.001	0.000	**0.772**	0.000	0.009	0.000
Assiniboine	0.012	0.093	0.006	0.051	0.082	0.141	0.278	**0.285**	0.052
Sioux	0.015	0.152	0.004	0.012	0.036	0.202	0.113	**0.461**	0.006
Pawnee	0.077	0.092	0.004	0.054	0.065	**0.314**	0.129	0.232	0.032
Wichita	0.087	0.133	0.006	0.044	0.041	**0.339**	0.088	0.248	0.014
Caddo	0.029	0.013	0.000	0.014	0.243	0.132	**0.250**	0.181	0.138

Eastern

The Eastern culture area was one of the less densely populated regions in North America south of the subarctic. Kroeber (1939) gives a total population of 426,400 living on 61,328,000 km², with an average density of 6.95 per 100 km². The density was about 45 percent higher in the Southeast than in the Northeast.

Six language families were represented in the area, most having a continuous geographic distribution. The liguistic areas tend to coincide with genetic language areas; bilingualism was rare; lines of social communication tended to stay within linguistic boundaries; most contacts with outside tribes were not friendly (Sherzer 1973).

Two trees for four tribes located in the Eastern culture area are shown in Figure 7.12. The upper tree (A) is based on the observed gene frequencies and the lower tree (B) is based on gene frequencies adjusted to remove some of the non-Indian contribution to the gene pools. Ojibwa (*28*) and Chippewa (*28a*)—considered members of the same tribe by most ethnographers—do not join together at a proximal node. Note also that the two proximate nodes are both rather remote from the present populations, considerably more remote than the closest node in all of the other trees for individual culture areas (compare Figures 7.8–7.11 and 7.13).

Kinietz (1940, p. 317) wrote of the Chippewa: "The make-up of this tribe is very confusing." And Swanton (1952, p. 263) reported that ". . . the present population of Chippewa includes thousands of mixed bloods, partly representing mixtures with other tribes and partly mixtures with whites." The amount of European mixture in the four tribes was estimated using A^2, B, and $r(cde)$ frequencies and Bernstein's equation. Gene frequencies for the assumed parental

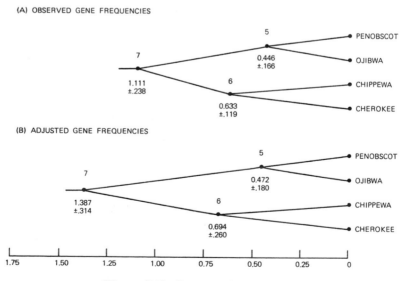

Figure 7.12. Eastern culture area tree.

Table 7.17. *Gene Frequency Classification of Eastern Groups by Culture Areas*

Group	Arctic	Sub-arctic	North-west Coast	Plateau	Cali-fornia	Plains	South-west	North-east	South-east
Northeast Chippewa	0.011	0.113	0.003	0.014	0.051	0.179	0.165	**0.452**	0.012
Southeast Cherokee	0.006	0.002	0.000	0.020	0.251	0.006	0.172	0.014	**0.527**

white population were taken from Reed (1968). The following percentages of mixture were obtained: Ojibwa, 25.5; Chippewa, 15.6; Penobscot, 44.2; and Cherokee, 9.3. The ABO and rhesus frequencies were adjusted by removing A^2, B, and r. The MN locus gene frequencies were calculated using the above estimates of the degree of admixture (m) in the equation $q_{unmixed} = (q_{mixed} - mq_{white})/(1 - m)$. The adjustment changed the depth of the nodes but not the topography of the tree.

Because the discriminant function analysis (Table 7.17) considered separately the Ojibwa as the single representative of the Northeast and the Cherokee as the one sample from the Southeast culture area, the two tribes are necessarily classified with highest probability in their own areas. It is of interest that the next higher probabilities place Ojibwa with the Plains area $(P = 0.18)$ and the Southwest $(P = 0.16)$ and connect the Cherokee to California $(P = 0.25)$, reflecting a high affinity to Diegueño and to the Southwest $(P = 0.17)$.

The area center for the Southeast (point 28 in Figure 7.6) is the most centrally located in mean distance from the other eight area centers, and that of the Northeast (point 33 in Figure 7.6) is the fourth most distant among the nine areas.

Southwest

The Southwest, the final culture area to be considered here, is the best known by gene frequencies in the continent and is represented here by 18 tribes: five Pueblos (Hano, Zuni, Laguna, Jemez, and Isleta), six circumpueblo Apacheans (San Carlos, Chiricahua, and Mescalero Apache; Western, Central, and Eastern Navajo), and seven groups from the desert and sierra subarea (Huichol, Cora, Tarahumara, Yuma, Maricopa, Pima, and Papago). Linguistic diversity is characteristic of the area, with five language families—Athapaskan, Hokan, Penutian, Aztec–Tanoan, and Keresan—represented in the culture area and in the sample. The Southwest as a whole is not a linguistic area, but the Pueblo region within the Southwest is a linguistic area (Sherzer 1973). Archaeologic and ethnologic evidence shows that, except for the Apacheans, linguistic diversity in the area is old (Ford *et al.* 1972, Spencer and Jennings 1977).

Kroeber's (1939) demographic data were restricted to that part of the Southwest culture area within the United States where the total population numbered

103,000 and occupied 9,671,000 km², with an average density of 10.70 per 100 km², which would rank the culture area as the third most densely peopled in North America. However, the 33,800 Pueblo Indians occupied only 446,000 km² as oases in the southwestern desert, with a density reaching 75.70 per 100 km², by far the highest in the continent. The desert Yumans and Apacheans lived in small, open bands, the river Yumans in large tribes, and the Pueblos in villages with populations in the 100s to the 1000s.

The 18 southwestern tribes divide into three major branches on the area logit tree (Figure 7.13). The six Apacheans cluster together in the upper main branch, with Jemez meeting Chiricahua Apache at the closest node in the tree, both then joining Eastern Navajo at the next node, and with Hano connecting to the branch at its most remote node. Jemez had a close alliance with the Apache living west of the Rio Grande from about 1600–1690 A.D. (Schroeder 1968) and with the

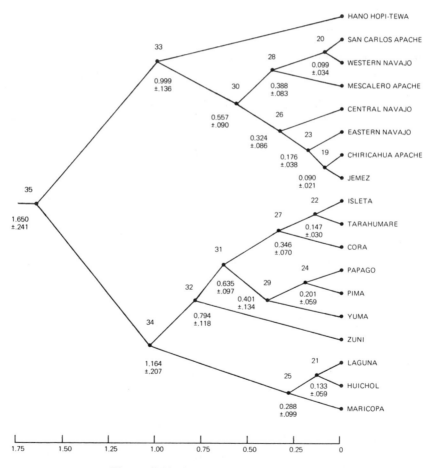

Figure 7.13. Southwest culture area tree.

Eastern Navajo from the time of the Pueblo Revolt to the time of the Navajo captivity at Bosque Redondo, that is, from 1696 to 1868 (Bandelier 1890, Hester 1962, Underhill 1967).

Hano was founded on First Mesa in Hopiland about 1700 A.D. by Tanos, or southern Tewa, originally from the Galisteo basin south of Santa Fe (Reed 1943). Dozier (1966), without giving quantitative details, reports that the contemporary Hano Hopi–Tewa are biologically more Hopi than Tewa, although the Tewa language is still intact. The Tanoan contribution to the modern Hano gene pool cannot be estimated accurately because gene frequencies of the parental Tanos (and the modern Rio Grande Tewa in general) are unknown. If we pool available gene frequency data from Towa Jemez (itself highly mixed with Apacheans) and Tiwa Isleta, the ABO frequencies give an estimate that Hano is more than 100 percent Hopi (see Boyd 1939 for an explanation of such results) and the MN frequencies that they are about 50 percent Hopi. The available demographic information indicates that the modern Hano are some 90 percent Hopi and only 10 percent Tewa (Fewkes 1894; census data summarized in Dozier 1966). In considering Hopi affinities in this chapter it is important to note that other Shoshonean-speaking peoples were not included in the analyses. [After the already massive computer runs were completed, I sometimes regretted that I had not enlarged the total series by including Ute gene frequencies of Matson and Piper (1947) as a Shoshonean representative of the omitted Great Basin culture area.]

In the middle branch of the southwestern tree (Figure 7.13) Pima and Papago meet at a close node, later joined by the Yuma. Isleta shows closest relation with the Tarahumara, and Zuni has the most remote connection in that main branch. Keresan Laguna joins Huichol and then Maricopa in the third major branch.

Multiple affinities of southwestern peoples is indicated on the 53-group map (Figure 7.6), where Navajo (point 21) and Jemez (44) approach the Subarctic cluster in a left-upward extension, Mescalero Apache (22) approaches the Plains area center in a right-upward extension, Zuni (41) approaches the California area center in a right-downward extension, and Hopi Hopi–Tewa (43) goes off to the periphery in a left-downward, long, narrow, extension. In the 53-group logit tree (Figure 7.3), Mescalero Apache joins Assiniboine in a close node (number 64) followed next by Central Navajo (79); Hano connects remotely to Nootka (node 96), whereas Zuni associates with Cherokee–Diegueño more closely at node 78.

Only six of the 14 Southwestern tribes show highest probable affinity for their own culture area in the stepwise discriminant analysis (Table 7.18). Central Navajo and Jemez (because of strong Apache and Navajo mixture, as explained above) classify in the Sub-arctic, whereas Zuni, consistent with other analyses in this study, shows Californian affinity, and Mescalero Apache, as expected on ethnohistorical grounds (M.E. Opler and Opler 1950, M. Opler 1971), is similar in gene frequencies to groups in the southern plains. Here Hopi–Tewa joins Ojibwa in the Northeast with the highest probability in the table ($P = 0.80$), and Maricopa–Pima–Papago relate to the Southeast (that is, to Cherokee) with

Table 7.18. *Gene Frequency Classification of Southwest Groups by Culture Areas*

Group	Arctic	Sub-arctic	North-west Coast	Plateau	Cali-fornia	Plains	South-west	North-east	South-east
Yuma	0.001	0.007	0.000	0.002	0.298	0.014	**0.404**	0.169	0.104
Maricopa	0.003	0.003	0.001	0.041	0.168	0.005	0.216	0.014	**0.549**
Papago	0.013	0.006	0.001	0.061	0.189	0.019	0.207	0.025	**0.479**
Pima	0.010	0.018	0.005	0.116	0.146	0.028	0.284	0.047	**0.346**
Hopi–Tewa	0.000	0.020	0.000	0.000	0.017	0.003	0.158	**0.801**	0.000
Central Navajo	0.001	**0.472**	0.153	0.115	0.004	0.033	0.089	0.131	0.003
Zuni	0.067	0.000	0.000	0.002	**0.461**	0.002	0.031	0.002	0.434
Laguna	0.001	0.014	0.003	0.023	0.184	0.009	**0.467**	0.076	0.223
Isleta	0.002	0.029	0.002	0.012	0.166	0.028	**0.451**	0.222	0.088
Jemez	0.003	**0.316**	0.116	0.236	0.011	0.053	0.134	0.117	0.014
Mescalero Apache	0.053	0.112	0.004	0.047	0.010	**0.536**	0.040	0.183	0.005
Tarahumara	0.002	0.035	0.006	0.039	0.149	0.023	**0.461**	0.132	0.154
Cora	0.002	0.034	0.015	0.098	0.114	0.012	**0.430**	0.065	0.230
Huichol	0.001	0.042	0.016	0.058	0.110	0.010	**0.515**	0.093	0.157

probabilities from 0.35 to 0.55. Yuma, Laguna, Tarahumara, Cora, and Huichol in this table relate most strongly to their own southwestern culture area.

A one-way analysis of variance establishes that only four of the nine gene frequencies significantly discriminate the 53 groups by culture area: A, $R^1 + r'$, $R^2 + r''$ ($P < 0.025$), and O($P < 0.01$).

Comparison of within- and between-group genetic distances for culture areas reveals that the within-area mean genetic distances are larger than the mean between-area genetic distance in three of the nine areas: the Subarctic, Plains, and Southwest.

A summary of the probabilistic classification of the 53 groups by culture area is given in Table 7.19. Across the main diagonal of the table the gene frequencies assign groups to their present culture area in 31 of 53 cases (58.5 percent), a

Table 7.19. *Number of Groups Classified by Gene Frequencies into each Culture Area*

Area	Percent[a]	Arctic	Sub-arctic	North-west Coast	Plateau	Cali-fornia	Plains	South-west	North-east	South-east
Arctic	83	**10**	0	0	0	0	1	0	0	1
Subarctic	44	0	**4**	1	1	0	1	1	1	0
Northwest Coast	50	0	1	**2**	1	0	0	0	0	0
Plateau	50	0	0	1	**1**	0	0	0	0	0
California	50	0	0	0	0	**1**	0	0	0	1
Plains	62	0	0	0	0	0	**5**	1	2	0
Southwest	43	0	2	0	0	1	1	**6**	1	3
Northeast	100	0	0	0	0	0	0	0	**1**	0
Southeast	100	0	0	0	0	0	0	0	0	**1**

[a] Percentage of groups belonging to culture area classified into own area.

Table 7.20. *Proportion of Total Dispersion Removed by Gene Frequency Variables by Culture Area, Language Phyla, and Family*

Analysis by	Proportion removed by variable	Cumulative proportion of total
(a) Culture area		
$R^2 + r''$	0.60337	0.60337
O	0.26277	0.86608
$R^0 + r$	0.10969	0.97577
B	0.02423	1.00000
(b) Language phylum		
O	0.57891	0.57891
$R^1 + r'$	0.30922	0.88813
M	0.07631	0.96444
$R^0 + r$	0.02923	0.99367
B	0.00633	1.00000
(c) Language family		
$R^1 + r'$	0.47494	0.47494
O	0.22766	0.70260
A	0.15392	0.85652
$R^z + r^y$	0.06536	0.92188
$R^2 + r''$	0.04709	0.96897
M	0.03103	1.00000

proportion less than the 64.7 percent (excluding the two one-member phyla) of 51 groups classified by the gene frequencies into their own language phylum (see Table 7.10). Note that the Plains—the most consolidated North American culture area—shows a fairly high proportion of classification into its current area (62 percent), whereas the genetically heterogeneous Southwest (43 percent) and Subarctic (44 percent) show relatively low proportions of classification to their own areas. Examination of Tables 7.3, 7.4, and 7.12 suggests that division of the Subarctic into two (Eastern and Western of Kroeber 1939) or three (Yukon, Mackenzie, and Eastern of Driver 1969) subdivision would increase the proportion assigned to current own culture subarea. The Arctic shows the highest concordance of group affiliation to their culture area.

The proportion of the total dispersion of the 53 tribes removed by specific gene frequency variables differs in kind, in amount, and in order when the dispersion is classified by (a) culture area, (b) language phylum, and (c) language family (Table 7.20).

These results support the conclusion that the native population of North America consists of many small, partially isolated subpopulations, or neighborhoods in Wright's (1969) sense, and that local differentiation of the red blood cellular antigen gene frequencies in native North America has occurred largely by random genetic drift, even in the case of polymorphisms possibly maintained by overdominance or other similar genetic systems (Nei 1965, Nei and Imaizumi 1966a, b).

Summary and Conclusions

Gene frequencies of red blood cellular antigens are the raw materials used to measure the genetic affinities of 53 North American Indian tribes. The gene frequencies for each group are transformed into genetic distances that are zero for populations identical in gene frequency and increasingly larger (up to the limit of 2) the more dissimilar the populations become. The genetic distances are used to make genetic trees and maps.

The only complete, unabridged way to show genetic affinities among 53 populations is to present a matrix containing the 1378 distances between all possible pairs taken two at a time. With care and perseverance one can locate the two most closely related, the two most distantly related, and the degree of relationship for each of the remaining 1376 pairs. Clearly, for more than a few populations, a matrix of genetic distances is hard to interpret at a glance.

A genetic tree reduces this complexity by connecting the 53 tribes to one stem, using 52 intermediate nodes or points of bifurcation, a total of 105 points. A genetic map reduces the intricacy to only 53 points. In both tree and map, the net relations of each tribe to the other 52 are shown at a glance. Generally, the map summary is less distorted than the tree, especially when more than 10 or 15 tribes are considered. A major defect of the genetic tree is that it neglects tribal admixture. An advantage is that the tree more accurately pictures the closest genetic ties, the pairs of tribes joined at the most recent nodes, a picture often more blurred in the map.

American anthropologists long ago recognized that biologic, linguistic, and cultural inheritance might not be closely correlated. Boas especially (1911, pp. 6–14) pointed to well-known cases where biology was constant but language and culture changed, where language was constant but physical type changed, or where biology and language persisted but culture changed. Among North American Indian tribes there is significant, but not high, correlation among biology, language, and culture. The association is not perfect, even for the Eskimo language family and culture area; the association is weakest for the Algonquian languages and the Southwest culture area. In general, gene frequency affinities are a little stronger by language family than by culture area; overall, genetic distances classify tribes into their present culture area in only 58.5 percent of the 53 cases and languages into their families in 64.7 percent.

Only a small part of the total Indian, Eskimo, and Aleut biologic diversity was sampled by the 53 groups. The results represent biologic relationships at the time the samples were taken and may differ from the relations that held before 1600 A.D. Most of the groups sampled had some non-Indian mixture; many of the samples were collected excluding individuals with known genes of European origin, especially A^2, B, and $r(cde)$. In most cases Indian–Indian mixture was a more important determiner of current affinities than was gene flow from non-Indian sources.

The form of genetic trees and maps depends on the history of genetic isolation, gene exchange, genetic drift, and probably some adaptation by selection to local and district environments. For example, the Indian biologic data reflect the recency of the Athapaskan migration to the Southwest, ending about 500 years ago, followed by admixture with the Pueblos and, in the case of the Mescalero Apache, with the Plains tribes.

Frequencies of the rhesus chromosomes and of the O and A genes are the best indicators of North American Indian genetic similarities and dissimilarities.

The genetic evidence is consistent with long occupation (say 2000 to 10,000 years) in most culture areas by representatives of two or more language families with opportunity for some local gene exchange and some regional adaptation through natural selection (for example, in the areas of high group A^1 frequencies), but with random genetic drift an important agency of local differentiation in gene frequencies. In a few cases, for instance, Zuni and the Keres in the Southwest and Nootka versus the Na-Dené tribes in the Northwest Coast area, members of language families remained biologically distinct from neighboring groups over long periods of time. The northern Athapaskans and Algonquians tended to mix freely.

The best summary of this chapter is found in Figures 7.2, 7.3, 7.4, and 7.6, and in Tables 7.10 and 7.19; the best digest of the chapter is in Tables 7.2–7.9 and 7.11–7.18.

On a world scale, the North American Indians, Eskimos, and Aleuts are a genetically distinct geographic race or breeding population, related most closely to the Mongoloid peoples of eastern Asia.

References

Ahrengot, V., and K. Eldon. (1952). Distribution of the ABO-MN and Rh Types among Eskimos in South-West Greenland. *Nature (London)* 169:1065.

Alfred, B.M., T.D. Stout, J. Birkbeck, M. Lee, and N.L. Petrakis. (1969). Blood Groups, Red Cell Enzymes, and Cerumen Types of the Ahonsat (Nootka) Indians. *Am. J. Phys. Anthropol.* 31(3):391–398.

Alfred, B.M., T.D. Stout, R.B. Lowry, M.Lee, and J. Birkbeck. (1972a). Blood Groups and Red Cell Enzymes of the Ross River (Northern Tuchone) and Upper Liard (Slave) Indians. *Am. J. Phys. Anthropol.* 36(2):161–164.

Alfred, B.M., T.D. Stout, M. Lee, R. Tipton, N.L. Petrakis, and J. Birkbeck. (1972b). Blood Groups of Six Indian Bands of Northeastern British Columbia. *Am. J. Phys. Anthropol.* 36(2):157–160.

Allegro, L. M. Lewis, and H. Kaita. (1972). Unpublished data cited in manuscript by M. Lewis and B. Chown, in preparation for *Handbook of North American Indians*, Vol. 3.

Allen, F.H. Jr. (1959). Summary of Blood Group Phenotypes in Some Aboriginal Americans. *Am. J. Phys. Anthropol.* 17(1):86.

Allen, F.H. Jr., and P.A. Corcoran. (1960). Blood Groups of the Penobscot Indians. *Am. J. Phys. Anthropol.* 18(2):109–114.

Balakrishnan, V. (1974). Use of Genetic Distance in Hybrid Analysis. *In* N.E. Morton, Ed., *Genetic Structures of Populations*. Honolulu: Univ. Hawaii Press.

Bandelier, A.H. (1890). Final Report of Investigations among the Indians of the Southwestern United States, Part I. *Pap. Archaeol. Inst. Am.* Am. Ser., No. III.

Blumberg, B.S., J.R. Martin, F.H. Allen, J.L. Weiner, E. Vitagliano, and A. Cooke. (1964). Blood Groups of the Naskapi and Montagnais Indians of Schefferville, Quebec. *Hum. Biol.* 36(3):263–272.

Boas, F. (1911). *Handbook of American Indian Languages, Part 1.* Washington, D.C.: Bureau of American Ethnology, Bulletin 40.

Bodmer, W.F., and L.L. Cavalli-Sforza. (1976). *Genetics, Evolution, and Man.* San Francisco: W.H. Freeman, *xv* + 782 pp.

Boyd, W.C. (1939). Blood Groups of American Indians. *Am. J. Phys. Anthropol.* 25:215–235.

Brown, K.S., B.L. Hanna, A.A. Dahlberg, and H.H. Strandskov. (1958). The Distribution of Blood Group Alleles among Indians of Southwest North America. *Am. J. Hum. Genet.* 10(2):175–195.

Cavalli-Sforza, L.L. (1973). Analytic Review: Some Current Problems in Human Population Genetics. *Am. J. Hum. Genet.* 25(1):82–104.

Cavalli-Sforza, L.L., and W.F. Bodmer. (1971). *The Genetics of Human Populations.* San Francisco: W.H. Freeman.

Cavalli-Sforza, L.L., and A.W.F. Edwards. (1967). Phylogenetic Analysis: Models and Estimation Procedures. *Am. J. Hum. Genet.* 19, 22(3) Pt. 1:24–49/233–257.

Chakraborty, R. (1975). Estimation of Race Mixture—A New Method. *Am. J. Phys. Anthropol.* 42(3):507–511.

Chown, B., and M. Lewis. (1952). Unpublished data cited in manuscript by M. Lewis and B. Chown, in preparation for *Handbook of North American Indians*, Vol. 3.

Chown, B., and M. Lewis. (1953). The ABO, MNSs, P, Rh, Lutheran, Kell Lewis, Duffy, and Kidd Blood Groups and the Secretor Status of the Blackfoot Indians of Alberta, Canada. *Am. J. Phys. Anthropol.* 11(3):369–383.

Chown, B., and M. Lewis. (1955a). The Blood Group and Secretor Genes of the Stoney and Sarcee Indians of Alberta, Canada. *Am. J. Phys. Anthropol.* 13(2):181–189.

Chown, B., and M. Lewis. (1955b). The Inheritance of the Blood Group and Secretor Genes in the Blood Indians of Alberta, Canada. *Am. J. Phys. Anthropol.* 13(2):473–478.

Chown, B., and M. Lewis. (1956). The Blood Group Genes of the Cree Indians and the Eskimos of the Ungava District of Canada. *Am. J. Phys. Anthropol.* 14(2):215–224.

Chown, B., and M. Lewis. (1959). The Blood Group Genes of the Copper Eskimo. *Am. J. Phys. Anthropol.* 17(1):13–18.

Chown, B., and M. Lewis. (1962a). The Blood Group and Secretor Genes of the Eskimo on Southampton Island. *Bull. Nat. Mus. Can.* 80:181–190.

Chown, B., and M. Lewis. (1962b). The Blood Groups and Secretor Status of Three Small Communities in Alaska. *Oceania* 32(3):211–218.

Corcoran, P.A., F.H. Allen, A.C. Allison, and B.S. Blumberg. (1959). Blood Groups of Alaskan Indians and Eskimos. *Am. J. Phys. Anthropol.* 17(3):187–193.

Corcoran, P.A., D.L. Rabin, and F.H. Allen Jr. (1962). Blood Groups of 237 Navajo School Children at Piñon Boarding School, Piñon, Arizona, 1961. *Am. J. Phys. Anthropol.* 20(3):389–390.

Cordova, M.S., R. Lisker, and A. Loria. (1967). Studies on Several Genetic Hematological Traits of the Mexican Population. XII. Distribution of Blood Group Antigens on Twelve Indian Tribes. *Am. J. Phys. Anthropol.* 26(1):55–56.

Crawford, M.H., and P.L. Workman (Eds.). (1973). *Methods and Theories of Anthropological Genetics.* Albuquerque: Univ. New Mexico Press.

Dixon, W.J. (Ed.). (1973). *BMD: Biomedical Computer Programs.* Berkeley: Univ. California Press.

Dozier, E.P. (1966). *Hano, A Tewa Indian Community in Arizona*. New York: Holt, Rinehart & Winston.

Driver, H.E. (1969). *Indians of North America*, 2nd ed. Chicago: Univ. Chicago Press.

Driver, H.E., and W.C. Massey. (1957). Comparative Studies of North American Indians. *Trans. Am. Phil. Soc.* 47:165–456.

Drucker, P. (1955). *Cultures of the North Pacific Coast*. New York: American Museum of Natural History.

Eggan, F. (1950). *Social Organization of the Western Pueblos*. Chicago: Univ. Chicago Press.

Elston, R.C. (1971). The Estimation of Admixture in Racial Hybrids. *Ann. Hum. Genet.* 35(1):9–17.

Fernet, P., W.S. Mortensen, A. Langaney, and J. Robert. (1977). Hémotypologie du Scoresdysund (Est Groenland). *Bull Soc. Anthropol. Paris*, Ser. 12, 8(suppl.):177–189.

Fewkes, J.W. (1894). The Kinship of a Tanoan-Speaking Community in Tusayan. *Am. Anthropol.* (Old Ser.) 7:162–167.

Ford, R.I., A.H. Schroeder, and S.L. Peckham. (1972). Three Perspectives on Puebloan Prehistory. *In* A. Ortez, Ed., *New Perspectives on the Pueblos*. Albuquerque: Univ. New Mexico Press, pp. 19–39.

Garn, S.M. (1971). *Human Races*, 3rd ed. Springfield, Ill.: Charles C. Thomas.

Gershowitz, H. (1959). The Diego Factor among Asiatic Indians, Apaches, and West African Negroes. *Am. J. Phys. Anthropol.* 17(3):195–200.

Giblett, E.R. (1969). *Genetic Markers in Human Blood*. Oxford: Blackwell Scientific Publications.

Gifford, E.W. (1926). Californian Indian Types. *Nat. Hist.* 26:50–60.

Goodman, M.M. (1974). Genetic Distances: Measuring Dissimilarity among Populations. *Yearbook Phys. Anthropol.* 17:1–38.

Gower, J.C. (1972). Measures of Taxonomic Distance and Their Analysis. *In* J.S. Weiner and J. Huizinga, Eds., *The Assessment of Population Affinities in Man*. Oxford: Clarendon Press, p. 1–24.

Gray, M.P., and W.S. Laughlin. (1960). Blood Groups of Caddoan Indians of Oklahoma. *Am. J. Hum. Genet.* 12(1):86–94.

Gürtler, H. (1971). Personal communication; cited in Mourant *et al.* (1976).

Harpending, H., and T. Jenkins. (1973). Genetic Distance among Southern African Populations. *In* M.H. Crawford and P.L. Workman, Eds., *Methods and Theories of Anthropological Genetics*. Albuquerque: Univ. New Mexico Press, pp. 177–199.

Harpending, H., and T. Jenkins. (1974). !Kung Population Structure. *In* J.F. Crow and C.F. Denniston, Eds., *Genetic Distance*. New York: Plenum Press, pp. 137–161.

Harrington, J.P. (1945). The Sounds and Structure of the Aztecan Languages. *In* E.L. Hewett and B.P. Dutton, Eds., *The Pueblo Indian World*. Albuquerque: Univ. New Mexico Press, pp. 157–162.

Haas, M.R. (1964). Athapaskan, Tlingit, Yuchi, and Siouan. *Proc. Int. Cong. Americanists* 35(2):495–500.

Hester, J.J. (1962). Early Navajo Migrations and Acculturation in the Southwest. Santa Fe: Museum of New Mexico Papers in Anthropology, No. 6.

Hiorns, R.W., and G.A. Harrison. (1977). The Combined Effects of Selection and Migration in Human Evolution. *Man* (London) 12:438–445.

Hulse, F.S. (1957). Linguistic Barriers to Gene-Flow. The Blood-Groups of the Yakima, Okanagon, and Swinomish Indians. *Am. J. Phys. Anthropol.* 15:235–246.

Hulse, F.S. (1960). Ripples on a Gene Pool: The Shifting Frequencies of Blood-Type Alleles among the Indians of the Hupa Reservation. *Am. J. Phys. Anthropol.* 18(2):141–152.

Jenness, D. (1955). *The Indians of Canada*, 3rd ed. Ottawa: National Museum of Canada, Bulletin 65.

Kinietz, W.V. (1940). The Indians of the Western Great Lakes, 1615–1760. Ann Arbor: Occas. Pap. Mus. Anthropol. Univ. Michigan, No. 10.

Kraus, B.S., and C.B. White. (1956). Micro-evolution in Human Population: A Study of Social Endogamy and Blood Type Distributions among the Western Apache. *Am. Anthropologist* 58(6): 1017–1043.

Kroeber, A.L. (1939). *Cultural and Natural Areas of Native North America*. Berkeley: University of California Publications in American Archaeology and Ethnology, No. 38, pp. xii, 1–242.

Kruskal, J.B. (1964). Nonmetric Multidimensional Scaling: A Numerical Method. *Psychometrika* 29: 115–129.

Landar, H. (1973). The Tribes and Languages of North America: A Checklist. *Curr. Trends Ling.* 10: 1253–1446.

Lange, C.H. (1953). A Reappraisal of Evidence of Plains Influences among the Rio Grande Pueblos. *Southwestern J. Anthropol.* 9: 212–230.

Laughlin, W.S. (1950). Blood Groups, Morphology, and Population Size of the Eskimos. *Cold Spring Harbor Symp. Quant. Biol.* 15: 165–173.

Laughlin, W.S. (1951). The Alaska Gateway Viewed from the Aleutian Islands. *In* W.S. Laughlin, Ed., *The Physical Anthropology of the American Indian*. New York: Viking Fund, pp. 98–126.

Laughlin, W.S. (1957a). Personal communication to Mourant, Kopeć and Domaniewska-Sobczak, in Mourant *et al.* (1976).

Laughlin, W.S. (1957b). Blood Groups of the Anaktuvuk Eskimos, Alaska. *Univ. Alaska Anthropol. Pap.* 6: 5–15.

Laughlin, W.S. (1963a). Origins of Mongoloids and Eskimo–Aleut Cultures: Synthesis. Abs., 28th Annu. Mtg. Soc. Am. Archaeol.

Laughlin, W.S. (1963b). Eskimos and Aleuts: Their Origins and Evolution. *Science* 142: 633–645.

Laughlin, W.S. (1975). Aleuts: Ecosystem, Holocene History, and Siberian Origin. *Science* 189: 507–515.

Lewis, M., and B. Chown. (1952). Unpublished data cited in manuscript by M. Lewis and B. Chown, in preparation for *Handbook of North American Indians*, Vol. 3.

Lewis, M.V., J.A. Hildres, H. Kaita, and B. Chown. (1961). The Blood Groups of the Kutchin Indians at Old Crow, Yukon Territory. *Am. J. Phys. Anthropol.* 19(4): 383–389.

Lewontin, R.C. (1972). The Apportionment of Human Diversity. *Evol. Biol.* 6: 381–398.

Matson, G.A., and C.L. Piper. (1947). Distribution of the Blood Groups, M–N, Rh Types and Secretors among the Ute Indians of Utah. *Am. J. Phys. Anthropol.* 5: 357–368.

Matson, G.A., and H.F. Schrader. (1933). Blood Groupings among the "Blackfeet" and "Blood" Tribes of the American Indians. *J. Immunol.* 25(2): 155–163.

Matson, G.A., E.A. Koch, and P. Levine. (1954). A Study of the Hereditary Blood Factors among the Chippewa Indians of Minnesota. *Am. J. Phys. Anthropol.* 12(3): 413–426.

Matson, G.A., T.A. Burch, H.F. Polesky, J. Swanson, H.E. Sutton, and A. Robinson. (1968). Distribution of Hereditary Factors in Blood of Indians of the Gila River, Arizona. *Am. J. Phys. Anthropol.* 29(3): 311–337.

Morton, N.E. (1973). Population Structure of Micronesia. *In* M.H. Crawford and P.L. Workman, Eds., *Methods and Theories of Anthropological Genetics*. Albuquerque: Univ. New Mexico Press, pp. 333–366.

Morton, N.E., I. Roisenberg, R. Lew, and S. Yee. (1971). Pingelap and Mokil Atolls: Genealogy. *Am. J. Hum. Genet.* 23: 350–360.

Morton, N.E., J.N. Hurd, and G.F. Little. (1973). Pingelap and Mokil Atolls: A Problem in Population Structure. *In* M.H. Crawford and P.L. Workman, Eds., *Methods and Theories of Anthropological Genetics.* Albuquerque: Univ. New Mexico Press, pp. 315–332.

Morton, N.E., J.M. Lalouel, and L.L. Cavalli-Sforza. (1974). Controversial Issues in Human Population Genetics. *Am. J. Hum. Genet.* 26:259–271.

Mourant, A.E. (1954). *The Distribution of the Human Blood Groups.* Oxford: Blackwell Scientific Publications.

Mourant, A.E., A.C. Kopeć, and K. Domaniewska-Sobczak. (1958). *The ABO Blood Groups. Comprehensive Tables and Maps of World Distribution.* Oxford: Blackwell Scientific Publications.

Mourant, A.E., A.C. Kopeć, and K. Domaniewska-Sobczak. (1976). *The Distribution of the Human Blood Groups and Other Polymorphisms,* 2nd ed. London: Oxford Univ. Press.

Murdock, G.P., and T.J. O'Leary. (1975). *Ethnographic Bibliography of North America* (5 Vols.), 4th ed. New Haven: Human Relations Area Files Press.

Nei, M. (1965). Variation and Covariation of Gene Frequencies in Subdivided Populations. *Evolution* 19:256–258.

Nei, M., and Y. Imaizumi. (1966a). Genetic Structure of Human Populations I. Local Differentiation of Blood Group Gene Frequencies in Japan. *Heredity* 21:9–35.

Nei, M., and Y. Imaizumi. (1966b). Genetic Structure of Human Populations II. Differentiation of Blood Group Gene Frequencies among Isolated Populations. *Heredity* 21:183–190.

Nei, M., and A.K. Roychoudhury. (1972) Gene Differences Between Caucasian, Negro, and Japanese Populations. *Science* 177:434–436.

Nie, N., D.H. Bent, and C.H. Hull. (1970). *SPSS: Statistical Package for the Social Sciences.* New York: McGraw-Hill.

Oliver, S.C. (1962). *Ecology and Cultural Continuity as Contributing Factors in the Social Organization of the Plains Indians.* University of California Publications in American Archaeology and Ethnology, Vol. 48, No. 1.

Opler, M.K. (1971). Plains and Pueblo Influence in Mescalero Apache Culture. *In* M.D. Zamora, J.M. Mahar, and H. Otenstein, Eds., *Themes in Culture.* Quezon City: Kayumanggi, pp. 73–112.

Opler, M.E., and C.H. Opler. (1950). Mescalero Apache History in the Southwest. *New Mexico Hist. Rev.* 25:1–36.

Pantin, A.M., and R. Kallsen. (1953). The Blood Groups of the Diegueño Indians. *Am. J. Phys. Anthropol.* 11(1):91–96.

Parsons, E.C. (1925). *The Pueblo of Pecos.* New Haven: Yale Univ. Press.

Parsons, E.C. (1939). *Pueblo Indian Religion* (2 Vols.). Chicago: Univ. Chicago Press.

Pollitzer, W.S., R.C. Hartmann, H. Moore, R. Rosenfeld, H. Smith, S. Hakim, P.J. Schmidt, and W.C. Leyshon. (1962). Blood Types of the Cherokee Indians. *Am. J. Phys. Anthropol.* 20:33–43.

Post, R.H., J.V. Neel, and W.J. Schull. (1968). Tabulations of Phenotype and Gene Frequencies for 11 Different Genetic Systems Studied in the American Indian. *In Biomedical Challenges Presented by the American Indian.* Washington, D.C.: Pan American Health Organization Scientific Publication No. 165, pp. 141–185.

Race, R.R., and R. Sanger. (1975). *Blood Groups in Man,* 6th ed. Oxford: Blackwell Scientific Publications.

Reed, E.K. (1943). The Origins of Hano Pueblo. *El Palacio* 50(4):73–76.

Reed, T.E. (1968). Distributions and Tests of Independence of Seven Blood Group Systems in a Large Multi-racial Sample from California. *Am. J. Hum. Genet.* 20:142–150.

Rodriguez, H., E. Rodriguez, A. Lorice, and R. Lisker. (1963). Studies on Several Genetic Hematological Traits of the Mexican Population. V. Distribution of Blood Group Antigens in Nahuas, Tarahumaras, Tarascos and Mixtecos. *Hum. Biol.* 35:350–360.

Rohlf, J.F., J. Kispaugh, D. Kirk, and M. Schum. (1972). *NT-SYS: Numerical Taxonomy of Multivariate Statistical Programs.* Albuquerque, N.M.: Univ. New Mexico. [Computer printout.]

Rokala, D.A. (1971). *The Anthropological Genetics and Demography of the Southwestern Ojibwa in the Greater Leech Lake–Chippewa National Forest Area.* Ph.D. Dissertation, University of Minnesota. Ann Arbor, Mich.: University Microfilms.

Rychkov, Yu.G., and V.A. Sheremetyeva. (1972a). Population Genetics of Aleuts in Commander Islands in Connection with Peoples History and Population of Ancient Beringean Land. *Vopr. Antropol.* 40:45–70. [In Russian.]

Rychkov, Yu.G., and V.A. Sheremetyeva. (1972b). Population Genetics of Peoples in the North of the Pacific Basin: The Problems of Their History and Adaptation. Communication III. Populations of Asian Eskimo and Chukchi of Bering Sea Coast. *Vopr. Antropol.* 42:3–30. [In Russian.]

Sapir, E. (1929). Central and North American Languages. *Encyclopaedia Britannica* (14th ed.) 5:138–141.

Schroeder, A.A. (1968). Shifting for Survival in the Spanish Southwest. *New Mexico Hist. Rev.* 43:291–310.

Sherzer, J. (1973). Areal Linguistics in North America. *Curr. Trends Ling.* 10:749–795.

Sneath, P.H.A., and R.R. Sokal. (1973). *Numerical Taxonomy. The Principles and Practice of Numerical Classification.* San Francisco: W.H. Freeman.

Spencer, R.F., and J.D. Jennings, Eds. (1977). *The Native Americans: Ethnology and Backgrounds of the North American Indians, 2nd Edition.* New York: Harper and Row.

Spuhler, J.N. (1951). Some Genetic Variations in American Indians. *In* W.S. Laughlin, Ed., *Papers on the Physical Anthropology of the American Indian.* New York: Viking Fund, pp. 177–202.

Spuhler, J.N. (1972). Genetic, Linguistic, and Geographical Distances in Native North America. *In* J.S. Weiner and J. Huizinga, Eds., *The Assessment of Population Affinities in Man.* Oxford: Clarendon Press, pp. 72–95.

Spuhler, J.N. (1973). Anthropological Genetics: An Overview. *In* M.H. Crawford and P.L. Workman, Eds., *Methods and Theories of Anthropological Genetics.* Albuquerque: Univ. New Mexico Press, pp. 423–451.

Spuhler, J.N., P. Gluckman, and M. Pori. (1972). A Method for Graphing a Phylogenetic Tree Based on a Logit Transformation of Gene Frequencies. (Manuscript.)

Swanton, J.R. (1952). The Indian Tribes of North America. Washington, D.C.: Bureau of American Ethnology, Bulletin 145.

Szathmary, E.J.E., and T.E. Reed. (1972). Caucasian Admixture in Two Ojibwa Indian Communities in Ontario. *Hum. Biol.* 44:655–671.

Tatsuoka, M.M. (1971). *Multivariate Analysis: Techniques for Educational and Psychological Research.* New York: John Wiley and Sons.

Thomas, J.W., M.A. Stuckey, H.S. Robinson, J.P. Gafton, D.O. Anderson, and J.N. Bell. (1964). Blood Groups of the Haida Indians. *Am. J. Phys. Anthropol.* 22(2):189–192.

Thompson, E.A. (1973). The Icelandic Admixture Problem. *Ann. Hum. Genet.* 37(1):69–80.

Underhill, R.M. (1967). *The Navajos,* Rev. ed. Norman: Univ. Oklahoma Press.

Voegelin, C.F., and F.M. Voegelin. (1966). *Map of North American Indian Languages.* Seattle: Publications of the American Ethnological Society, No. 20.

Ward, R.H., and J.V. Neel. (1970). A Comparison of a Genetic Network with Ethno-history and Migration Matrices, a New Index of Genetic Isolation. *Am. J. Hum. Genet.* 22(5):562–573.

Willey, G.R. (1966). *An Introduction to American Archaeology.* Vol. 1, *North and Middle America.* Englewood Cliffs, N.J.: Prentice-Hall.

Workman, P.L., and J.D. Niswander. (1970). Population Studies on Southwestern Indian Tribes. II. Local Genetic Differentiation in the Papago. *Am. J. Hum. Genet.* 22(1):24–49.

Workman, P.L., J.D. Niswander, K. S. Brown, and W.C. Leyshon. (1974). Population Studies on Southwestern Indian Tribes, IV. The Zuni. *Am. J. Phys. Anthropol.* 41(1): 119–132.

Wright, S. (1968). *Evolution and the Genetics of Populations.* Vol. 1, *Genetic and Biometric Foundations.* Chicago: Univ. Chicago Press.

Wright, S. (1969). *Evolution and the Genetics of Populations.* Vol. 2, *The Theory of Gene Frequencies.* Chicago: Univ. Chicago Press.

Wright, S. (1977). *Evolution and the Genetics of Populations.* Vol. 3, *Experimental Results and Evolutionary Deductions.* Chicago: Univ. Chicago Press, 613 pp.

Yeung, A.T., and J.N. Spuhler. (1975). Blood Group Antigens in Navajo and Pueblo Indians. (Manuscript.)

Chapter 8

Blood Groups of Siberians, Eskimos Subarctic and Northwest Coast Indians: The Problem of Origins and Genetic Relationships

Emöke J.E. Szathmary

Introduction

Anthropologists are in general agreement that North American Indians and Eskimos represent branches of the Mongoloid family tree. What disagreement exists concerns the nature of the relationship between these branches. Taylor (1968), Levin (1963), and McGhee (1972), in reviewing hypotheses of Eskimo origins, listed a variety of theories that perhaps can be summarized under two or three broad themes.

One of the oldest hypotheses claimed that Eskimo originated in the central arctic, perhaps between Hudson Bay and the Mackenzie River. From this region groups gradually migrated to the arctic coast and from there to Alaska. Ultimately the advancing Eskimoid groups disrupted the indigenous population distribution along the North Pacific shore and Bering Sea rim. As a consequence, Indian populations of that area, in whose music, myths and selected cultural practices Siberian origins were seen, became separated from their Siberian relatives by an "Eskimo wedge."

The most current hypothesis of Eskimo origin, one for which there is archaeologic and linguistic support, is that modern Eskimo and Aleut evolved *in situ* in the Bering Sea area from a series of ancestral populations that occupied the shore from Hokkaido to Umnak Island in the Aleutian chain. After the submergence of Beringia, Eskimo spread north along the Alaskan coast, eventually migrating across Canada and into Greenland (Laughlin 1963), while the ancestral Aleut turned southward to occupy the Aleutian chain. Indians in this hypothesis are considered descendants of populations that moved into the New World from the far east perhaps as early as 40,000 years ago (Chard 1963) or, as is more commonly held, from the interior of Siberia at least 25,000 years ago (Chard 1963).

The first of these hypotheses, although not now in vogue, suggested that northwest coast Indian populations had a close relationship with people of northeastern Siberia. Jochelson (1928), among others (Levin 1963), claimed that physical, cultural, and linguistic evidence indicated that the Chukchi, Koryak, Kamchadal, Chuvants, Yukagir, and Gilyak were Siberian Americanoids; that is, people whose ancestors had reentered Siberia from the New World. Biologic connection between these people and northwest coast Indian populations would be close and either's connection with Eskimo, more remote.

The modern hypothesis suggests the reverse relationship. Laughlin (1962, 1963) has presented both osteologic and archaeologic evidence in support of a close connection between Chukchi and Eskimo. Levin (1963) has categorically stated that the Bering Sea physical type (Eskimo, Chukchi) and Kamchatka type (Koryak) are close to each other and cannot be shown to be morphologically similar to Indians of northwestern North America. On linguistic grounds, Swadesh (1964) attests to the same argument.

Both the major themes, then, agree in suggesting distinctiveness of Eskimos from Indians. They contradict each other, however, in the postulated relationships of Indians, and Eskimos, with northeastern Siberians.

In addition to these theories, there is at least one hypothesis suggesting that Indians and Eskimos are not uniformly as distinct as has been claimed. About a dozen years ago a number of archaeologists argued that the culture of the Dorset people, who in Canada and Greenland preceded the Thule populations from which modern Eskimo are derived, showed boreal influence. Meldgaard (1962) has particularly argued that artifacts and cultural motifs in Dorset that are appropriate to forest environments are most likely to have originated in the area bounded by the Great Lakes, James Bay, and Newfoundland. Extending this into the biologic sphere, gene flow from Indians into the Dorset Eskimo was postulated and employed occasionally to explain the distinctiveness of certain Eskimo groups from others (e.g., Chown and Lewis 1959, on the Dorset residuum in Copper Eskimo).

It would appear, then, that there is a number of contradictory hypotheses about the nature of the relationship between Eskimos, Indians, and Siberians. Much of the evidence offered through the years for or against the claimed biologic relationships has been osteologic or morphologic. However, genetic marker data, principally blood groups, were also employed by Laughlin (1966) to emphasize the distinctiveness of Eskimos and Indians. Mourant (1960) and Lewis and Chown (1960) used ABO, MN, and Rh gene frequency data to determine relationships with Siberians and even Polynesians.

The unavoidable deficiency of these studies was the lack of essential information about the genetic traits of Siberians. Until recently, only ABO and MN data had been published. A more serious flaw, particularly in the assessment of relationships between Indians and Eskimos, was that all conclusions were based on single loci. Until 1972 there had not been any published attempt to analyze genetic marker data with techniques that used all the information provided by all available loci. McGhee (1972) addressed himself to the question of Copper Eskimo relationships and did not concern himself with the large problem of Eskimo ethnogenesis. Szathmary (1975), using genetic distance analysis, did. The preliminary conclusions (using data from five blood group systems) suggested great genetic separation of Eskimo and Athapascan-speaking Indians, a closer relationship of Eskimo and Algonkian speakers, and closest relationship (based on ABO and MN data only) between Eskimo and the Asiatic Chukchi.

Given the paucity of information, whatever genetic comparisons were made in the past were subject to considerable error and should properly be viewed as trial attempts. However, with the publication of Zolotareva's and Bashlay's Yakut data in the English-language literature (Mourant *et al.* 1976) and Russian-language papers on blood groups of three groups from the east Sayan mountains (Rychkov *et al.* 1969) and the Aleuts of the Commander Islands (Rychkov and Sheremetyeva 1972), five Siberian populations have become available for study.

The purpose of this chapter, therefore, is twofold. First, to present the blood group gene frequency estimates available for North American Indians, Eskimos, Siberians, and isolates, such as Aleut and Ainu. Second, to analyze this data by means of distance statistics to establish support for any one of the following hypotheses:

1. *Eskimo wedge hypothesis:* Northwest Coast people are closer to Siberians than are the Eskimo; there is great genetic distance between Eskimos and Indians.

2. *Modern hypothesis:* There is great genetic distance between Eskimos and Indians, the former being closer to Mongoloids than to Indians, the latter being closer to Siberians than to Eskimos.

3. *Dorset hypothesis:* Eskimos are the descendants of (and, therefore, are closest to) northeastern subarctic Indian populations.

Selection of Populations

Fitch and Neel (1969) and Neel (1976) have stressed the genetic distinctiveness of South American Indians from their North American counterparts. Because prevailing hypotheses involve comparisons between arctic and subarctic Mongoloids, no attempt has been made to include South American Indians in this analysis. For similar reasons, North American Indians resident in the subarctic and northwest coast are alone considered.

Fitch and Neel (1969), among others (e.g., Ward *et al.* 1975), have restricted their analyses of native populations of the Western Hemisphere to those groups that met criteria of (1) minimal sample size (200 persons); (2) and/or ethnic

" purity," i.e., non-Indian admixture less than 5 percent as judged by the ABO system.

It was not possible to meet these standards in collecting data on arctic and subarctic groups. Because Siberian genetic data are scarce, it is not yet possible to judge what characteristics in these groups suggest European admixture. It seemed illogical to attempt to exclude North American groups on this basis and yet to include Siberians that may or may not have a recent European contribution to their ancestry.

With respect to sample size, some populations essential for the purposes of this study fell well below the desired number of 200 persons (e.g., Tlingit, $N = 79$). When several samples of the same tribe were available, phenotypes were pooled prior to gene frequency calculation for the entire tribe. Other "tribes" were grouped on a regional basis (e.g., South Alaskan Eskimo). Because Spuhler (1972) was able to demonstrate that genetic distances within linguistic groups were significantly less than between linguistic groups, linguistic relationship was also used as a criterion for grouping small samples into single large aggregates (e.g., "Northern Algonkian"). By these manipulations, most populations in this study consist of a sample of 100 individuals or more, the Tlingit and Aleut excepted. A map indicating the location of the populations used in this study is shown in Figure 8.1.

Blood group data of Eskimo were grouped as follows:

1. South Alaskan Eskimo [Yupik speakers (Dumond 1965)]
   a. Kodiak Island: Old Harbor, Karluk, and Kaguyak (Denniston 1966)
   b. Koniag Isolate: groups "X," "Y," "Z" (Chown and Lewis 1961)
2. North Alaskan Eskimo [Inyupik speakers (Dumond 1965)]
   a. Wainwright, Pt. Barrow, and Anaktuvuk Pass samples (Corcoran et al. 1959)
3. Central Eskimo (districts of Mackenzie and Keewatin, Northwest Territories)
   a. Copper (Chown and Lewis 1959)
   b. Aivilik (Chown and Lewis 1962)
4. Eastern Eskimo (north shore of Hudson Strait, Ungava Bay, and east shore of Hudson Bay)
   a. Okomiut (Chown and Lewis 1962)
   b. Fort Chimo (data of Auger, in Simpson et al. 1976)
   c. Hudson Bay (Chown and Lewis 1956)
5. West Greenland Eskimo
   a. Thule
   b. Augpilagtok
   c. Southwest Greenland (all three data of Gürtler, Bilberg, and Tingsgard in Simpson et al. 1976)
6. East Greenland Eskimo
   a. Angmagssalik (data of Gürtler et al. in Simpson et al. 1976)
   b. Scoresbysund (Fernet et al. 1971)

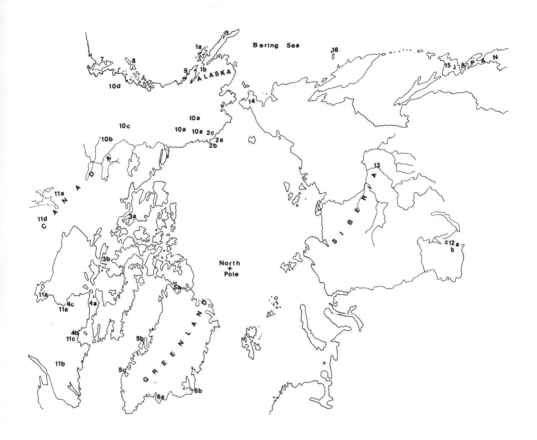

Figure 8.1. Location of populations. (1a) Kodiak Island = Old Harbor, Karluk, Kaguyak Eskimo. (1b) Koniag Isolates = Anchorage Bay Eskimo. (2a) Wainwright. (2b) Point Barrow. (2c) Anaktuvuk Pass Eskimo. (3a) Copper Eskimo. (3b) Aivilik Eskimo. (4a) Okomiut. (4b) Fort Chimo. (4c) Hudson Bay Eskimo. (5a) Thule. (5b) Augpilagtok. (5c) Southwest Greenland Eskimo. (6a) Angmagssalik. (6b) Scoresbysund Eskimo. (7) Nootka, Ahousat, British Columbia. (8) Haida, Masset, and Skidegate, British Columbia. (9) Tlingit, Sitka, and Mt. Edgecombe, Alaska. (10) Northern Athapascan speakers: (10a) Kutchin—Fort Yukon and Arctic Village, Alaska; Old Crow, Yukon Territory. (10b) Slave—Upper Liard River, Yukon Territory. (10c) Tuchone—Ross River, Yukon Territory. (10d) Chilcotin—Williams Lake, British Columbia. (11) Northern Algonkian: (11a) Cree—Post-de-la-Baleine, Quebec; James Bay; Northern Manitoba. (11b) Montagnais—Schefferville, Quebec. (11c) Naskapi—around Fort Chimo, Quebec. (11d) Ojibwa—Pikangikum, Ontario. (12) East Sayan groups: (12a) Tuvin—Tuvinskaya ASSR. (12b) Todzhin—Tuvinskaya ASSR. (12c) Tofalar—north slope of East Sayan ridge, Siberia. (13) Yakut, northeastern Siberia. (14) Chukchi, Chukotka Peninsula. (15) Ainu, Hokkaido Island. (16) Aleut, Commander Islands.

North American Indian populations were grouped as follows:

1. Nootka (Alfred *et al.* 1969)
2. Haida (Thomas *et al.* 1964)
3. Tlingit (Concoran *et al.* 1959)
4. Northern Athapascan
    a. Kutchin
        i. Fort Yukon and Arctic Village, Alaska (Corcoran *et al.* 1959)
        ii. Old Crow, Yukon Territory (Lewis *et al.* 1961)
    b. Slave (Alfred *et al.* 1972)
    c. Tuchone (Alfred *et al.* 1972)
    d. Chilcotin (Alfred *et al.* 1970)
5. Northern Algonkian
    a. Cree
        i. Poste-de-la-Baleine, Quebec (Auger *et al.* 1975)
        ii. James Bay (Chown and Lewis 1956)
        iii. Manitoba (Lucciola *et al.* 1974)
    b. Montagnais (Blumberg *et al.* 1964)
    c. Naskapi (Blumberg *et al.* 1964)
    d. Northern Ojibwa (Szathmary *et al.* 1975)

Siberian populations include:

1. Populations of the east Sayan Mountains
    a. Tuvin (Rychkov *et al.* 1969)
    b. Todzhin (Rychkov *et al.* 1969)
    c. Tofalar (Rychkov *et al.* 1969)
2. Yakut (Mourant *et al.* 1976)
3. Chukchi (Rychkov and Sheremetyeva 1972)
4. "Siberian": a large number of indigenous populations for which ABO and MN data have been published (Levin 1958, 1959, Rychkov 1965, Zolotareva 1965)

Two isolate North Pacific groups include:

1. Ainu
    a. Districts of Shizunai and Piratori, Hokkaido (Simmons *et al.* 1953)
    b. Niikapu, Hokkaido (data of Omoto and Harada in Simpson *et al.* 1976)
2. Aleut (Rychkov and Sheremetyeva 1972)

To represent Caucasian and Negroid blood group gene frequencies, the data of Cavalli-Sforza and Bodmer (1971) were employed. Their Mongoloid frequencies were used only in those instances where their sample sizes exceeded the size of a Japanese (Lewis *et al.* 1957) and/or a Korean sample (Won *et al.* 1960).

Calculation of Gene Frequencies

Because gene frequencies estimated by maximum likelihood methods are the most suitable for distance analysis, Reed's and Schull's (1968) program MAXLIK was used to compute all gene frequencies. The exceptions to this included the Negro, Caucasian, and Mongoloid data mentioned above and that of the Chukchi (Rychkov and Sheremetyeva 1972), for which no phenotype data were provided. The latter group was not included in the "total" Siberian sample.

The gene frequencies for 25 blood group genes from eight systems (ABO, Rh, MNSs, Duffy, P, Kell, Diego, and Kidd) are shown for total Eskimos, total Indians, total Siberians, Mongoloids, Caucasians, Negroes, and Ainu in Table 8.1. The Duffy, Diego, and Kidd frequencies of Siberians are average values of the respective alleles in Eskimos, Indians, Mongoloids, and Aleut, as no testing for these systems was done in Siberians.

The gene frequencies for 26 blood group genes for the above eight systems of Eskimos, Indians, Siberians, and Aleut (calculated by geographic region or linguistic group) are shown in Tables 8.2a and 8.2b. Total Eskimo frequencies have been provided as estimates of the incidence of the Duffy and Kidd alleles in West Greenland Eskimos and the Diego alleles in West and East Greenland Eskimos. The Kidd frequencies in the Haida are total Indian values, whereas the Kidd frequencies in the Aleut are average Mongoloid values (Mongoloid, Eskimo, Indian). Unfortunately, testing for these systems was not done in these populations.

The ABO, Rh, and MN gene frequencies of tribal populations of minimum sample size of 100 or more (excepting Tlingit, Aleut, and Chukchi) are shown in Tables 8.3 and 8.4. The Yakut Rh frequencies should be viewed with caution, as for these MAXLIK would not converge. The presented Yakut R^z, R^1, and R^2 frequencies are within 0.1 percent of published values, whereas the r', R^0, and r frequencies are within 1.7 percent.

Choice of the Distance Statistic

Within the past 20 years several statistics that use gene frequency data from multiple loci have been developed to measure the amount of divergence between any two populations. By inference, the smaller the computed distance between a pair of populations, the closer the genetic relationship between them.

Three of the more commonly used distance statistics, Cavalli-Sforza's and Edward's (1967) chord distance E and Balakrishnan's and Sanghvi's (1968) B and G measures, have been shown to be highly correlated (Sanghvi and Balakrishnan 1972). A fourth measure, Nei's (1972) standard distance, has also been shown to correlate positively with E (Rothhammer *et al.* 1977). This suggests that it would make little difference which measure was selected for the analysis if relative ranking of population pairwise distances was the only criterion. However, because the precision of the distance estimates is also a consideration, Nei's

E.J.E. Szathmary

Table 8.1. Blood Group Gene Frequencies of Seven Population Aggregates

			Populations				
Allele	Eskimo	North American Indian[a]	Siberian	Mongoloid	Caucasoid	Negroid	Ainu
A	0.2948	0.1199	0.1873	0.1864	0.2786	0.1780	0.2667
B	0.0603	0.0062	0.1750	0.1700	0.0612	0.1143	0.2118
O	0.6449	0.8739	0.6377	0.6436	0.6602	0.7077	0.5215
N^b	5061	2250	4698	817	3459	858	312
R^1	0.5560	0.2949	0.2257	0.6275	0.4234	0.0287	0.5786
R^2	0.3859	0.5656	0.2991	0.3174	0.1670	0.0427	0.1852
R^z	0.0000	0.0288	0.1567	0.0059	0.0008	0.0000	0.0020
R^0	0.0078	0.0242	0.2086	0.0491	0.0186	0.7395	0.0135
r	0.0503	0.0768	0.0702	0.0000	0.3820	0.1184	0.0249
r'	0.0000	0.0070	0.0123	0.0000	0.0052	0.0707	0.0000
r''	0.0000	0.0026	0.0274	0.0000	0.0029	0.0000	0.1958
N^b	1402	2206	418	532	8297	644	248
MS	0.1329	0.2071	0.2730	0.0232	0.2371	0.0925	0.0217
Ms	0.6129	0.5250	0.2291	0.4975	0.3054	0.4880	0.3827
NS	0.0399	0.0268	0.3335	0.0632	0.0709	0.0436	0.1810
Ns	0.2143	0.2411	0.1644	0.4161	0.3866	0.3759	0.4146
N^b	1739	2100	239	145	1000	205	298
Fy^a	0.7557	0.7883	0.6408[c]	0.9496	0.4208	0.0607	1.0000
Fy^b	0.2443	0.2117	0.3592	0.0504	0.5492	0.0000	0.0000
Fy	0.0000	0.0000	0.0000	0.0000	0.0300	0.9393	0.0000
N^b	1625	2143	—	394	1944	365	523
P^1	0.3043	0.3495	0.2816	0.1677	0.5161	0.8911	0.1468
P^2	0.6957	0.6505	0.7184	0.8323	0.4839	0.1089	0.8532
N^b	3901	1938	839	293	1166	900	522
K	0.0071	0.0056	0.0188	0.0000	0.0462	0.0029	0.0000
k	0.9929	0.9944	0.9812	1.0000	0.9538	0.9971	1.0000
N^b	3732	2146	241	7200	9875	1205	523
Di^a	0.0010	0.0294	0.0434[c]	0.0312	0.0000	0.0000	0.0388
Di^b	0.9990	0.9706	0.9566	0.9688	1.0000	1.0000	0.9612
N^b	1010	1746	—	277	2600	—	77
Jk^a	0.6056	0.5166	0.4775[c]	0.3103	0.7640	0.7818	0.3312
Jk^b	0.3944	0.4834	0.5225	0.6897	0.2360	0.2182	0.6688
N^b	1106	1421	—	103	4275	105	313

[a] Based on characteristics of Subarctic and Northwest Coast populations.
[b] N, sample size. For sources see text.
[c] Estimate based on pooled data from non-Siberian Mongoloid populations. See text.

Table 8.2a. *Blood Group Gene Frequencies of Eskimo Populations*

Allele	South Alaskan	North Alaskan	Central Arctic	Eastern Arctic	West Greenland	East Greenland
A	0.2464	0.3854	0.3240	0.2596	0.2559	0.4165
B	0.0831	0.0921	0.0248	0.0092	0.0601	0.0800
O	0.6702	0.5225	0.6512	0.7312	0.6840	0.5035
N^a	563	230	429	436	2418	985
R^1	0.5532	0.4543	0.5536	0.5493	0.5746	0.6537
R^2	0.2376	0.5283	0.4114	0.4025	0.3068	0.3081
R^z	0.0000	0.0000	0.0000	0.0000	0.0011	0.0008
R^0	0.0305	0.0029	0.0175	0.0002	0.0046	0.0000
r	0.1672	0.0145	0.0175	0.0480	0.0958	0.0374
r'	0.0000	0.0000	0.0000	0.0000	0.0171	0.0000
r''	0.0115	0.0000	0.0000	0.0000	0.0000	0.0000
N^a	564	230	429	436	2408	981
MS	0.2297	0.1139	0.1640	0.0878	0.2134	0.1182
Ms	0.3773	0.7187	0.6332	0.6193	0.5399	0.6643
NS	0.0933	0.0112	0.0074	0.0122	0.1334	0.0160
Ns	0.2997	0.1661	0.1954	0.2807	0.1133	0.2015
N^a	257	230	429	425	152	246
Fy^a	0.6553	0.8526	0.8946	0.8230	0.7557[b]	0.8169
Fy^b	0.3447	0.1474	0.1054	0.1770	0.2443	0.1831
N^a	564	230	90	383	—	358
P^1	0.3952	0.1479	0.3900	0.2756	0.4073	0.1536
P^2	0.6048	0.8521	0.6100	0.7244	0.5927	0.8464
N^a	257	230	86	383	1961	984
K	0.0161	0.0022	0.0000	0.0105	0.0088	0.0000
k	0.9839	0.9978	1.0000	0.9895	0.9912	1.0000
N^a	564	230	94	431	1428	985
Di^a	0.0000	0.0044	0.0000	0.0000	0.0010[b]	0.0010[b]
Di^b	1.0000	0.9956	1.0000	1.0000	0.9990	0.9990
N^a	257	230	429	94		—
Jk^a	0.5263	0.4483	0.6286	0.7280	0.6056[b]	0.7221
Jk^b	0.4737	0.5517	0.3714	0.2720	0.3944	0.2779
N^a	205	230	87	338	—	246
M	0.6483	0.8217	0.7972	0.7080	0.7296	0.8970
N	0.3517	0.1793	0.2028	0.2920	0.2704	0.1030
N^a	563	230	429	435	2417	985

[a] N, sample size. For sources see text.
[b] Estimates based on frequencies in all Eskimos.

Table 8.2b. *Blood Group Gene Frequencies of Subarctic and Northwest Coast Indian Populations of North America and the Aleut of the Commander Islands*

Allele	Nootka	Haida	Tlingit	Northern Athapascan	Northern Algonkian	Commander Islands Aleut
			Indian			
A	0.0076	0.1038	0.0789	0.1416	0.1429	0.3292
B	0.0000	0.0092	0.0321	0.0015	0.0074	0.0253
O	0.9924	0.8870	0.8890	0.8569	0.8498	0.6455
N^a	198	437	79	650	886	60
R^1	0.0739	0.3258	0.2524	0.2695	0.3515	0.5034
R^2	0.8416	0.5386	0.6165	0.5847	0.4991	0.2617
R^z	0.0498	0.0026	0.0134	0.0160	0.0473	0.1466
R^0	0.0347	0.0140	0.0000	0.0380	0.0190	0.0493
r	0.0000	0.1190	0.0640	0.0918	0.0618	0.0390
r'	0.0000	0.0000	0.0000	0.0000	0.0213	0.0000
r''	0.0000	0.0000	0.0537	0.0000	0.0000	0.0000
N^a	198	437	79	641	851	60
MS	0.0276	0.1537	0.1219	0.1124	0.3808	0.3005
Ms	0.3714	0.6495	0.7389	0.6301	0.3750	0.4054
NS	0.0507	0.0096	0.0286	0.0430	0.0300	0.0720
Ns	0.5503	0.1872	0.1106	0.2145	0.2143	0.2221
N^a	198	437	79	600	786	51
Fy^a	0.7341	0.8210	1.0000	0.8236	0.7529	0.0697
Fy^b	0.2659	0.1790	0.0000	0.1764	0.2471	0.9303
N^a	198	437	79	643	786	52
P^1	0.2086	0.3818	0.1633	0.2693	0.4853	0.2521
P^2	0.7914	0.6182	0.8367	0.7307	0.5147	0.7479
N^a	198	437	30	635	638	59
K	0.0282	0.0115	0.0063	0.0008	0.0006	0.0084
k	0.9718	0.9885	0.9937	0.992	0.9994	0.9916
N^a	198	437	79	645	787	60
Di^a	0.0382	0.0023	0.0000	0.0037	0.0813	0.1120
Di^b	0.9618	0.9977	1.0000	0.9963	0.9187	0.8880
N^a	160	437	79	538	532	52
Jk^a	0.3644	0.5166[b]	0.5361	0.5393	0.5531	0.4775[c]
Jk^b	0.6356	0.4834	0.4639	0.4607	0.4469	0.5225
N^a	198	—	79	523	621	—
M	0.3990	0.8032	0.8608	0.7425	0.7223	0.7059
N	0.6010	0.1968	0.1392	0.2575	0.2777	0.2941
N^a	198	437	79	600	886	51

[a] N, sample size. For sources see text.
[b] Estimates based on frequencies in all Indians.
[c] Estimates based on frequencies in all Mongoloids.

Table 8.3. *Blood Group Gene Frequencies of Ten Eskimo Tribal Populations*

Allele	Wainwright	Copper	Aivilik	Fort Chimo	Thule	Kodiak Island	Koniag Isolates	Augpilagtok	Scoresbysund	Angmagssalik
A	0.3578	0.2926	0.4253	0.2110	0.0895	0.2767	0.2115	0.2273	0.3558	0.4391
B	0.0900	0.0333	0.0000	0.0127	0.0000	0.0636	0.1071	0.1027	0.0143	0.1030
O	0.5522	0.6740	0.5747	0.7763	0.9105	0.6597	0.6814	0.6700	0.6299	0.4579
N^a	111	320	109	278	152	306	257	152	246	739
R^1	0.4775	0.4984	0.7156	0.5396	0.7252	0.6449	0.4436	0.5428	0.6509	0.6474
R^2	0.5135	0.4906	0.1789	0.3885	0.2748	0.2199	0.2641	0.3963	0.2250	0.3483
R^z	0.0000	0.0000	0.0000	0.0000	0.0000	0.0000	0.0000	0.0000	0.0219	0.0010
R^0	0.0047	0.0078	0.0562	0.0366	0.0000	0.0181	0.0429	0.0341	0.0232	0.0007
r'	0.0043	0.0031	0.0493	0.0353	0.0000	0.1171	0.2295	0.0268	0.0789	0.0026
r''	0.0000	0.0000	0.0000	0.0000	0.0000	0.0000	0.0000	0.0000	0.0000	0.0000
r'''	0.0000	0.0000	0.0000	0.0000	0.0000	0.0000	0.0199	0.0000	0.0000	0.0000
N^a	111	320	109	278	151	307	257	152	243	738
M	0.8243	0.8422	0.6651	0.7050	0.6118	0.6830	0.6070	0.7533	0.7805	0.9350
N	0.1757	0.1578	0.3349	0.2950	0.3882	0.3170	0.3930	0.2467	0.2195	0.0650
N^a	111	320	109	278	152	306	257	152	246	739

[a] N, population size.

Table 8.4. *Blood Group Gene Frequencies of North American Indian and Siberian Tribal Populations*

Allele	Indian				Siberian		
	Kutchin	Chilcotin	Cree	Naskapi	East Sayan	Yakut	Chukchi
A	0.0401	0.3057	0.1198	0.1809	0.1996	0.1737	0.1578
B	0.0000	0.0000	0.0072	0.0033	0.1252	0.1995	0.1233
O	0.9599	0.6943	0.8730	0.8158	0.6752	0.6267	0.7188
N^a	280	195	555	152	490	501	—[b]
R^1	0.2706	0.2113	0.3634	0.4375	0.0617	0.3571^c	0.4720
R^2	0.6527	0.5877	0.4832	0.5197	0.2223	0.3897	0.2110
R^z	0.0133	0.0225	0.0370	0.0000	0.2749	0.0480	0.1410
R^0	0.0186	0.0905	0.0000	0.0092	0.3240	0.1047	0.1750
r	0.0448	0.0880	0.0929	0.0346	0.0522	0.0773	0.0000
r'	0.0000	0.0000	0.0235	0.0000	0.0000	0.0232	0.0000
r''	0.0000	0.0000	0.0000	0.0000	0.0648	0.0000	0.0000
N^a	280	186	519	152	153	265	—[b]
M	0.8464	0.5270	0.7720	0.9276	0.5355	0.7295	0.6007
N	0.1536	0.4730	0.2280	0.0724	0.4645	0.2705	0.3993
N^a	280	185	454	152	479	501	—[b]

[a] N, population size.
[b] Sample size not given. See text.
[c] Rh frequencies would not converge. See text.

(1972) statistic for which the variance can be computed is the most appropriate. In this chapter, for comparative purposes, both E and D measures are employed.

Two successive sets of genetic distances were obtained with the data presented in Table 8.1 and Tables 8.2a and 8.2b (excluding MN data), respectively. The first set, shown in Table 8.5, lists the distances between major population aggregates (i.e., Caucasians, Mongoloids, Negroids, Eskimos, Indians, Ainu, and Siberians). The second set, shown in Tables 8.6a and 8.6b, lists the distances between smaller population groupings.

Relationship Between Major Population Aggregates

The pairwise genetic distances E and D between major populations are highly correlated, Spearman's rank correlation coefficient being $r = +0.95$ ($t_{19} = 12.30$). If one is concerned with the magnitude of the distances only, it is clear that both measures show that Eskimos and North American Indians are more similar to each other in blood group genetic characteristics than to any other population. Both measures agree that Eskimos are more similar to Mongoloids than are North American Indians and accord identical ranks to the two Eskimo–Mongoloid and two Indian–Mongoloid pairwise distances, respectively. Similarly, both E and D show that, of the seven populations compared, Negroes are the most divergent and give identical ranks to the 12 Negroid–other population pairwise distances.

Table 8.5. *Genetic Distances Between Seven Population Aggregates*[a]

Populations	E	D
Eskimo–Indian	0.308	0.022 ± 0.013
–Siberian	0.564	0.054 ± 0.035
–Mongoloid	0.420	0.037 ± 0.018
–Caucasoid	0.473	0.069 ± 0.032
–Negroid	1.150	0.340 ± 0.208
–Ainu	0.591	0.055 ± 0.021
Indian–Siberian	0.504	0.046 ± 0.026
–Mongoloid	0.499	0.050 ± 0.021
–Caucasoid	0.538	0.088 ± 0.040
–Negroid	1.130	0.330 ± 0.199
–Ainu	0.663	0.079 ± 0.035
Siberian–Mongoloid	0.624	0.073 ± 0.038
–Caucasoid	0.599	0.082 ± 0.037
–Negroid	1.115	0.326 ± 0.187
–Ainu	0.685	0.076 ± 0.037
Mongoloid–Caucasoid	0.769	0.156 ± 0.069
–Negroid	1.203	0.448 ± 0.247
–Ainu	0.381	0.011 ± 0.006
Caucasoid–Negroid	1.016	0.249 ± 0.168
–Ainu	0.838	0.169 ± 0.077
Negroid–Ainu	1.242	0.478 ± 0.262

[a] Based on frequencies of 25 blood group genes.

If, however, one is concerned with the precision of the distance measures and wishes to decide objectively whether significant differentiation has occurred between these populations, a perusal of the D measures is necessary. Of the 21 distances in Table 8.5, 12 are not significantly different from zero. Only the Negroid–other, Siberian–Eskimo, Siberian–Indian, Siberian–Mongoloid, and Mongoloid–Ainu population distances are greater than zero.

Significance tests comparing all the D distances in Table 8.5 indicate that only the Eskimo–Ainu and Mongoloid–Ainu and Mongoloid–Ainu and Caucasian–Ainu measures are statistically different. In both cases, the Mongoloid–Ainu distance is significantly smaller than either the Eskimo–Ainu distance or the Caucasian–Ainu distance. None of the other distances can be shown to differ at the 5 percent level of significance. It is therefore impossible to determine to which group—Siberian, Mongoloid, or any other—Eskimos and Indians have the greatest affinity.

The most probable explanation for these findings is that too few loci have been examined for the derivation of meaningful estimates of genetic diversity. Nei and his associates have shown that the interlocus variance contributed by random genetic drift is very large and is much larger than the sampling variance even when as few as 20 individuals per locus are compared (Li and Nei 1975). It is to minimize this large interlocus variance that information from many systems is desirable (Nei and Roychoudhoury 1974a), certainly many more than the eight considered here.

Table 8.6a. Genetic Distances (E) Between 11 Small Population Aggregates[a]

Population	Eskimo						Indian					Asian	
	South Alaska	North Alaska	Central Arctic	East Arctic	West Greenland	East Greenland	Nootka	Haida	Tlingit	North Athapascan	North Algonkian	Aleut	Ainu
North Alaska	0.444												
Central Arctic	0.376	0.271											
Eastern Arctic	0.368	0.310	0.200										
West Greenland	0.239	0.376	0.295	0.309									
East Greenland	0.404	0.244	0.260	0.230	0.334								
Nootka	0.691	0.643	0.663	0.612	0.704	0.747							
Haida	0.368	0.370	0.290	0.280	0.327	0.427	0.510						
Tlingit	0.604	0.439	0.449	0.455	0.526	0.519	0.617	0.387					
North Athapascan	0.399	0.358	0.312	0.293	0.355	0.418	0.451	0.157	0.379				
North Algonkian	0.403	0.495	0.395	0.429	0.358	0.517	0.558	0.313	0.556	0.331			
Aleut	0.578	0.711	0.726	0.704	0.625	0.694	0.790	0.700	0.927	0.673	0.583		
Ainu	0.609	0.560	0.609	0.634	0.632	0.591	0.800	0.685	0.569	0.655	0.737	0.942	
Siberian	0.470	0.642	0.649	0.674	0.506	0.668	0.679	0.603	0.720	0.552	0.497	0.573	0.685

[a] Based on the frequencies of 24 blood group genes.

Table 8.5b. *Genetic Distances (D) Between 11 Small Population Aggregates*[a]

Population	Eskimo						Indian					Asian	
	South Alaska	North Alaska	Central Arctic	East Arctic	West Greenland	East Greenland	Nootka	Haida	Tlingit	North Atha-pascan	North Algonkian	Aleut	Ainu
North Alaska	0.051 ±0.021												
Central Arctic	0.027 ±0.013	0.053 ±0.012											
East Arctic	0.028 ±0.012	0.029 ±0.016	0.008 ±0.003										
West Greenland	0.011 ±0.007	0.036 ±0.015	0.009 ±0.004	0.014 ±0.007									
East Greenland	0.042 ±0.016	0.023 ±0.016	0.018 ±0.010	0.013 ±0.008	0.026 ±0.013								
Nootka	0.098 ±0.063	0.088 ±0.045	0.104 ±0.045	0.091 ±0.047	0.111 ±0.060	0.145 ±0.074							
Haida	0.035 ±0.018	0.035 ±0.023	0.021 ±0.012	0.023 ±0.011	0.023 ±0.013	0.057 ±0.030	0.049 ±0.026						
Tlingit	0.075 ±0.032	0.029 ±0.020	0.036 ±0.017	0.054 ±0.015	0.050 ±0.022	0.059 ±0.033	0.057 ±0.032	0.016 ±0.008					
North Athapascan	0.041 ±0.024	0.026 ±0.017	0.024 ±0.013	0.021 ±0.013	0.029 ±0.018	0.050 ±0.030	0.038 ±0.020	0.004 ±0.002	0.009 ±0.004				
North Algonkian	0.027 ±0.014	0.065 ±0.032	0.031 ±0.014	0.038 ±0.017	0.024 ±0.013	0.077 ±0.033	0.064 ±0.031	0.017 ±0.013	0.050 ±0.025	0.025 ±0.016			
Aleut	0.088 ±0.074	0.158 ±0.132	0.167 ±0.148	0.148 ±0.142	0.121 ±0.102	0.150 ±0.119	0.186 ±0.107	0.158 ±0.122	0.226 ±0.183	0.156 ±0.123	0.133 ±0.102		
Ainu	0.035 ±0.024	0.072 ±0.043	0.071 ±0.037	0.071 ±0.039	0.047 ±0.027	0.086 ±0.044	0.084 ±0.046	0.060 ±0.035	0.082 ±0.042	0.054 ±0.032	0.046 ±0.021	0.101 ±0.073	
Siberian	0.059 ±0.023	0.046 ±0.027	0.060 ±0.026	0.066 ±0.033	0.066 ±0.025	0.060 ±0.034	0.117 ±0.076	0.082 ±0.035	0.076 ±0.042	0.075 ±0.036	0.101 ±0.038	0.221 ±0.192	0.076 ±0.037

[a] Based on frequencies of 24 blood group genes.

What is the rationale, then, for continuing the analysis? There is good reason to think that genetic distances obtained between subpopulations of a single species will not be significantly different if there has been gene flow between the subpopulations since initial separation. Chakraborty and Nei (1974) have shown that genetic differentiation in such groups becomes appreciable only when the migration rate between them is very small, on the order of 10^{-4} or 10^{-3}. This holds for populations of equal or unequal sizes, when gene flow is either unidirectional or two way. Nei's and Roychoudhoury's (1974b) analysis of divergence between Caucasians, Negroids, and Mongoloids, using data from 56 systems, corroborates this theoretical expectation, for no significant differences between the pairwise distances are found.

With groups more recently separated than these, even if 30 or more systems are employed, it is questionable whether significant differences in genetic distances will be detected. It is most unlikely, therefore, that the question of Eskimo and Indian origins and affinities as indicated by genetic markers (certainly the blood groups) can ever be resolved within the limits of statistical significance. What similarities or differences are found will have to be measured against similarities or differences indicated by nonbiologic criteria, which are themselves subject to an unknown amount of error.

Obviously, this can be a problem. As discussed in the introduction to this chapter, the close relationship of North American Indians and Eskimos is unexpected. However, finding that Eskimos are ranked "closer" to Mongoloids than are American Indians is in keeping with the pattern of relationship expected from the survey of the literature. Anthropologists have stressed the "Mongoloidness" of the Eskimo and have suggested an intermediate position between Mongoloids and Caucasians for North American Indians both on genetic (Neel 1976) and morphologic grounds (Montagu 1960). Although the D and E distances do not agree on the rank positions, both measures show that Indians are "closer" to Siberians than are Eskimos, a relationship predicted by the "modern" theory of Eskimo origin.

Because some relative positions are in accordance with expectations, whereas the Eskimo–Indian closeness is not, it can be argued that the former relationships are correct and the latter are not. For example, the Eskimo–Indian relationship could be a consequence of recent Caucasian gene flow into the two populations. Unfortunately, this issue is not a simple one to resolve. First, the Caucasian–Eskimo and Caucasian–Indian distances are not significantly different. Second, assuming that K and r were absent in the Eskimo prior to European contact an admixture estimate of 0.143 ± 0.014 was obtained. However, a similar mean admixture estimate cannot be calculated for Indians because the single-locus admixture estimates are significantly different (for M_B, M_r, and M_K, $\chi^2 = 19.23$). One cannot determine, therefore, whether the two groups have had similar or very different amounts of Caucasian admixture.

Fortunately, there is some evidence suggesting that European gene flow, whatever its magnitude, has not yet appreciably distorted the pattern of aboriginal genetic relationships in North America. Spuhler's (1972) studies (using a distance statistic other than E or D) on 20 indigenous North American populations have shown that linguistic within-group distances are smaller than between-group distances. He claimed that the groups compared showed less than 5 percent non-Indian admixture. However, judgment of that depended on the presence or absence of low-frequency alleles (A^2, B, K, Lu^a, or Mi^a, Js^a, V, V^w, and Wr) that are low in non-Indian populations themselves. A reexamination of Spuhler's data indicates that eight of his 20 groups do have more than 5 percent admixture. This suggests strongly that aboriginal relationships in North America can still be identified in spite of varying amounts of gene flow from European populations. With reference to the question at hand, the closeness of Eskimos and Indians relative to other populations probably reflects phyletic closeness and is not necessarily convergence produced by Caucasian gene flow.

Relationship Between Linguistic or Geographic Population Aggregates

There is historic, demographic, and genetic evidence showing that all native populations have not undergone similar amounts of admixture with Europeans. By breaking the "total" Indian and "total" Eskimo populations into sub-populations, a more precise pattern of relationship between these groups may be obtained.

Tables 8.6a and 8.6b show the genetic distances between 14 subpopulations measured by E and D, respectively. Again the distances are significantly correlated ($r_s = +0.94$, $t_{89} = 26.31$). Tests were done on all Eskimo–Indian, Eskimo–Aleut, Eskimo–Siberian, Indian–Aleut, Indian–Siberian, and Siberian–Aleut distances. None was significant at the 5 percent level. Northwest Coast Indians (Nootka, Haida, Tlingit) are as similar to Siberians as is any other group; similarly, subarctic Indians (Northern Athapascan, Northern Algonkian) are as similar to Eskimo as is any other group. It is therefore impossible to determine statistically which groups have greater affinity.

Two-dimensional representations of the relationships between these groups can be obtained by the construction of dendrograms. All such methods consider only the magnitudes of the distances and thus provide schemas of relative affinities only. Figure 8.2 shows the genetic network based on E distances. Given certain assumptions, such "minimum string" networks may be viewed as evolutionary trees representing phylogenetic lines of descent (Cavalli-Sforza and Edwards 1967). Figure 8.3 shows the dendrogram based on the D measures (henceforth called the D tree) constructed according to the method of Sneath and Sokal (1973) described in Nei (1975). Dendrograms such as this show current genetic similarities and are not meant to suggest the route of evolutionary descent.

Because the trees are constructed according to different principles they cannot be directly compared. There are, however, some striking similarities: Each has an identifiable Eskimo cluster and a Na-Dené language phylum cluster (Tlingit,

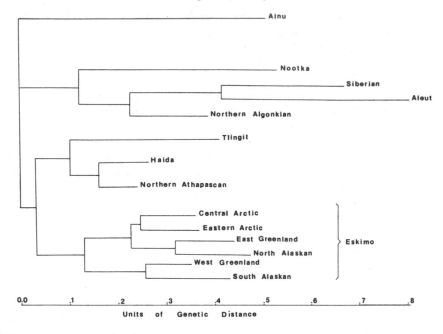

Figure 8.2. "Minimum-string" tree depicting the relationship between 14 subpopulations. Based on the frequencies of 24 genes at the ABO, Rh, MNSs, Diego, Duffy, Kell, Kidd, and P loci.

Haida, Northern Athapascan; in each the Aleut are considerably separated from Eskimo.

On linguistic grounds, Aleut would be expected to group with Eskimo (Dumond 1965). The fact that it does not must reflect the genetic composition of the Aleut of the Commander Islands. These people are the descendants of a very heterogeneous founding population that included Aleut, Kodiak Eskimo, Indians from Sitka, Kamchadal from Kamchatka, and other Siberian indigenes, as well as persons of Baltic origin (Jochelson 1928). Although the language of the group and its nucleus were Aleut, these factors alone led to its being called Aleut. Genetically speaking they are decided not Eskimoid, and their position in the dendrograms indicates this fact.

The Eskimoid cluster in the minimum-string tree includes all six of the Eskimo groups compared, in contrast to the D tree, in which North Alaskan Eskimo cluster with the Na-Dené. With the exception of this difference, both dendrograms show two basic divisions in Eskimo that are linguistically expected: the division of Eskimo into Yupik and Inyupik speakers. Unfortunately the clustering of the groups within these divisions is imperfect, for West Greenlanders are closer to South Alaskan Yupik speakers than to any of their more immediate linguistic or geographic cognates. As in the Aleut case, there are grounds for thinking that

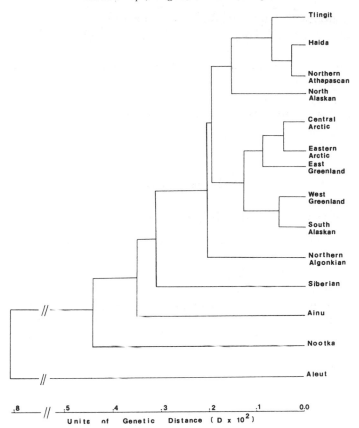

Figure 8.3. Dendrogram of the relationship between 14 populations determined by the standard distance D. Based on frequencies of 24 genes at ABO, Rh, MNSs, Diego, Duffy, Kell, Kidd, and P loci.

this odd association reflects genetic heterogeneity in West Greenlanders and South Alaskans. Both these subpopulations happen to be the Eskimo groups that have been exposed to the longest period of European contact and have had the greatest amount of European admixture. For example, the single-locus admixture estimates range from 0.19 (M_K) to 0.25 (M_r) in West Greenlanders and from 0.35 (M_K) to 0.44 (M_r) in South Alaskans. In all other Eskimo groups M_r (a more precise measure than M_K) is less than 13 percent. European admixture, then, certainly seems to affect the pattern of relationship, but in detail only and not in overall topography.

The separation of North American Indians into various branches in both trees suggests either than there have been successive waves of immigration into the New World or that tremendous differentiation has occurred if only one group entered North America. The former hypothesis is more likely to be correct. Of the Indian groups examined, the Na-Dené appear to have been very close to the groups that gave rise to Eskimo—far closer than any other Indian or the Siberian population.

There is no support in the dendrograms for closeness of Siberians (phylogenetic or otherwise) and Northwest Coast Indians (Nootka, Haida, Tlingit), nor is there support for the hypothesis that Eskimo are an offshoot of subarctic Indians (Northern Athapascan, Northern Algonkian).

Discussion

Blood Group Characteristics of Eskimos, Indians, and Siberians

Tables 8.3 and 8.4 show that after all the serologic work that has been done in the past only ABO, Rh, and MN data are uniformly available for all the tribal populations studied in this chapter. These gene frequencies do not represent the entire genome of any one of these groups and analysis done on them cannot be expected to show true population affinities. Of necessity, then, population groupings above the tribal level had to be done to maximize both sample sizes and the number of systems available for analysis. In some instances it was necessary to ignore particular findings in the calculation of gene frequencies. The two simplifications worth mentioning are the groupings of C^w phenotypes with C and of A_2 with A. The former was done because none of the Greenland or Hudson Bay Eskimo, none of the Nootka, Slave, Tuchone, Chilcotin, or James Bay Cree, or the Yakut had been tested with anti-C^w, the latter (A_2 with A) because none of the Siberians had been grouped with anti-A_1.

C^w is a rare allele in most Eskimos and Indians. Of the populations which had been tested for its presence, only the Fort Chimo, Kodiak Island, and Koniag Eskimo and the Northwest Coast Haida (J.W. Thomas, Correction of Haida Rh Phenotype, personal communication 1976) had the allele. C^w also occurs in the Aleut and in the populations of the eastern Sayans.

In the Bering Sea area the R^{1w} (C^wDe) frequency ranges from 0.0052 (Karluk and Koniag isolate "X") to 0.0500 in the Aleut. The frequencies in the east Sayans range from 0.0 to 0.0330; in one group (Todzhin) R^{zw} (C^wDE) occurs instead of R^{1w}, with a frequency of 0.0540.

It is most unfortunate that the Yakut were not tested with anti-C^w. If one can generalize from the presence of the gene in the east Sayan area it is possible that Siberians have higher frequencies of this gene than Caucasians. In such a case, the occurrence of the allele in Eskimo and Indians need not be taken as evidence of European admixture as has been suggested (e.g., Simpson et al. 1976). A similar argument may be made for the occurrence of R^0, which in North American studies is most commonly thought to be evidence of Negro admixture. R^0 occurs in all the Siberians tested (Table 8.4), in all groups with frequencies greater than 10 percent.

Relationship Between Eskimos, Indians, and Siberians

In the assessment of population affinities one consistent question is whether it is valid to rule out selection as a force responsible for unexpected convergences or divergences in the gene frequencies of any two groups. Cavalli-Sforza and Edwards (1967) have pointed out that directional selection acting uniformly over a geographic area will cause a uniform shift in the gene frequencies of populations in that area. This process would then slow down the rate of differentiation of related groups. Differential selection, operating in different ways in different areas or times, would increase the rate of diversification.

It is worth noting that efforts to demonstrate clinal variation in blood group gene frequencies of arctic and subarctic groups have not been particularly successful Simpson *et al.* (1976) postulated a north to south decline in O and N frequencies, yet the gene frequencies of Indians presented in this chapter (Tables 8.2, 8.3, and 8.4) uniformly show higher O frequencies than do the Eskimos to the north of them. Similarly, there is no clear north to south decline in N frequency in these populations.

The linguistic connections within the Na-Dené phylum embrace populations living in the subarctic as well as on the northwest Pacific coast. Similarly Eskimo occupy territory that ranges climatically from relatively benign (Kodiak Island) to very harsh across the high arctic. If selection operates differently on blood group characters in the climatic zones occupied by these Indian and Eskimo populations, it has not had a large enough effect to distort the pattern of relationship within each of these groups, for clusterings expected on the basis of non-biologic criteria are observable.

Conclusions

Regarding the questions posed at the beginning of this paper, answers can only be as accurate as the data themselves. To date far fewer Eskimos, Indians, and Siberians have had their serum protein and red cell enzyme characteristics determined than have had their blood groups assessed. The best samples yet of their respective genomes are their blood group genes. The evidence from 24 such genes indicates, however, that the question of Eskimo and Indian origins and affinities cannot be resolved within the limits of statistical significance. Other criteria must therefore be used to judge the relative merits or demerits of the relationship patterns established from the magnitudes of the distance measures alone.

Unfortunately this approach is also beset with problems. The data in this chapter do not clearly support either the Eskimo wedge hypothesis, the modern hypothesis, or the Dorset Indian hypothesis. The most conclusive finding is that North American Indians and Eskimos, contrary to common opinion, are closely related. However, the relationship is not with groups in the eastern subarctic (Dorset hypothesis) but with groups in the western subarctic and northwest coast

(Na-Dené). Insofar as Northern Athapascans occupy the drainage between the MacKenzie River and Hudson Bay, the Eskimo wedge theory, which suggests Eskimos originated in this area, would appear plausible. However, the Eskimo wedge theory postulated closeness of Siberians to Northwest Coast people, and this clearly is not the case. Genetic distance analysis, then, raises as many questions as it attempts to solve. Because the approach has proved useful in other situations (e.g. Ward and Neel 1970), it is not likely that erroneous results would be obtained in the Indian–Eskimo case. Perhaps it would be judicious to reexamine existing theories formulated from the archaeologic and ethnographic record. Concurrent with this, it would behoove physical anthropologists to turn their attention to both Siberia and the western subarctic. Very little systematic genetic work has been conducted in these areas, yet well-designed studies could do much to eliminate the problems posed by insufficient number of systems tested and small sample sizes. Northeastern Siberia, Alaska, and the Yukon and Northwest Territories are critically important regions, and clearly, to solve the problem of Eskimo and Indian ethnogenesis, work must begin there.

Acknowledgment

I am grateful to W.J. Schull and R. Ward for providing the computer programs used in the distance analyses and construction of the dendrograms.

References

Alfred, B.M., T.D. Stout, J. Birkbeck, M. Lee, and N.L. Petrakis. (1969). Blood Groups, Red Cell Enzymes and Cerumen Types of the Ahousat (Nootka) Indians. *Am. J. Phys. Anthropol.* 31:391–398.

Alfred, B.M., T.D. Stout, M. Lee, J. Birkbeck, and N.L. Petrakis. (1970). Blood Groups, Phophoglucomutase and Cerumen Types of the Anaham (Chilcotin) Indians. *Am. J. Phys. Anthropol.* 32:329–338.

Alfred, B.M., T.D. Stout, M. Lee, R. Tipton, N.L. Petrakis, and J. Birkbeck. (1972). Blood Groups and Red Cell Enzymes of the Ross River (Northern Tuchone), and Upper Liard (Slave) Indians. *Am. J. Phys. Anthropol.* 36:161–164.

Balakrishnan, V., and L.D. Sanghvi. (1968). Distance Between Populations on the Basis of Attribute Data. *Biometrics* 24:859–865.

Blumberg, B.S., J.R. Martin, F.H. Allen, J.L. Weiner, E.M. Vitagliano, and A. Cook. (1964). Blood Groups of the Naskapi and Montagnais Indians of Schefferville, Quebec. *Hum. Biol.* 36:263–272.

Cavalli-Sforza, L.L., and W.F. Bodmer. (1971). *The Genetics of Human Populations.* San Francisco: W.H. Freeman.

Cavalli-Sforza, L.L., and A.W.F. Edwards. (1967). Phylogenetic Analysis: Models and Estimation Procedures. *Am. J. Hum. Genet.* 19:233–257.

Chakraborty, R., and M. Nei. (1974). Dynamics of Gene Differentiation Between Incompletely Isolated Populations of Unequal Sizes. *Theoret. Pop. Biol.* 5:460–469.

Chard, C.S. (1963). The Old World Roots: Review and Speculations. *Anthropol. Pap. Univ. Alaska.* 10:115–122

Chown, B., and M. Lewis. (1956). The Blood Group Genes of the Cree Indians and the Eskimo of the Ungava District of Canada. *Am. J. Phys. Anthropol.* 14:215–224.

Chown, B., and M. Lewis. (1959). The Blood Group Genes of the Copper Eskimo. *Am. J. Phys. Anthropol.* 17:13–18.

Chown, B., and M. Lewis. (1961). The Blood Groups and Secretor Status of Three Small Communities in Alaska. *Oceania* 32:211–218.

Chown, B., and M. Lewis. (1962). The Blood Group and Secretor Genes of the Eskimo on Southampton Island. *Natl. Mus. Can. Bull.* 180:181–190.

Corcoran, P.A., F.H. Allen, A.C. Allison, and B.S. Blumberg. (1959). Blood Groups of Alaskan Eskimos and Indians. *Am. J. Phys. Anthropol.* 17:187–193.

Denniston, C. (1966). The Blood Groups of Three Konyag Isolates. *Arctic Anthropol.* 3:195–205.

Dumond, D.E. (1965). On Eskaleutian Linguistics, Archaeology and Prehistory. *Am. Anthropologist* 67:1231–1257.

Fernet, P., W.A. Mortensen, A. Langaney, and J. Robert. (1971). Hémotypologie du Scoresbysund (Est Greenland). Cahier du C.R.A. No. 11–12. *Bull. Mem. Soc. Anthropol. Paris* T. 8e Ser., pp. 177–185.

Fitch, W.M., and J.V. Neel. (1969). The Phylogenetic Relationships of Some Indian Tribes of Central and South America. *Am. J. Hum. Genet.* 21:384–397.

Jochelson, W. (1928). *Peoples of Asiatic Russia.* New York: The American Museum of Natural History.

Laughlin, W.S. (1962). Genetic Problems and New Evidence in the Anthropology of the Eskimo Aleut Stock. *In* J.M. Campbell, Ed., *Cultural Relationships Between the Arctic and Temperate Zones of North America.* Arctic Inst. North Am. Tech. Pap. 11, pp. 100–112.

Laughlin, W.S. (1963). Eskimos and Aleuts: Their Origins and Evolution. *Science* 142:633–645.

Laughlin, W.S. (1966). Genetical and Anthropological Characteristics of Arctic Populations. *In* P.T. Baker and J.S. Weiner, Eds., *The Biology of Human Adaptability.* London: Oxford Univ. Press, pp. 469–496.

Levin, M.G. (1958). Blood Groups among the Chukchi and Eskimo. Reprinted from *Soviet. Etnogr.* 5:113–116 in *Arctic Anthropol.* 1:87–91.

Levin, M.G. (1959). New Material on the Blood Groups among the Eskimos and Lamuts. Reprinted from *Soviet. Etnogr.* 3:98–99 in *Arctic Anthropol.* 1:91–92.

Levin, M.G. (1963). *Ethnic Origins of the Peoples of Northeastern Siberia.* H.N. Michael, Ed., Arctic Inst. North Am. Translations from Russian Sources, No. 3. Toronto: Univ. Toronto Press.

Lewis, M., and B. Chown. (1960). The Blood Groups of the Eskimo: An Analysis and Tentative Hypothesis. *Proc. Alaska Sci. Conf.* 11:10–19.

Lewis, M., J.A. Hildes, H. Kaita, and B. Chown. (1961). The Blood Groups of the Kutchin at Old Crow, Yukon Territory. *Am. J. Phys. Anthropol.* 19:383–389.

Lewis, M., H. Kaita, and B. Chown. (1957). The Blood Groups of a Japanese Population. *Am. J. Hum. Genet.* 9:274–283.

Li, W.H., and M. Nei. (1975). Drift Variance of Heterozygosity and Genetic Distance in Transient States. *Genet. Res. Cambridge* 25:229–248.

Lucciola, L., H. Kaita, J. Anderson, and S. Emery. (1974). The Blood Groups and Red Cell Enzymes of a Sample of Cree Indians. *Can. J. Genet. Cytol.* 16:691–695.

McGhee, R. (1972). *Copper Eskimo Prehistory*. Ottawa: National Museum of Canada, Publications in Archaeology, No. 2.

Meldgaard, J. (1962). On the Formative Period of the Dorset Culture. In J.M. Campbell, Ed., *Cultural Relationships Between the Arctic and Temperate Zones of North America*. Arctic Inst. North Am. Tech. Pap. 11, pp. 92–95.

Montagu, A. (1960). *An Introduction to Physical Anthropology*. Springfield, Ill.: Charles C Thomas.

Mourant, A.E. (1960). The Relationship Between the Blood Groups of Eskimos and American Indians and Those of the Peoples of Eastern Asia. *Proc. Alaska Sci. Conf.* 11: 1–9.

Mourant, A.E., A.C. Kopeć, and K. Domaniewska-Sobczak. (1976). *Distribution of Human Blood Groups and Other Polymorphisms*. London: Oxford Univ. Press.

Neel, J.V. (1976). Applications of Multiple Variable Analysis to Questions of Amerindian Relationship. Paper presented at Annual Meeting, American Association for the Advancement of Science, February 18, 1976.

Nei, M. (1972). Genetic Distance Between Populations. *Am. Naturalist.* 106: 283–292.

Nei, M. (1975). *Molecular Population Genetics and Evolution*. New York: American Elsevier.

Nei, M., and A.K. Roychoudhoury. (1974a). Sampling Variances of Heterozygosity and Genetic Distance. *Genetics* 76: 379–390.

Nei, M., and A.K. Roychoudhoury. (1974b). Genetic Variation Within and Between the Three Major Races of Man, Caucasoids, Negroids, and Mongoloids. *Am. J. Hum. Genet.* 26: 421–443.

Reed, T.E., and W.J. Schull. (1968). A General Maximum Likelihood Estimation Program. *Am. J. Hum. Genet.* 20: 579–580.

Rothhammer, F., R. Chakraborty, and E. Llop. (1977). A Collation of Gene and Dermatoglyphic Diversity at Various Levels of Population Differentiation. *Am. J. Phys. Anthropol.* 46: 51–60.

Rychkov, Y.G. (1965). Peculiarities of Serological Differentiation in Siberian Peoples. *Vopr. Antropol.* 21: 18–34. [In Russian.]

Rychkov, Y.G., and V.A. Sheremetyeva. (1972). Population Genetics of Commander Island. *Vopr. Antropol.* 40: 45–70. [In Russian.]

Rychkov, Y.G., I.V. Perevozchikov, V.A. Sheremetyeva, T.V. Volkova, and A.S. Bashley. (1969). On the Population Genetics of Native Peoples of Siberia, Eastern Sayans. *Vopr. Antropol.* 31: 3–32. [In Russian.]

Sanghvi, L.D., and V. Balakrishnan. (1972). Comparison of Different Measures of Genetic Distance Between Human Populations. In J.S. Weiner and J. Huizinga, Eds., *The Assessment of Population Affinities in Man*. London: Oxford Univ. Press, pp. 25–36.

Simmons, R.T., J.J. Graydon, N.M. Semple, and S. Kodama. (1953). A Collaborative Genetical Survey in Ainu: Hidaka, Island of Hokkaido. *Am. J. Phys. Anthropol.* 11: 47–82.

Simpson, N.E., A.W. Eriksson, and W. Lehmann. (1976). Part I. Genetic Markers in Blood. In F.A. Milan, Ed., *Circumpolar Peoples*. Cambridge: Cambridge Univ. Press. (In Press.)

Sneath, P.H.A., and R.R. Sokal. (1973). *Numerical Taxonomy*. San Francisco: W.H. Freeman.

Spuhler, J.N. (1972). Genetic, Linguistic and Geographical Distances in Native North America. In J.S. Weiner and J. Huizinga, Eds., *The Assessment of Population Affinities in Man*. London: Oxford Univ. Press, pp. 72–95.

Swadesh, M. (1962). Linguistic Relations across Bering Strait. *Am. Anthropologist* 64: 262–291.

Swadesh, M. (1964). Linguistic Overview. *In* J.D. Jennings and E. Norbeck, Eds., *Prehistoric Man in the New World.* Chicago: Univ. Chicago Press. pp. 527–556.

Szathmary, E.J.E. (1975). Genetic Relationships of Eskimo Populations. *Am. J. Phys. Anthropol.* 42:333.

Szathmary, E.J.E., J.F. Mohn, H. Gershowitz, R.M. Lambert, and T.E. Reed. (1975). The Northern and Southeastern Ojibwa: Blood Group Systems and the Causes of Genetic Divergence. *Hum. Biol.* 47:351–368.

Taylor, W.E. (1968). An Archaeological Overview of Eskimo Economy. *In* V.F. Valentine and F.G. Vallee, Eds., *Eskimo of the Canadian Arctic.* Toronto: McClelland and Stewart.

Thomas, J.W., M.A. Stuckey, H.S. Robinson, J.P. Gofton, D.O. Anderson, and J.N. Bell. (1964). Blood Groups of the Haida Indians. *Am. J. Phys. Anthropol.* 22:189–192.

Ward, R.H., and J.V. Neel. (1970). Gene Frequencies and Microdifferentiation among the Makiritare Indians. IV. A Comparison of a Genetic Network with Ethnohistory and Migration Matrices; a New Index of Genetic Migration. *Am. J. Hum. Genet.* 22: 538–561.

Ward, R.H., H. Gershowitz, M. Layrisse, and J.V. Neel. (1975). The Genetic Structure of a Tribal Population, the Yanomamo Indians, XI. Gene Frequencies for 10 Blood Groups and the ABH-Le Secretor Traits in the Yanomamo and Their Neighbours: The Uniqueness of the Tribe. *Am. J. Hum. Genet.* 27:1–32.

Won, C.D., H.S. Shin, S.W. Kim, J. Swanson, and G.A. Matson. (1960). Distribution of Hereditary Blood Factors among Koreans Residing in Seoul, Korea. *Am. J. Phys. Anthropol.* 18:115–124.

Zolotareva, I.M. (1965). Blood Group Distribution of the Peoples of Northern Siberia. *Arctic Anthropol.* 3:26–33.

Chapter 9

Genetic Differentiation in Australia and its Bearing on the Origin of the First Americans[1]

Robert L. Kirk

Introduction

A decade and a half ago Simmons (1962) comprehensively reviewed the population genetic data for peoples in the Pacific, and he discussed in particular the relationship between American Indians and Polynesians. Simmons was restricted by the limited amount of data available at that time, most of which were for the distribution of alleles in the ABO, MNS, and Rh blood group systems, with meager information for some of the other blood group systems.

During the intervening period, from the publication of Simmons' review to the present, population genetics has become a more exact discipline. Information is available now for many more genetic marker systems, and sophisticated approaches have been developed for interpreting the data in terms of genetic relatedness between populations. In addition new evidence from archaeology, linguistics, and anthropometrics has added greater breadth to the study of human origins and migrations in the Pacific area, topics so lucidly summarized by Howells (1973) in *The Pacific Islanders,* and by Bellwood (1974). My purpose in this chapter is to complement much that has been written recently in these areas of study by updating the population genetic data and applying the concepts of genetic distance analysis to the problem of microevolution in the western Pacific and Australasian region and also to add to the genetic evidence on the relationship between the American Indians and other Pacific and circumpacific peoples.

---

[1] This paper is dedicated to the late Dr. Roy Simmons—a pioneer in the study of gene distribution in Pacific peoples.

Genetic Markers

Most recent genetic studies of human populations have concerned themselves with the distribution of alleles at loci controlling traits detectable in samples of blood: cell antigens, serum proteins, and enzyme systems in cells or serum. Although other traits showing simple Mendelian inheritance can be studied, relatively few have been documented for a sufficient number of populations to make worthwhile detailed discussion of them here.

Red Cell Antigens

The distribution of red cell antigens in Australia has been reviewed by Kirk (1965) and in New Guinea and other southwest Pacific Islands by Simmons and Booth (1971). Simmons (1966) commented on the general features of blood group distribution in the Pacific as a whole, and a comprehensive summary of data for Japan and neighboring territories by Akaishi and Kudo (1975) has appeared. There are certain characteristics of the blood group distributions that are well known: the absence of the gene A_2, except in rare instances; the absence of the rhesus negative gene (*cde*) over the greater part of the region; a clinal variation of the M antigen from higher values in the north; the low values of the S antigen in many places and its complete absence in Australian Aborigines; complete absence of Kidd and Lutheran (a+) reactions in nearly all populations; and nearly 100 percent positive reactions for Fy(a), but with relatively few tests for Fy(b). A complete tabulation of blood group data has been published recently by Mourant *et al* (1976).

Of special interest is the distribution of Diego (a+) reactions. This was reviewed earlier by Layrisse and Wilbert (1960) but additional information has made possible the preparation of Table 9.1. Di(a+) reactions have not been observed in Australia, New Guinea, Melanesia, Micronesia, or Polynesia. Di(a) therefore appears to be a specific Mongoloid marker that has been carried into the Americas and into a few other populations subject to gene flow from neighboring Mongoloid groups (Kirk *et al*. 1962). Di^a values fluctuate from zero to over 40 percent in different American Indian populations (Post *et al*. 1968), but the mean value based on 112 series is 3.7 percent. The Kidd blood group system is polymorphic in those populations in which it has been tested, and for the other known blood group systems data either are not available, are restricted to a few populations only, or are of variable quality. This latter is true for the P system and especially for the Lewis system. In most surveys Lewis results are not given for each ABO blood group, and the ABH secretor status also is not available. In addition tests in most cases have been carried out only with anti-Le(a). Where careful study has been made, as by Vos and Comley (1967) for Australian Aborigines, it is clear that interpretations made on the basis of results for Caucasians are not adequate.

Some studies have revealed the presence of unusual red cell antigen variants, and some of these reach polymorphic frequencies over part of the range. A unique

Table. 9.1. *Percentage Frequency of the Di^a gene in selected Pacific and Circumpacific Peoples*

Locality	Di^a (percent)	Locality	Di^a (percent)
Japan		Canada	
Ainu	4.32	Eskimo (Igloolik)	1.00^a
Japanese	4.91	Others	0.00
Ryukyuans	3.28	American Indians	
Taiwan		Chippewa	5.87
Atayal	1.84	Cree	6.20
Thailand		Blackfoot	2.31
Bangkok	3.80	United States	
Northern Thais	3.70	Athabascan	<0.2
Maeo and Yeo	1.53	Apache	1.26
India		Tuscarora	4.50
Oraons	2.43	Central America	
Bengalis (low caste)	1.81	Maya	3.70
Malaysia		Venezuela	
Malays	1.25	Warrau (Guayo)	1.87
Senoi	0.00	Warrau (Winikina)	0.00
Land Dyak	2.49	Guahibo	7.53
Sea Dyak	0.00	Piaroa	6.46
Singapore		Brazil	
Chinese	2.33	Yanomama	0.00
China		Caingang	9.08
Cantonnese	2.54	Colombia	
New Guinea		Noanama	0.00
Various localities	0.00	Peru	
Australia		Aymara	9.03
Aborigines—various	0.00	Quechua	12.84
localities		Chile	
New Hebrides		Mapuche	2.13
Banks Islands	0.00		
Micronesia	0.00		
Chamorro	4.11		
Polynesia			
Tokelau Island	0.00		
Marquesas	0.00		
Easter Island	0.00		
Cook Islands	0.00		
New Zealand			
Maoris	0.00		

^{a} An aberrant result: Two Di(a+) persons only out of nearly 1500 Eskimos tested in Alaska and Canada.

variant of the Rh E antigen was reported by Vos and Kirk (1962) in high frequency among Aborigines in the center of Australia, but it was absent in the northeast of the continent. Booth *et al.* (1972a) similarly has plotted the distribution in New Guinea of a variant of the N antigen. Of greater interest is Booth's *et al.* (1970) discovery of a high frequency of persons negative for the Gerbich blood group antigen in many parts of New Guinea. Extensive studies of the distribution of the *Ge*$^-$ allele in New Guinea have been summarized by Booth *et al.* (1972b). He believes this allele was introduced by Austronesian language speakers who pushed up the Markham Valley and into other areas of coastal New Guinea. However, there is not a one-to-one correlation with language, so that other factors, such as malaria, may well have affected its distribution. The matter is further complicated by the discovery of another example of anti-Ge(a) which gives different results in some cases: further testing is in progress and the final analysis should be of very great interest.

Space does not permit me to discuss in more detail some other aspects of red cell antigen distributions in the Pacific and circumpacific peoples. What is clear, however, is the need for (a) greater standardization of antisera used by different investigators, and (b) the importance of screening pregnancy sera in local populations to find antisera against specific alleles that do not occur in Caucasians or Black Americans. The discovery of Di(a), E^T, N^A, and Ge$^-$ demonstrate the wealth of information on local genetic differentiation that can be uncovered if this is done—although, incidentally, anti-E^T was found as a naturally occurring antibody in a 70-year-old male.

Serum Proteins

The world distribution of haptoglobins and transferrins up to 1968 was summarized by Kirk (1968, 1969a). Additional data for New Guinea are given by Simmons and Booth (1971) and for Japan and neighboring populations by Omoto (1975a). The latter review also includes data on the Gc types, immunoglobulin allotypes, serum cholinesterase loci E_1 and E_2, protease inhibitor, and the low-density lipoprotein types, Ag(x) and Ag(y). Little information is available for other parts of the Pacific for the last four loci.

Several of the better documented serum protein systems are of great importance for understanding genetic differentiation and these will require more extended discussion.

Transferrin

I have pointed out elsewhere (Kirk, 1976) the sharp division between the distribution of the two transferrin alleles, TfD_1 and TfD_{Chi}. TfD_1 occurs in all Australian Aborigine populations, in nearly all New Guinea populations, and in many other parts of Melanesia. By contrast, TfD_{Chi} is found, with somewhat lower frequency, in Japanese, Chinese, Malays, Micronesians, and Indonesians, and in Ainu and some but not all American Indian populations. It is not found in Polynesia, Melanesia, New Guinea, or Australia. It seems that TfD_{Chi} is a Mongoloid marker of some antiquity. Some years ago I estimated (Kirk 1969b) that the D_{Chi} mutation could have occurred in a Mongoloid precursor population some 33,000 to 80,000 years ago. Similarly I estimated the D_1 mutation to have occurred 100,000 or more years ago.

Several other transferrin variants with more restricted distribution in Pacific and Australian populations have been described. In Cape York a TfB gene occurs in one Australian Aborigine population with a frequency of 6 percent (Kirk et al.

1963a). Another B allele, Tf^B_{Lae}, was found in my laboratory by Lai (1963) and it is localized to populations in New Britain and neighboring areas of New Guinea. Another D variant, TfD^{Manus} is restricted to populations in the Admiralty Islands (Malcolm *et al.* 1972).

The Group-Specific Component

The world distribution of Gc has been reviewed by Cleve (1967). Detailed figures for Japan are given by Omoto (1975a) and for a number of localities in Australia by Kirk (1965) and for New Guinea by Kitchin *et al.* (1972). In Australia and New Guinea there is a variant of Gc discovered first in an Australian Aborigine population (Cleve *et al.* 1963) but now found widespread in New Guinea and other parts of Melanesia. It has not been detected in other parts of south and southeast Asia (Kirk *et al.* 1963b) or in Japan (Kirk 1969a). Like the transferrin allele TfD_1, Gc^{Ab} appears to be a very old mutation, and it is interesting to note that, as in the case of TfD_1, a possibly identical allele occurs also in Black African populations (McDermid and Cleve 1972).

The Immunoglobulin Allotypes

Three main genetic loci controlling the immunoglobulin allotypes are currently providing information of interest to population geneticists. The first is the complex allelic series controlling changes in the structure of the "heavy" chains of the IgG molecules in the serum. These are the Gm groups, and we will deal with them here as though they were from a single locus: It is more likely that they represent at least three separate but closely linked cistrons, but this is a complication that can be ignored for the moment. What is important is that the products of the various alleles occur together in complexes called allogroups. The second is a smaller number of alleles controlling changes in the structure of the kappa class of "light" chains of the IgG molecules. These alleles belong to what was known formerly as the Inv groups but are now renamed as the Km groups. Finally, the most recently discovered, and least studied, is a series of alleles controlling the structure of the IgA molecules, which are present in exocrine secretions as well as serum and designated as the A_2 m groups.

The first study of the Gm and Km groups in Asia and Australasia was published in 1963 by Vos *et al.* (1963) using the limited range of testing reagents available at that time. More detailed studies for New Guinea and Australia were published by Friedlaender and Steinberg (1970), Steinberg *et al.* (1972), Giles *et al.* (1965), and Steinberg and Kirk (1970). Testing with a more comprehensive battery of reagents has been carried out in the same areas by Curtain and his collaborators (1972, 1976). For the western Pacific and east and southeast Asian areas Schanfield's and Steinberg's laboratories have been most active and important contributions have been made also by Matsumoto's laboratory in Japan (see Omoto 1975a for references).

Gm and Km data for selected populations in the Pacific and circumpacific area have been summarized in Table 9.2 and the essential features of the Gm data

Table 9.2. *Gammaglobulin Allogroup Frequencies in Selected Pacific and Circumpacific Peoples*

Population	a:g za:g 1, 21 1, 17, 21	ax:g zax:g 1, 2, 21 1, 2, 17, 21	a:b⁰b³st za:b⁰b³st 1, 5, 13, 14 1, 10, 13, 15, 16, 17	a:fb a:fb⁰,¹,³,⁴ 1, 3, 5, 13, 14 1, 3, 10, 15	x:g 2, 21	KM 1 InV 1	A₂M 1
Ainu	52.2	0	27.3	2.6	17.9	21.8	
Japanese	47.0	17.5	27.6	7.8	0	30.6	
Korean	50.7	21.3	13.8	14.1	0	30.5	
Formosa							
Atayal	15.5	2.5	0	82.1	0	18.4	
North Chinese	41.3	13.8	9.0	35.9	0	31.1	
South Chinese	24.6	10.6	7.4	57.4	0	33.2	
Ryukus							
Okinawan	47.8	16.2	30.2	5.7	0	24.4	
Malaya							
Senoi	0.6	0	0.6	98.9	0	31.5	
Moluccans	12.4	1.4	3.5	82.3	0	12.1	
New Guinea							
Motu	1.2	0	15.8	83.0	0	20.3	26.0
Eastern Highlands	75.4	3.7	15.3	5.6	0	1.8	85.0
Australia							
Central	79.7	15.2	0	0	0	—	94.0
North	74.2	9.1	16.7	0	0	36.6	86.0
Micronesia							
Pingelap	5.3	1.1	0	92.7	0	25.3	
New Hebrides							
Tongariki	48.5	11.9	0	39.6	0	24.5	
America							
Eskimo	87.9	0	12.1	0	0	53.2	29.0
Ojibwa	86.0	7.1	6.9	0	0	28.6	
Athabascan	62.3	19.8	15.6	2.4	0	—	
Papago	92.5	2.5	1.5	0	0	22.5	
Peruvian	75.0	16.0	8.0	3.0	0	—	
Xavante	78.0	22.0	0	0	0	37.0	

are shown diagrammatically in Figure 9.1. The Km alleles do not vary in such a striking manner as those in the Gm groups: the A_2m groups have not been documented so extensively, but sufficient is known already to indicate marked differences in the frequency of A_2 m alleles in various populations (M.S. Schanfield, personal communication 1976).

Some of the most detailed studies of the Gm, Km, and A_2 m groups have been carried out in New Guinea and Australia, and these have been summarized recently by Curtain *et al.* (1976). Briefly, among the Papuan speakers in New Guinea, the allogroup Gm(za:n:b) combined with A_2m(1) is common, whereas, by contrast, among the Melanesian speakers the allogroup Gm(fa:n:b) with either A_2m(1) or A_2m(2) is common. It is suggested that these latter combinations are Melanesian markers, but there is not a completely clear-cut separation along language lines in many areas because of gene flow between later and earlier

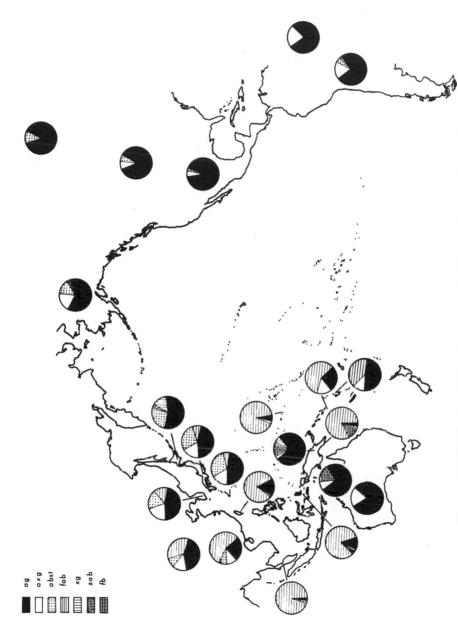

Figure 9.1. Distribution of Gm allogroups in Pacific and circumpacific peoples.

populations. This is particularly true among Melanesian speakers in the Markham Valley, who have considerable frequencies of the Gm(za:b) allogroup, and among the Papuan speakers in New Britain, where the Gm(fa:b) allogroup is present with frequencies comparable to that among the Melanesian speakers.

Australia also is heterogeneous for the distribution of the Gm allogroups. In the center of the continent only Gm (za:n − :g) and (zax:n − :g), mainly with $A_2 m(1)$, appear to be present, whereas in the north there is Gm(za:n:b) with $A_2 m(1)$ in addition.

Curtain and his colleagues interpret their data as indicating that the earliest populations in the Australia–New Guinea region possessed the allogroups Gm(za:n − :g) and Gm(zax:n − :g) with very high frequencies of $A_2 m(1)$, and today these people are represented by the desert groups in the center of Australia and in the Western Highlands of New Guinea. In the Western Highlands, however, Gm(za:n:b) and Gm(za:n − :b) also occur and these allogroups may have been introduced by a later, but still pre-Melanesian group of people moving into New Guinea. These allogroups are found today also among Papuan-speaking populations on the south coast of New Guinea (both in the Asmat of Irian Jaya and the Fly River delta of Papua New Guinea) and one of these, Gm(za:n:b), was carried into northern populations in Australia.

Finally, Melanesian-speaking peoples could have introduced the Gm(fa:n:b) allogroup, which occurs with both the $A_2 m(1)$ and the $A_2 m(2)$ alleles. Curtain et al. (1976) comment that it is strange that the Gm(fa:n:b) complex was not introduced into Australia. This may, however, be an apparent rather than real absence of the allogroup Gm(fa:n:b) in the north of Australia, because not many populations have been tested adequately. Moreover, our own genetic distance studies, utilizing information from a larger number of loci does indicate closer affinities with Melanesian-speaking groups than with Papuan speakers.

In the broader context of the circumpacific populations Curtain et al. (1976) comment that Schanfield believes the allogroup Gm(za:n − :g) in combination with $A_2 m(2)$, because of its wide distribution in Asia and among American Indians, is a very "old" allogroup. Unfortunately, most surveys have been carried out without the benefit of testing for the n factor, a deficiency which makes difficult a critical evaluation of Schanfield's interesting hypothesis.

Leukocyte Antigens

One of the most rapidly developing areas of research in human genetics is concerned with unraveling the complexities of the genetic control of antigenic structures on the surface of tissue cells, particularly the lymphocytes. Of great medical significance because of the role such antigens play in the rejection of "foreign" tissue, they also are being invoked as significant factors in conferring resistance or susceptibility to a variety of diseases.

At least four loci, all relatively closely located on a portion of chromosome 6, are involved. Only two of these, HLA-A and HLA-B (formerly HLA first and second loci) have been studied in enough populations to enable useful comparisons to be made, although the next few years no doubt will see the present situation changed drastically. The best recent review of work on the HLA groups is in the volumes *Histocompatability Testing 1972* (Dausset and Colombani 1973) and *Histocompatibility Testing 1975* (Kissmeyer-Nielsen 1976).

The HLA-A and HLA-B loci each contain a large number of alleles. The cell surface antigens they control are recognizable by antisera, some of which are not monospecific, so that testing must be carried out with a large number of antisera to enable cross-reactivity to be evaluated. To obtain comparability of results in different parts of the world carefully selected antisera have been made available to collaborating laboratories, and this has enabled us to get a preliminary view of the distribution of alleles at the HLA-A and -B loci for Pacific and circumpacific peoples. Table 9.3 summarizes these results, some of which for HLA-A are reproduced in diagrammatic form in Figure 9.2.

It is important to note the figures in the Table 9.3 column headed "O." This indicates the difference between the summated frequencies of all other alleles detected at the locus in question and unity: In some cases it is a measure of the lack of specificity of testing reagents developed in one population when used to define antigens present in another. Further research on the production of "locally" defined antisera therefore is of great importance, and its potential success can be illustrated with reference to Chinese populations. Initial testing of Chinese in Hong Kong and Singapore, mainly originating in provinces in the south of China, revealed 16 percent "O" at the HLA-A locus and 25 percent "O" at the HLA-B locus. It was concluded that Chinese probably possessed cell surface antigens not present in Europeans and therefore not detectable with antisera derived from European sources. Careful screening of pregnant Chinese women in Singapore finally yielded antisera that did recognize at least some specifically Chinese cell surface antigens (Simons *et al.* 1975): In this case an antiserum with similar specificity was also obtained from a Chinese woman in California (Payne *et al.* 1975). Similar screening of pregnancy sera from other "local" populations will almost certainly reveal specific alleles with highly restricted distributions. Their detection is of importance for understanding transplantation reactions in such populations but also will be of great significance in population genetic studies.

Because only limited studies of leukocyte antigens have been carried out so far in Australia and New Guinea, it is possible to outline only the most striking features of the distributions. In both areas there is a restricted number of alleles present, and the frequency of "O" alleles is very low. In New Guinea and Australia four alleles (A_2, A_9, A_{10}, A_{11}) at the HLA-A locus have been detected, and at the HLA-B locus four alleles (B_{13}, Bw_{40}, Bw_{15}, Bw_{22}) are present in Australia and in the New Guinea highlands, with the addition of B_{18} in both Takia and Waskia populations of Karkar Island. Both New Guinea and Australian populations are broadly similar in the distribution of allele frequencies, except for the additional allele on Karkar, but at the HLA-A locus the highland New Guinea populations have a very high frequency of A_9, whereas among the Karkar

Table 9.3. Human Leukocyte Antigen Gene Frequencies in Pacific and Circumpacific Peoples

Population	No. tested	A_1	A_2	A_3	A_9	A_{10}	A_{11}	A_{28}	"O"	B_5	B_7	B_8	B_{12}	B_{13}	Bw_{35}	Bw_{40}	B_{14}	Bw_{15}	Bw_{16}	Bw_{17}	B_{18}	Bw_{22}	B_{27}	"O"a
										Locus HLA-A →				Locus HLA-B										
Ainu	127	0	33	2	32	20	4	4	5	13	1	0	4	0	0	12	0	28	—	1	2	2	4	33
Japanese	120	0	28	2	37	14	5	2	12	23	5	0	7	1	0	20	0	7	—	1	0	9	0	27
Southern Chinese	155	1	33	1	16	4	29	0	16	5	1	0	1	11	3	19	0	18	—	8	1	6	2	25
Filipinos	144	0	11	1	46	5	22	1	14	5	4	0	2	3	12	22	0	18	—	4	4	2	3	21
Northern Vietnamese	217	2	25	1	11	2	15	1	43	5	8	0	4	3	3	3	0	9	—	4	0	4	1	56
Koreans	150	3	24	2	14	8	11	8	30	13	3	1	11	5	7	11	1	10	—	7	0	8	4	19
Malays	104	3	12	3	43	1	17	1	20	7	6	0	4	6	10	5	0	27	—	6	0	2	2	25
Moluccans	139	1	12	0	55	3	6	0	23	4	3	0	1	8	24	9	0	19	—	0	7	1	6	18
New Guineans Eastern Highlands	214	0	0	0	76	4	5	0	15	0	0	0	0	8	0	23	0	29	—	0	0	29	7	4
New Guineans Waskia	152	0	8	0	24	11	59	0	0	0	0	0	0	15	0	25	0	7	—	0	11	26	10	6
New Guineans Takia	116	0	2	0	23	9	61	0	5	5	0	0	0	19	0	17	0	9	—	0	13	22	6	9
Australians Waljbiri	172	0	13	0	33	41	10	0	3	0	0	0	0	21	0	27	0	22	2	0	0	32	0	0
Fijians	76	0	8	0	59	10	13	0	10	1	0	0	0	4	0	21	0	19	—	1	4	25	2	23
Samoans	—	5	27	2	27	3	9	2	25	0	1	0	1	4	1	43	0	2	—	1	—	27	1	19
Tokelauans	142	0	21	0	46	8	13	0	12	2	0	3	4	5	2	67	0	0	—	0	—	13	0	4
Americans Eskimo	236	3	19	3	64	0	0	—	11	12	6	3	0	0	1	51	0	6	—	0	0	12	—	10
Papago	101	0	51	1	37	1	1	0	9	5	1	1	2	0	16	24	0	1	18	1	1	0	6	24
Zuni	85	0	51	0	31	0	1	3	14	9	0	1	0	0	28	19	0	5	17	1	0	1	8	11
Pima	62	0	49	2	49	0	0	3	0	3	2	0	0	0	8	20	2	0	14	0	0	0	10	41
Ixil	79	2	34	0	20	0	7	25	12	9	0	0	3	0	25	15	1	0	21	1	3	0	0	22
Maya	58	0	49	0	20	1	0	19	11	1	0	1	0	0	69	8	2	0	18	1	0	0	0	0
Quechua	90	1	59	1	20	0	1	10	8	4	4	1	2	1	36	7	2	19	11	1	2	1	1	8
Whites, USA	250	16	27	13	9	9	7	6	13	6	12	9	16	2	16	5	4	4	—	4	6	2	2	12
New Zealanders Whites	492	19	32	17	8	5	5	1	13	3	23	13	21	1	6	7	3	3	—	3	—	3	3	11

a "O" = blank.

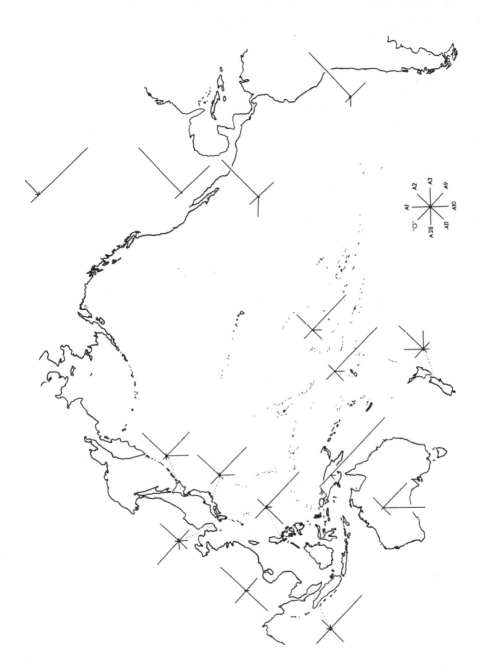

Figure 9.2. Relative frequencies of alleles at the HLA-A locus. The New Zealand figures are for Caucasians.

Islanders the most frequent allele is A_{11}. The Waljbiri from central Australia have frequencies more similar to the New Guinea highlanders.

Although information for circumpacific peoples is more limited, certain clines and specific allele distributions are already evident. As indicated in Figure 9.2 the most prominent allele at the HLA-A locus in many Pacific populations and Eskimos is A_9. A_2 is important in Chinese, Japanese, and Ainu and is the most frequent allele in American Indians. A_{10}, which is absent in American Indians, has modest frequencies in Ainu, Japanese, and other Pacific Islanders but is the most frequent allele in the Waljbiri of central Australia. However, A_{11}, also absent in most American Indian populations, is present in the Pacific, particularly among Malays, Filipinos, and southern Chinese: It is less frequent among the Waljbiri and the Eastern Highlanders of New Guinea. It is important to note that the Caucasian alleles A_1 and A_3 are virtually absent in all Asian, Pacific, and American aboriginal populations.

Certain alleles at the HLA-B locus are of significance in the whole region when contrasted with those for Caucasians. Bw_{40} is prominent in Eskimos and Pacific Island populations but is largely replaced in American Indians by Bw_{35} and Bw_{16}. Bw_{15} also is present with reasonable frequencies in Ainu, Chinese, Malays, and some American Indians and achieves its highest frequencies among the Eastern Highlanders of New Guinea and the Australian Aborigines. The latter two groups also have the highest frequencies of Bw_{22}. This allele is present in some Pacific Island populations, and with reduced frequencies in Chinese and Malays; Eskimos have a value similar to that for some Pacific Islanders. Finally, the characteristic Caucasian alleles B_7, B_8, and B_{12}, are either absent or have only a low frequency in the populations of the region in which we are interested here.

Red Cell Enzymes

More than 20 loci controlling enzymes detectable in red blood cells are routinely screened in our laboratory. Similar information has been published for other parts of the Pacific by Ishimoto (1975), by Lie-Injo (1976), and in more restricted surveys by a number of other authors (see for example Omoto 1973).

A number of the loci are monomorphic over the whole area or show the presence only of sporadic rare alleles. Among these are the loci controlling:

1. Adenylate kinase. This is monomorphic also in Eskimo and American Indian populations but is polymorphic in Europeans, Black Africans, western Asians, and Asiatic Indians.

2. Superoxide dismutase. Monomorphic in all parts of the world except for parts of Scandinavia and populations influenced by migration from that region.

3. Lactate dehydrogenase. Polymorphic in India and in one population only in the Pacific region; sporadic variants elsewhere.

4. Glutamic oxaloacetic transaminase. Monomorphic except for sporadic variants in most populations. Appears to be just polymorphic in Japanese, and

GOT^2 has a frequency of 2 percent in Athabascan Indians. It could become a valuable marker in American Indians.

5. Isocitrate dehydrogenase. Sporadic variants only except for polymorphism in some populations in New Guinea.

6. Peptidase B. Polymorphic in Black Africans; monomorphic elsewhere except for some populations in Irian Jaya and Australia. Some areas have not been surveyed. Other peptidase systems may be polymorphic but data are scanty and unreliable.

7. Phosphohexose isomerase. Sporadic variants only except among Asiatic Indians, where frequencies approximate 1 percent.

The other loci controlling red cell enzymes are either universally polymorphic or reveal geographically restricted polymorphism, which is of value in throwing light on human migration routes in the Australian and western Pacific region. Briefly, the salient features of the distribution of alleles in these systems are as follows:

1. Red cell acid phosphatase. Effectively only two alleles, p^a and p^b, are present in the whole region, with p^b fixed in some populations. The European allele p^c and the Black African allele p^R have been detected in a few individuals, their presence possibly resulting from admixture or from independent mutation. The p^a allele most frequently is in the range of 10 to 25 percent, the only striking exception being the value of 70 percent reported for a small series of Easter Islanders (Thorsby *et al.* 1973). It is of interest that other high values reported are for Eskimo (45 percent at Igloolik, with 1 percent p^c; McAlpine *et al.* 1974) and Athabascan Indians (67 percent; Scott *et al.* 1966). By contrast p^a in the Yanomama is only 8 percent (Weitkamp and Neel 1972) and our own study of the Noanama in Columbia gives $p^a = 22$ percent (Kirk *et al.* 1974).

2. 6-Phosphogluconate dehydrogenase. Two alleles present in most populations, with sporadic cases of other alleles. In the western Pacific values for PGD^C are about 5 percent, but with higher values in Island Melanesia, and the highest value of 25 percent in the Polynesian outliers (Rennell and Bellona, unpublished results). Values for PGD^C in the Americas are significantly lower than elsewhere, being zero in the Yanomama and Athabascan, 3 percent in the Aymara, 2 percent in Maya, and 1 percent in Noanama and also in Igloolik Eskimo.

3. Phosphoglucomutase (locus 1). Polymorphic in all but a few populations. The distribution has been reviewed recently by Blake and Omoto (1975). In addition to the universally polymorphic alleles PGM_1^1 and PGM_1^2, three less frequent alleles are of great interest in southeast Asia and the western Pacific. The populations in which these alleles have been detected are shown in Figure 9.3. All three occur in the south and southeast Asian area and in the northwestern Pacific except among the Ainu. Elsewhere PGM_1^6 has been found only at two localities, but PGM_1^3 and PGM_1^7 have a wide distribution in the southwest Pacific. PGM_1^7 appears to be a Melanesian gene or a more recent introduction. None of these three genes has been detected in the Americas so far.

Figure 9.3. Presence or absence of specific markers at the PGM_1 locus in the western Pacific and east Asia. ▲, PGM_1^3; ⊙, PGM_1^6; ■, PGM_1^7; *, no variants detected.

4. Phosphoglucomutase (locus 2). Monomorphic in nearly all parts of the world except among Black Africans and some parts of Australia and in the south-west Pacific. Blake and Omoto (1975) have reviewed the distribution of the specific alleles PGM_2^9 and PGM_2^{10}. The only other populations in which one or both occur are the Banks and Torres Islands and the Solomon Islands. Among the Waljbiri of central Australia the PGM_2^3 allele appears to be the same as that detected in Black Africans.

5. Glutamic pyruvic transaminase. Universally polymorphic for two alleles, GPT^1 and GPT^2, with sporadic occurrence of other alleles. There appears to be a cline of decreasing GPT^1 frequencies from north to south in the western Pacific, but with very high values in the highlands of New Guinea and in central Australia, but lower values in Melanesian and Polynesian populations (Blake 1976). In American Indians few studies have been reported, but the Aymara have 41

percent (Van der Does *et al.* 1973), the Athabascan 56 percent (Duncan *et al.* 1944), and Igloolik Eskimo 62 percent (McAlpine *et al.* 1974).

6. Esterase D. Universally polymorphic for two alleles, EsD^1 and EsD^2, with sporadic rarer alleles. A similar picture to that for GPT, with decreasing values for EsD^1 from north to south in the Pacific except for higher values in New Guinea and Australia (Blake 1976). Mestriner *et al.* (1976) have published EsD values for eight South American Indian groups which show a range for the EsD^1 allele from 36 to 100 percent.

7. Malate dehydrogenase. Monomorphic in nearly all parts of the world except for New Guinea and parts of Island Melanesia and Micronesia. The allele in New Guinea and Island Melanesia (Blake *et al.* 1970) is different from that found in the Western Caroline Islands (Blake 1978).

8. Phosphoglycerate kinase. A sex-linked locus, monomorphic in nearly all parts of the world; polymorphic in New Guinea, Island Melanesia, and Micronesia. Two different variant alleles are present in New Guinea, but the more common of these, PGK^2, is a marker for other parts of the Western Pacific and some coastal areas of New Guinea. No information is available for the Americas except for the Igloolik Eskimo, where no variants were detected.

9. Glucose-6-phosphate dehydrogenase. Another sex-linked locus, monomorphic in populations except those exposed to endemic malaria. This is true also for the region at present being considered. High frequencies of the Gd^{B-} allele, controlling G6PD deficiency of the European type, occurs in the Solomons, New Hebrides, and coastal areas of New Guinea, all highly malarious, but not elsewhere.

10. Adenosine deaminase. Probably universally polymorphic, but information is available for only a limited number of populations in the region. Adenosine deaminase distribution in the Pacific has been reviewed by Omoto (1972, 1973), and Vergnes *et al.* (1976) have found the ADA^1 allele to be 100 percent in a small series of Siriono and Aymara. We consider results based on samples collected in the field under less than ideal conditions to be unreliable.

11. Carbonic anhydrase. Two separate loci are involved, *CAI* and *CAII*. Recently, new techniques have made screening for these loci relatively easy, but so far only a few populations have been examined. The original work of Tashian *et al.* (1963) and of Lie-Injo (1967) suggests that at least one variant occurs in southwest Pacific populations. Our unpublished observations however indicate that these variants are rare or absent in most populations in the area.

12. Glyoxalase. Work in our own laboratory and in that of Omoto and collaborators in Tokyo suggests this is universally polymorphic, with high values of the GLO^2 allele in western Pacific populations. No information is available yet for the Americas.

13. Other enzymes. Several other enzyme systems have been examined in a few populations, but there are not enough results to allow meaningful analysis. These include: aldolase, glutathione reductase, glyceraldehyde-3-phosphate dehydrogenase, hexokinase, phosphofructose kinase, aconitase, pyruvate kinase, diaphorase, and uridine monophosphate kinase. The latter may be universally

polymorphic and should be tested on many more populations. Unfortunately the enzyme is not particularly stable, which makes the use of stored samples difficult.

Hemoglobin

Abnormal hemoglobins occur sporadically in all populations and in some achieve polymorphic frequency. In the region under consideration here only two hemoglobin variants, HbE and HbJ Tongariki, are of interest, although a number of other variants has been reported, especially in Japan (Yanase and Imamura 1975). HbE is widely distributed in southeast Asia with considerable frequencies, but it does not occur in Taiwan or further north, or in New Guinea, Australia, or other parts of the Pacific, nor does it occur in Eskimo or American Indians. HbJ Tongariki, in contrast, was detected first in the New Hebrides (Gajdusek *et al.* 1967) and has now been found in various parts of Island Melanesia and at a few coastal localities in New Guinea. However, it does not occur in the New Guinea highlands or in Australia, Micronesia, or Polynesia. HbJ Tongariki therefore appears to be a Melanesian marker of relatively recent origin, and its distribution may also have been favored in populations where malaria was endemic.

Genetic Distance and Population Diversity

I have pointed out already that specific genetic markers can provide valuable insights into the relatedness of populations in the Pacific and circumpacific region. For example, American Indians can manifest their Mongoloid affinities not only in morphologic traits but also through the presence of such alleles as Di^a and TfD^{Chi} and in the array of alleles in the Gm and HLA systems. Similarly in the southwest Pacific area Austronesian-speaking peoples reveal some genetic relationships among themselves through the presence of such alleles as $HbJ^{Tongariki}$, PGK^2, and PGM_2^{10}. The movement of these alleles into non-Austronesian populations can be some guide also to the extent of gene flow across the interface separating two major linguistic divisions.

There are serious limitations, however, to the amount of information that such gene markers can provide. In particular, genes with low frequencies can be lost readily through random drift and also when small populations subdivide, or when founding populations stem from a small number of migrants. For this reason it is necessary also to utilize the information provided by the frequencies of alleles in all systems, using the methods of genetic distance analysis.

During the last few years several approaches to the measurement of genetic distance have been propounded (see Crow and Denniston 1974, Morton 1973 for reviews). Our experience is that each of these measures gives essentially the same results for populations not long separated in time, where gene divergence has been maximized.

Differentiation in New Guinea and Australia

In both Australia and New Guinea several workers have calculated genetic distances and in some cases have attempted to correlate these with various measures of population diversity, particularly linguistic differences (Sanghvi *et al.* 1971, Balakrishnan *et al.* 1975, Kirk 1973, Parsons and White 1976, Sinnett *et al.* 1970, Rhoads and Friedlaender 1975, Littlewood 1972, Booth 1974, Booth and Taylor 1976). The most detailed analysis so far has been completed in my laboratory by Keats (1976) for 19 New Guinea populations, one Island Melanesian population and 12 Aboriginal populations in Australia. The dendrogram based on this analysis is shown in Figure 9.4. The most striking feature is the split

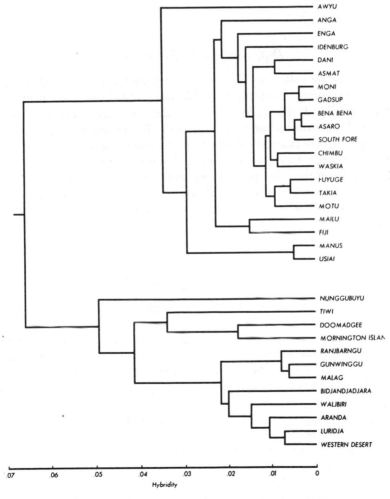

Figure 9.4. Dendrogram based on hybridity for 20 Melanesian and 12 Australian populations. (From Keats 1976.) $R = 0.8323$.

between the 20 Melanesian and Papuan populations and the 12 Australian Aboriginal populations. The genetic distance matrix shows also that the highland populations in Papua New Guinea and the Papuan populations of Irian Jaya are most remote from the central desert populations in Australia (Bidjandjadjara, Waljbiri, Aranda, Luridja, and Western Desert). This is in contrast to the evidence of the specific gene markers at the Gm and HLA loci, which suggests closest affinities between the highlands of New Guinea and central Australia. By contrast Keats' analysis shows that the Melanesian-speaking groups, such as Motu, Fiji, Manus, and Usiai, are more similar to northern Australians than to those in central Australia and suggests that the Melanesians were sea travelers who left their genetic imprint not only on parts of coastal New Guinea, e.g., in Motu, but also on a broad front across the north of Australia.

Differentiation in Micronesia

Morton and his colleagues have made a detailed study of genetic relatedness in eastern Micronesia. In one analysis (Morton and Lalouel 1973) comparison is made between dendrograms derived from estimates of hybridity based on either genetic markers, anthropometric measures, or (for a smaller number of populations) percentages of shared cognates in the languages. Both the phenotypic and the anthropometric assays agree that the Polynesian outlier, Kapingamarangi, stands apart from the other populations and that, within these, Pingelap is the most divergent. In addition, there is an appreciable east–west divergence within eastern Micronesia, a divergence that is demonstrated also by the linguistic analysis.

Differentiation Among Pacific Peoples

On a broader scale, covering Pacific peoples and their relationships with populations in other parts of the world, Imaizumi et al. (1973) and Omoto (1975b) have constructed dendrograms based on genetic distance matrices derived from gene frequency data. Omoto's dendrogram was constructed using data for 16 polymorphic loci. Under the constraints imposed in constructing a dendrogram of this type the three southwestern Pacific populations are related, followed by a grouping of Mongoloid populations (Ainu, Japanese, Chinese, and North American Indians), with the Caucasoid and Black Africans related together and most divergent from the other populations.

The principal components of the genetic distance matrices also can be reduced to a two-dimensional display to allow visualization of the genetic relationships between populations. Morton and Keats (1976) have done this for a number of Pacific populations contrasted with several major groupings from other parts of the world.

Table 9.4. *Genetic Loci Used in Distance Analysis*[a]

Blood groups	Leukocyte groups	Serum proteins	Red cell enzymes
● ABO	HLA-A	● Hp α	● Ac.Ph
● MNS	HLA-B	● Hp β	● 6PGD
● Rh		● Tf	● PGM1
P		Alb	● PGM2
Kell		Gc	GPT
Fy		Gm	MDH
Di		Km(Inv)	PGK
Jk			● AK
			● Hb α
			● Hb β
			ADA

[a] Those marked ● used in the 13 loci analysis.

A similar analysis has been carried out here for 27 Pacific and circumpacific populations using Nei's (1972) measure of genetic distance based on 13 polymorphic systems (Table 9.4). The dendrogram derived from the genetic distance matrix is shown in Figure 9.5, and the corresponding eigenvector diagram using the first two principal components is given in Figure 9.6.

Several striking features are evident. Geographic propinquity in general is reflected in the clustering of populations, and these clusters agree broadly with ethnic classifications based on other criteria. For example, the Mongoloid populations bordering the western Pacific form one cluster: Ainu, Japanese, Ryukyuan, Southern Chinese, Batak from Sumatra, Malays, and Atayal from Taiwan, with the aboriginal Senoi of Malaysia slightly removed. The Polynesian outliers Rennel and Bellona form a cluster with Samoa. Another cluster is comprised of the Austronesian-speaking Motu of the southern coast of Papua New Guinea, Fijians, Banks and Torres Islanders, Ulithi in Micronesia, and Tokelau Islanders. The Fore from the Eastern Highlands of Papua New Guinea and Pingelap in the eastern Caroline Islands are slightly removed from this latter cluster, and the Central and North Australian Aborigines lie between the Melanesian–Micronesian cluster and the Mongoloid–Polynesian cluster, although the North Australians are displaced to a more extreme position by the first principal component. Finally, the Central and South American Indian populations cluster together: Maya, Aymara, Noanama, with the Yanomama being more distinctive.

However, there are some strange bedfellows and some "loners." The Eskimo and Maori cluster together in the dendrogram, but Maori are close to the Maya and Aymara in the eigenvector diagram, and the Eskimo are between the Athabascan and the Mongoloid cluster. Athabascans, surprisingly, are very distinctive from the Mongoloids of the western Pacific, whereas the Easter Islanders are the most isolated on the eigenvector diagram. The position of the Easter Islanders may result from the relatively small number of families that contribute to the gene pool of the surviving "pure" Easter Islanders.

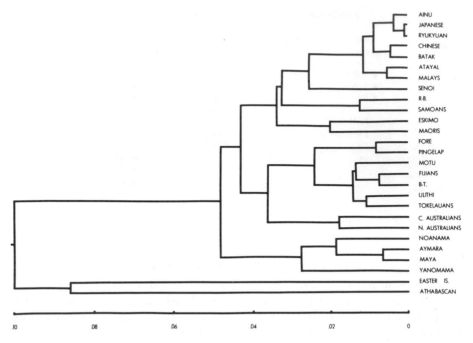

Figure 9.5. Dendogram based on Nei's genetic distances for 27 Pacific and circumpacific populations using data for 13 loci. B-T, Banks and Torres Islands; R-B, Rennell and Bellona Islands.

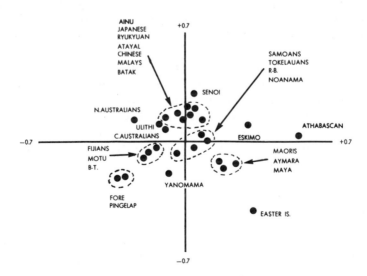

Figure 9.6. Eigenvector diagram for the first two principal components using the data for the populations shown in Figure 9.5. Horizontal axis, first component; vertical axis, second component. B-T, Banks and Torres Islands; R-B, Rennell and Bellona Islands.

The clustering of populations just outlined is based on data for 13 blood polymorphic systems. When data for 28 loci listed in Table 9.4 are included, a different picture emerges, sufficient data being available for 11 of the populations discussed above. The Eskimo now cluster with the Japanese, the Athabascan Indians being not far removed and the Aymara slightly further away. Southern Chinese cluster closely with Malays, and Samoans are fairly close; Pingelap and Banks and Torres Islands are a little further removed. The Central Australians cluster with the Fore of the Eastern Highlands of Papua New Guinea and both are some distance removed from the other nine populations.

In broad terms, therefore, this increase in the amount of genetic information shows that the American Indian groups and Eskimo now associate with a northern Asiatic Mongoloid group, represented by the Japanese. The southern Asian Mongoloid group is more closely related to a Polynesian series represented by the Samoans, and the New Guinea Highlanders and Australians are both more closely related to one another and, together, are a more distinct entity than when a restricted number of loci are employed.

Conclusions

Reviewing the archaeologic, linguistic, cultural, and physical anthropologic evidence bearing on the origin and evolution of Pacific populations, Bellwood (1974) sees an ancestral Australoid population with a time depth of at least 30,000 years occupying the Australia–New Guinea area (Sahul land): This early group may have extended to Santa Cruz in the Solomons and perhaps into New Caledonia and the New Hebrides. Bellwood argues that about 4000 years ago Austronesian-speaking people penetrated into this Melanesian area from a center probably in the Indonesian area, bringing with them pottery, pigs, and horticulture. Intermarriage with non-Austronesian speakers was followed by consolidation of Austronesian settlement in the islands as far east as the New Hebrides and New Caledonia. The Austronesian speakers who made Lapita pottery were coastal traders, and about 3000 years ago one or more groups of Lapita pottery makers moved further east to settle Fiji and western Polynesia. According to Bellwood, Fiji received later Melanesian settlers, but the Polynesians and possibly some of the central and eastern Micronesians are the descendants of the earlier Lapita-ware settlers. During the last 2500 years there has been complex population movement and interaction to produce the present-day pattern of peoples in the area. Bellwood does not accept the view that Micronesia was a stepping stone on the way to Polynesia, and he believes that Austronesians did not evolve in mainland southeast Asia. Finally, reviewing the arguments for American Indian influences in Polynesia, he agrees that most of the evidence supports a west–east source for the Polynesian peoples and culture but adopts the cautious position that one should not completely rule out any American contact, even though it may only have been on a small scale. He believes also that the Polynesians (and the Micronesians) have preserved many aspects of a Mongoloid phenotype. How does the genetic evidence fit with this reconstruction of Pacific prehistory? The following conclusions seem to be justified:

1. Australian, New Guinean, and Island Melanesian populations share a number of distinctive gene markers, which suggests a common substratum population of some considerable antiquity.

2. Marked divergence has taken place between Australian and New Guinean populations, reflecting the dichotomy apparent in the languages. However, Island Melanesians speaking Austronesian languages are genetically closer to the north Australian populations than are highland Papuan speakers, probably because of a later influence of the voyaging speakers of Austronesian on populations in the north of Australia.

3. Several specific markers are found widespread in Austronesian-speaking Melanesians (both from Island Melanesia and from New Guinea) and are found also in Micronesia and in Polynesian outliers in the western Pacific. Eastern Polynesia has not been studied in sufficient detail to allow definitive statements on their presence or absence in this area.

4. Some of these specific genetic markers, but not all, are found in southeast Asia and in Japan and the Ryukyuan Islands.

5. Certain clines are apparent for genes in the entire western Pacific area. These suggest that American Indians are most clearly aligned with populations in the northwest Pacific.

6. Despite the general view that Polynesians and Micronesians possess Mongoloid anthroposcopic traits, specific Mongoloid genetic markers have not penetrated far into the Pacific. Although complex population movements undoubtedly have been taking place in the southwestern Pacific during the last several thousand years, the penetration of the specific Mongoloid marker genes Di^a and $Tf D_{Chi}$ southward as far as Indonesia did not continue across into Australia, New Guinea, Island Melanesia, or, on present evidence, into Polynesia.

By contrast, genetic distances computed from allele frequencies in 13 blood genetic systems indicate a close relationship between the Noanama from Columbia and the Samoans, with the Yanomama not very far away from the general cluster of central Pacific populations. This analysis shows also that the Maori from New Zealand cluster with the Maya from Central America. These results support the proponents of a migratory movement from the west coast of the Americas into the Pacific. Results from the genetic distance analysis based on 28 blood genetic loci indicate, however, that the American Indians cluster more closely with the Japanese than they do with other Pacific populations, and that the Polynesians are genetically closest to the southern Chinese.

This change in the relationship between some of the groups underlines the importance of basing genetic relationships on the maximum number of genetic factors. It makes urgent the collection of such data before many of the populations being considered here become submerged in the wake of rapid technologic changes and political realignments taking place at the present time and which are likely to continue with accelerated speed in the future.

References

Akaishi, S., and T. Kudo. (1975). Blood Groups. *In* S. Watanabe, S. Kondo, and E. Matsunaga, Eds., *JIBP Synthesis.* Vol. 2, *Anthropological and Genetic Studies of the Japanese.* Tokyo: Univ. Tokyo Press, pp. 77–107.

Balakrishnan, V., L.D. Sanghvi, and R.L. Kirk. (1975). *Genetic Diversity among Australian Aborigines.* Canberra: Australian Institute of Aboriginal Studies.

Bellwood, P. (1974). The Prehistory of Oceania. *Curr. Anthropol.* 16:9–28.

Blake, N.M. (1976). Glutamic Pyruvic Transaminase and Esterase D types in the Asian–Pacific Area. *Hum. Genet.* 35:91–102.

Blake, N.M. (1978). Malate Dehydrogenase Types in the Asian–Pacific Area, and a Description of New Phenotypes. *Hum. Genet.* 43:69–80.

Blake, N.M., and K. Omoto. (1975). Phosphoglucomutase Types in the Asian–Pacific Area: A Critical Review Including New Phenotypes. *Ann. Hum. Genet.* 38:251–273.

Blake, N.M., R.L. Kirk, M.J. Simons, and M.P. Alpers. (1970). Genetic Variants of Soluble Malate Dehydrogenase in New Guinea Populations. *Humangenetik* 11:72–74.

Booth, P.B. (1974). Genetic Distances Between Certain New Guinea Populations Studied under the International Biological Programme. *Phil. Trans. Roy. Soc. London* Ser. B. 26B:257–267.

Booth, P.B., and H.W. Taylor. (1976). An Evaluation of Genetic Distance Analysis of Some New Guinea Populations. *In* R.L. Kirk and A.G. Thorne, Eds., *The Origin of the Australians.* Canberra: Australian Institute of Aboriginal Studies, pp. 415–430.

Booth, P.B., J.A. Albrey, J. Whittaker, and R. Sanger. (1970). Gerbich Blood Group System: A Useful Genetic Marker in Certain Melanesians of Papua New Guinea. *Nature (London)* 228:462.

Booth, P.B., R.W. Hornabrook, and L.A. Malcolm. (1972a). The Red Cell Antigen N^A in Melanesians: Family and Population Studies. *Hum. Biol. Oceania* 1:223–228.

Booth, P.B., L. Wark, K. McLoughlin, and R. Spark. (1972b). The Gerbich Blood Group System in New Guinea. I. The Sepik District. *Hum. Biol. Oceania* 1:215–272.

Cleve, H. (1967). Populationsgenetik des Gc-Systems. *Deutsche Ges. Anthropol.* 8:170–177.

Cleve, H., R.L. Kirk, and W.C. Parker. (1963). Two Genetic Variants of the Group Specific Component of Human Serum: Gc Chippewa and Gc Aborigine. *Am. J. Hum. Genet.* 15:368–379.

Crow, J.F., and Denniston, C. (Eds.). (1974). *Genetic Distance.* New York: Plenum Press.

Curtain, C.C., E. Van Loghem, H.H. Fudenberg, N.B. Tindale, R.T. Simmons, R.L. Doherty, and G. Vos. (1972). Distribution of the Immunoglobulin Markers at the IgGl, IgG2, IgG3, IgA$_2$ and K-Chain Loci in Australian Aborigines: Comparisons with New Guinea Populations. *Am. J. Hum. Genet.* 24:145–155.

Curtain, C.C., E. Van Loghem, and M.S. Schanfield. (1976). Immunoglobulin Markers as Indicators of Population Affinities in Australasia and the Western Pacific. *In* R.L. Kirk and A.G. Thorne, Eds., *The Origin of the Australians.* Canberra: Australian Institute of Aboriginal Studies, pp. 347–364.

Dausset, J., and J. Colombani. (Eds.). (1973). *Histocompatibility Testing 1972*, Copenhagen: Munksgaard.

Duncan, I.W., E.M. Scott, and R.C. Wright. (1974). Gene Frequencies of Erythrocytic Enzymes of Alaskan Eskimos and Athabascan Indians. *Am. J. Hum. Genet.* 25:244–246.

Friedlaender, J.S., and A.G. Steinberg. (1970). Anthropological Significance of Gamma Globulin (Gm and Inv) Antigens in Bougainville Island, Melanesia. *Nature (London)* 288:59–61.

Gajdusek, D.C., J. Guiart, R.L. Kirk, R.W. Carell, D. Irvine, P.M. Kynoch, and H. Lehmann. (1967). Haemoglobin J Tongariki (α 115 alanine → aspartic acid): The First New Haemoglobin Variant Found in a Pacific (Melanesian) Population. *J. Med. Genet.* 4:1–6.

Giles, E., E. Ogan, and A.G. Steinberg. (1965). Gamma Globulin Factors (Gm and Inv) in New Guinea: Anthropological Significance. *Science* 150:1159–1160.

Howells, W. (1973). *The Pacific Islanders*. London: Weidenfeld and Nicolson.

Imaizumi, Y., N.E. Morton, and J.M. Lalouel. (1973). Kinship and Race. *In* N.E. Morton, Ed., *Genetic Structure of Populations*. Honolulu: Univ. Press of Hawaii, pp. 228–233.

Ishimoto, G. (1975). Red Cell Enzymes. *In* S. Watanabe, S. Kondo, and E. Matsunaga, Eds., *JIBP Synthesis*. Vol. 2, *Anthropological and Genetic Studies of the Japanese*. Tokyo: Univ. Tokyo Press, pp. 109–139.

Keats, B. (1976). *Genetic Aspects of Growth and of Population Structure in Indigenous Peoples of Australia and New Guinea*. Ph.D. Thesis, Australian National University, Canberra.

Kirk, R.L. (1965). *Genetic Markers in Australian Aborigines*. Canberra: Australian Institute of Aboriginal Studies.

Kirk, R.L. (1968). *Haptoglobin Groups in Man*. Monographs in Human Genetics No. 4. Basel: Karger.

Kirk, R.L. (1969a). The World Distribution of Transferrin Variants and Some Unsolved Problems. *Acta Genet. Med. Gemellol.* 17:613–640.

Kirk, R.L. (1969b). Biochemical Polymorphism and the Evolution of Human Races. *Proc. 8th Intl. Cong. Anthropol. Ethnol. Sci.* 1:371.

Kirk, R.L. (1973). Genetic Studies of Cape York Populations. *In* R.L. Kirk, Ed., *The Human Biology of Aborigines in Cape York*. Canberra: Australian Institute of Aboriginal Studies, pp. 25–36.

Kirk, R.L. (1976). Serum Protein and Enzyme Markers as Indicators of Population Affinities in Australia and the Western Pacific. *In* R.L. Kirk and A.G. Thorne, Eds., *The Origin of the Australians*. Canberra: Australian Institute of Aboriginal Studies, pp. 329–346.

Kirk, R.L., L.Y.C. Lai, G.H. Vos, and L.P. Vidyarthi. (1962). A Genetical Study of the Oraons of the Chota Nagpur Plateau (Bihar, India). *Am. J. Phys. Anthropol.* 20:375–385.

Kirk, R.L., H. Cleve, and A.G. Bearn. (1963a). The Distribution of Gc Types in Sera from Australian Aborigines. *Am. J. Phys. Anthropol.* 21:215–223.

Kirk, R.L., H. Cleve, and A.G. Bearn. (1963b). The Distribution of the Group-Specific Component (Gc) in Selected Populations in South and South-East Asia and Oceania. *Acta Genet. Stat. Med. (Basel)* 13:140–149.

Kirk, R.L., E.M. McDermid, N.M. Blake, D.C. Gajdusek, W.C. Leyshon, and R. MacLennan. (1975). Blood Group, Serum Protein and Red Cell Enzyme Groups of Amerindian Populations in Colombia. *Am. J. Phys. Anthropol.* 41:301–316.

Kissmeyer-Nielsen, F. (Ed.). (1976). *Histocompatibility Testing 1975*. Copenhagen: Munksgaard.

Kitchin, F.D., A.G. Bearn, M. Alpers, and D.C. Gajdusek. (1972). Distribution of the

Inherited Serum Group Specific Protein (Gc) Phenotypes in New Guineans. *Am. J. Hum. Genet.* 24:486–594.

Lai, L.Y.C. (1963). A New Transferrin in New Guinea. *Nature (London)* 198:589.

Layrisse, M., and J. Wilbert. (1960). *El Antigeno del Systema Sanguineo Diego.* Caracas: La Fundacion Creole y la Fundacion Eugenio Mendoza.

Lie-Injo, L.E. (1967). Red Cell Carbonic Anhydrase Ic in Filipinos. *Am. J. Hum. Genet.* 19:130–132.

Lie-Injo, L.E. (1976). Genetic Relationships of Several Aboriginal Groups in South East Asia. *In* R.L. Kirk and A.G. Thorne, Eds., *The Origin of the Australians.* Canberra: Australian Institute of Aboriginal Studies, pp. 277–306.

Littlewood, R.A. (1972). *Physical Anthropology of the Eastern Highlands of New Guinea.* Seattle and London: Univ. Washington Press.

Malcolm, L.A., D.G. Woodfield, N.M. Blake, R.L. Kirk, and E.M. McDermid. (1972). The Distribution of Blood, Serum Protein and Enzyme Groups on Manus (Admiralty Islands, New Guinea). *Hum. Hered.* 22:305–322.

McAlpine, P.J., S.-H. Chen, D.W. Cox, J.B. Dossetor, E. Giblett, A.G. Steinberg, and N.E. Simpson. (1974). Genetic Markers in Blood in a Canadian Eskimo Population with a Comparison of Allele Frequencies in Circumpolar Populations. *Hum. Hered.* 24:114–142.

McDermid, E.M., and H. Cleve. (1972). A Comparison of the Fast Migrating Gc-variant of Australian Aborigines, New Guinea Indigines, South African Bantu and Black Americans. *Hum. Hered.* 22:249–253.

Mestriner, M.A., F.M. Salzano, J.V. Neel, and M. Ayres. (1976). Esterase D in South American Indians. *Am. J. Hum. Genet.* 28:257–261.

Morton, N.E. (Ed.). (1973). *Genetic Structure of Populations.* Honolulu: Univ. Press of Hawaii.

Morton, N.E., and B. Keats. (1976). Human Microdifferentiation in the Western Pacific. *In* R.L. Kirk and A.G. Thorne, Eds., *The Origin of the Australians.* Canberra: Australian Institute of Aboriginal Studies, pp. 379–400.

Morton, N.E., and J.M. Lalouel. (1973). Topology of Kinship in Micronesia. *Am. J. Hum. Genet.* 25:422–432.

Mourant, A.E., A.C. Kopeć, and K. Domaniewska-Sobczak. (1976). *The Distribution of Blood Groups and Other Polymorphisms.* London: Oxford Univ. Press.

Nei, M. (1972). Genetic Distance Between Populations. *Am. Naturalist.* 106:283–292.

Omoto, K. (1972). The Distribution of Red Cell Adenosine Deaminase Phenotypes in Oceania. *Jap. J. Hum. Genet.* 16:166–169.

Omoto, K. (1973). Polymorphic Traits in Peoples of Eastern Asia and the Pacific. *Israel J. Med. Sci.* 9:1195–1215.

Omoto, K. (1975a). Serum Protein Groups. *In* S. Watanabe, S. Kondo, and E. Matsunaga, Eds., *JIBP Synthesis.* Vol. 2, *Anthropological and Genetic Studies of the Japanese.* Tokyo: Univ. Tokyo Press, pp. 141–162.

Omoto, K. (1975b). Genetic Affinities of the Ainu as Assessed from Data on Polymorphic Traits. *In* S. Watanabe, S. Kondo, and E. Matsunaga, Eds., *JIBP Synthesis.* Vol. 2, *Anthropological and Genetic Studies of the Japanese.* Tokyo: Univ. of Tokyo Press, pp. 296–307.

Parsons, P.A., and N.G. White. (1976). Variability of Anthropometric Traits in Australian Aboriginals and Adjacent Populations: Its Bearing on the Biological Origin of the Australians. *In* R.L. Kirk and A.G. Thorne, Eds., *The Origin of the Australians.* Canberra: Australian Institute of Aboriginal Studies, pp. 227–244.

Payne, R., R. Radvany, and C. Grumet. (1975). A New Second Locus HL-A Antigen in Linkage Disequilibrium with HL-A2 in Cantonese Chinese. *Tiss. Antigens* 5:69–71.

Post, R.H., J.V. Neel, and J.W. Schull. (1968). Tabulation of Phenotype and Gene Frequencies for 11 Different Genetic Systems Studied in the American Indian. *In Biomedical Challenges Presented by the American Indian*. Washington, D.C.: Pan American Health Organization, pp. 141–185.

Rhoads, J.G., and J.S. Friedlaender. (1975). Language Boundaries and Biological Differentiation on Bougainville: Multivariate Analysis of Variance. *Proc. Natl. Acad. Sci. USA* 72:2247–2250.

Sanghvi, L.D., R.L. Kirk, and V. Balakrishnan. (1971). A Study of Genetic Distance Between Populations of Australian Aborigines. *Hum. Biol.* 43:445–458.

Scott, E.M., I.W. Duncan, V. Ekstrand, and R.C. Wright. (1966). Frequency of Polymorphic Types of Red Cell Enzymes and Serum Factors in Alaskan Indians and Eskimos. *Am. J. Hum. Genet.* 18:408–411.

Simmons, R.T. (1962). Blood Group Genes in Polynesians and Comparisons with Other Pacific Peoples. *Oceania* 32:198–210.

Simmons, R.T. (1966). The Blood Group Genetics of Easter Islanders (Pascuense), and Other Polynesians. *In* R. Heyerdahl and E.N. Ferdon, Eds., *Reports of the Norwegian Archaeological Expedition to Easter Island and the East Pacific*. Vol. II. Stockholm: Miscellaneous Papers, pp. 333–343.

Simmons, R.T., and P.B. Booth. (1971). *A Compendium of Melanesian Genetic Data*. Parts I–IV. Melbourne: Roneoed, Commonwealth Serum Laboratories (C.S.L. Nos. 546–549).

Simons, M.J., G.B. Wee, S.H. Chan, K. Shanmugaratnam, N.E. Day, and G. de Thé. (1975). Probable Identification of an HL-A Second Locus Antigen Associated with a High Risk of Nasopharyngeal Carcinoma. *Lancet* 1:142–143.

Sinnett, P., N.M. Blake, R.L. Kirk, R.J. Walsh, and L.Y.C. Lai. (1970). Blood, Serum Protein and Enzyme Groups among Enga-Speaking People of the Western Highlands, New Guinea with an Estimation of Genetic Distance Between Clans. *Arch. Phys. Anthropol. Oceania* 5:236–252.

Steinberg, A.G., and R.L. Kirk. (1970). Gm and Inv Types of Aborigines in the Northern Territory of Australia. *Archaeol. Phys. Anthropol. Oceania* 5:163–172.

Steinberg, A.G., D.C. Gajdusek, and M.P. Alpers. (1972). Genetic Studies in Relation to Kuru: The Distribution of Human Gamma Globulin Allotypes in New Guinea Populations. *Am. J. Hum. Genet.* 24:S95–110.

Tashian, R.E., C.C. Plato, and T.B. Shows. (1963). Inherited Variant of Erythrocyte Carbonic Anhydrase in Micronesians from Guam and Saipan. *Science* 140:53–54.

Thorsby, E., J. Colombani, J. Dausset, J. Figueroa, and A. Thorsby. (1973). HL-A, Blood Group and Serum Type Polymorphisms of Natives of Easter Island. *In* J. Dausett and J. Colombani, Eds., *Histocompatibility Testing 1972*. Copenhagen: Munksgaard, pp. 287–297.

Van der Does, J.A., J. D'Amaro, A. Van Leeuwan, P.M. Khan, L.F. Bernini, E. Van Loghem, L. Nijenhuis, G. Van der Steen, J.J. Van Rood, and P. Rubinstein. (1973). HL-A Typing in Chilean Aymara Indians. *In* J. Dausset and J. Colombani, Eds., *Histocompatibility Testing 1972*. Copenhagen: Munksgaard, pp. 391–396.

Vergnes, H., J.C. Quilici, and J. Constans. (1976). Serum and Red Cell Enzyme Polymorphisms in Six Amerindian Tribes. *Ann. Hum. Biol.* 3:577–585.

Vos, G.H., and P. Comley. (1967). Red Cell and Saliva Studies for the Evaluation of ABH and Lewis Factors among the Caucasian and Aboriginal Populations of Western Australia. *Acta Genet. (Basel)* 17:495–510.

Vos, G.H., and R.L. Kirk. (1962). Naturally Occurring Anti-E which Distinguishes a Variant of the E Antigen in Australian Aborigines. *Vox Sang* 7:22–32.

Vos, G.H., R.L. Kirk, and A.G. Steinberg. (1963). The Distribution of the Gamma Globulin Types Gm(a) Gm(b) Gm(x) and Gm-like in South and South East Asia and Australia. *Am. J. Hum. Genet.* 15:44–52.

Weitkamp, L.R., and J.V. Neel. (1972). The Genetic Structure of a Tribal Population, the Yanomama Indians IV. Erythrocyte Enzymes and Summary. *Ann. Hum. Genet.* 35:433–444.

Yanase, T., and T. Imamura. (1975). Haemoglobin Variants. *In* S. Watanabe, S. Kondo, and E. Matsunaga, Eds., *JIBP Synthesis*. Vol. 2, *Anthropological and Genetic Studies of the Japanese*. Tokyo: Univ. Tokyo Press, pp. 163–167.

Part 3

Adaptations

Chapter 10

Analytic Methods for Genetic and Adaptational Studies

William J. Schull and Francisco Rothhammer

Introduction

Our charge in this volume is to examine analytic methods and strategies for genetic and adaptational studies that may be useful in the study of the origins and affinities of the human populations of the New World. We propose to do this in a specific context, namely man's adaptation to the hypoxia of altitude. Extrapolation to other situations when warranted will, we trust, be relatively straightforward. We shall emphasize philosophy and general principles but will couch these frequently in terms of a particular study, that is, the Multinational Andean Genetic and Health Program which has been underway in the Departmento de Arica of Chile (see Schull and Rothhammer, 1977, for a full description of this program).

The Issue

Fifty millenia ago, but possibly more recently, human beings are thought to have made their way across the Bering Land Bridge to the Americas (Stewart 1973). Whether one or many migrations occurred or one or many routes were involved is unclear and, for our purposes, unimportant. These prehistoric migrants and their descendants presumably erupted into North America, filtered through Mexico and the isthmus, and ultimately found themselves in the southern continent. Twelve thousand years ago some of their descendants made their appearance in the highlands of South America near Ayacucho (MacNeish et al. 1970). We know virtually nothing of the physical appearance of these individuals and are not even certain whether they merely frequented the altiplano in search of game—only their tools have been found, which suggests this may be so—or resided there permanently. Moreover, we do not know to which of the numerous regions of the New World through which they passed they were best suited. We do know, however, that individuals accustomed to the levels of oxygen which obtain at or near sea level, when transported to these high altitudes, are apt to experience some one or combination of the following: shortness of breath, dizziness and marked palpitation upon exertion, exhilaration, sleeplessness, headache, the irregularity of respiratory rhythm known as Cheyne–Stokes breathing, impairment of mentation, diminished night visual acuity, accelerated pulse, as well as other less obvious signs (Buskirk 1971). Ultimately, however, within days or weeks at most, an accommodation is made, and one's normal activity level can again be achieved without undue physical stress or apprehension. We presume this sequence of events to be one that these early settlers also experienced.

This discomfiture experienced by lowlanders notwithstanding, in the 12 millenia that have intervened since his arrival, Andean man has obviously thrived; more than 10 million individuals now live in the Andes at altitudes of 3000 m or more. This increase in numbers undoubtedly reflects an ability, at least in part, to influence environmental circumstances through the domestication of a wide variety of frost-resistant plants and the discovery of means to store surpluses in the form of herds of alpaca and llama, for example. However, has man himself adapted? Has his genetic makeup been systematically altered as a consequence of the rigors of this environment? If so how is this fact to be demonstrated, for surely much of the genetic contribution to the adaptation of the contemporary Andean, the Aymara and Quechua, for example, must have occurred generations, possibly millenia ago, and so is inaccessible to study.

The Strategy

Functional adaptation to high-altitude hypoxia, as Frisancho (1975) has aptly noted, could occur through modifications in (1) pulmonary ventilation, (2) lung volume and pulmonary diffusing capacity, (3) transportation of oxygen in the blood, (4) diffusion of oxygen from blood to tissue, (5) utilization of oxygen at the tissue level, or some combination thereof. Perusal of the findings that typify Andean man suggests that all of these alternatives may be operative, but differentially (for a discussion of these findings, see Hurtado 1964, 1971). Proof of an adaptive change in any one entails the demonstration, first, that the variables associated with a particular pathway change systematically with changes in the selective force, here oxygen tension, and second, that a significant component of interindividual variation at any altitude in these variables stems from genetic variability. More simply put, the trait that responds must be one known to be under genetic control.

As previously stated, presumably much of the heritable contribution to the adjustment to hypoxia of the contemporary Andean occurred numerous generations ago and is now inaccessible to study. However, at a number of locations in Bolivia, Chile, and Peru, and possibly elsewhere, the recent past has seen a migration of these people to lower altitudes. This downward, frequently coastward, movement affords an opportunity to study changes in variability that may be indicative of a loss in adaptation, a deadaptation or relaxation of selection as it were. If the increase in variability that follows the relaxation of selection can be viewed as a mirror of the loss in variability through selection, human adaptation to hypoxia becomes a studiable phenomenon. It does not necessarily follow, however, that the two processes of which we write need proceed at the same rate; indeed, it is unlikely that they do, for presumably mutation is the only source of new genes through which variability may increase. Admixture, admittedly another source of new genes, would obviously complicate interpretation of data if, as seems likely, such admixture were not uniform over all differing oxygen tensions. If it were uniform, its effects might be treated as if they were analogous to new mutational events, merely orders of magnitude more common than true mutations. If it were nonuniform, as seems likely, the impact of admixture would depend upon when in the evolution of the contemporary biologic structure the adaptive change occurred. For example, if the latter changes were to have taken place before the subdivision of Andean peoples into Aymara, Quechua, and the like, the subsequent admixture of these "tribes" would presumably have little impact upon the adaptive changes because these changes would be present in all contemporary populations in the area. Parenthetically, it deserves note that if a region could be identified in the New World where a systematic invasion of hypoxic areas by an initially homogeneous group of people has occurred over historic times, this would provide a more direct study of the adaptive process. We are not aware, however, of such an area where the invasion has proceeded over a long enough period of time to effect changes that are both measurable and reasonable.

Selection of the Area and Study Population

Most previous studies of a population nature or the hypoxia of altitude have generally examined adaptation in terms of a dichotomy, e.g., altiplano versus coast (Baker 1969) or highlands versus lowlands (Harrison *et al.* 1969). This is obviously a defensible strategy but one that we believe to be fraught with possible pitfalls. Other sources of differences are too readily confounded, e.g., rural versus urban effects, or low population density versus high population density as such densities affect the distribution of potentially fatal contagious diseases, say. Ideally, a continuum is sought in the factor that has propelled and continues to propel the adaptation; this may be impracticable. At the very least, however, it should be possible to identify a locale in which at least three alternatives with respect to the variable of interest exist, here, for example, regions in which low, middle, and high oxygen tensions prevail. The Departamento de Arica of Chile, the most northerly of the Chilean areas, presents such an alternative.

Broadly speaking, the Departamento de Arica is divisible into three sharply differentiable, but adjacent, ecologic zones or niches—a coastal area, the sierra, and the altiplano. These differ pronouncedly in temperature, rainfall, and oxygen pressure. The coastal zone embraces four valleys, the Lluta, the Azapa, the Codpa, and the Camarones, and the terrace on which the present city of Arica sits. Although portions of these valleys are lush as a result of irrigation and the rivers that arise in the mountains, rainfall is sparse, and this appears to have been true throughout their human occupation (Bird 1943). Temperatures are moderate. It is a salubrious climate, moderated by the Pacific high-pressure center to the west and swept by the southeast trade winds. Partial oxygen pressure is about 150 mm Hg.

Eastward in the sierra, the terrain is rugged and highly irregular. At these heights partial oxygen pressure is approximately 100 mm Hg. Rain, although markedly seasonal, occurs with increasing itensity from 2500 m upward. Temperature, particularly the daily and annual minimums, are substantially lower than those of the coast, and the diurnal temperature cycle will occasionally exceed 30°F (as contrasted with 10°F on the coast).

The altiplano, still further eastward, is a high rolling plain; it is an austere land, alternately dry and wet. Temperatures commonly oscillate as much as 30°F in the course of 24 hr, and changes of as much as 50°F are frequent. Little shelter exists from the chilling, parching winds that blow through much of the year; there are virtually no trees and the coarse grass and scrubby growth that do occur rarely attain a meter's height. The humidity is generally less than 10 percent in the dry season, and the partial oxygen pressure is about 80 mm Hg.

Today the predominant indigenous people are the Aymara. The unique coastal culture that once existed disappeared literally centuries ago, as did a sierran one if the latter was not Incan or Aymara. Anecdotal as well as historical materials document the coastward migration of the latter people in the post-Colombian period. Today, their villages are scattered across the altiplano and the sierra; these range from tens of individuals to a few hundred. Until recently, the Aymara of Chile have been either peasant pastoralists deriving their livelihoods from their herds of llama and alpaca or subsistence agriculturalists, the former in the altiplano and the latter in the sierra. Several centuries ago, some of these people began to migrate into the coastal valleys, a process that has been markedly accelerated in the last several decades as a result of new roads and the growth of economic opportunities in the city of Arica. This coastward migration has taken two forms—a movement of altiplano people directly to the valleys and a "Brownian" process in which sierran families have moved to the coast to be replaced in the sierra by altiplano families. Both migratory routes, although accelerated of late, have doubtlessly existed for several generations.

Choice of Variables

We have previously cited five different pathways through which adaptive changes to hypoxia may occur. Some of these involve known biochemical parameters; others do not. To the extent that the former is true, these pathways should prove the more interesting, for the relationship between gene and product is more immediate. For example, the enzyme 2,3-diphosphoglycerate mutase (2,3-DPG mutase) is known to impinge directly upon the oxygen dissociation characteristics of hemoglobin, and thus upon the capacity to unload oxygen at a particular partial pressure in the tissues. Likewise, because the dissociation of oxygen from oxyhemoglobin depends in part upon the pH of the blood, an enzyme such as carbonic anhydrase, which affects the acidity of blood, can afford a basis of evolutionary change in oxygen transport. In both instances, although neither locus is polymorphic, at least as evidenced by populations under little oxygen stress, structural changes in 2,3-DPG mutase and carbonic anhydrase are known to be under simple genetic control. Changes in these variables or in similar ones presumably bear an uncomplicated interpretation in adaptational terms. That is to say, it is reasonable to presume that if the frequencies of the genes associated with such biochemical traits appear to respond directly to changes in oxygen tension, they reflect the selective process that we term adaptation.

Some pathways patently of importance are presently understood only in physiologic terms; generally the genetic contribution to variation in the latter is poorly understood. For example, pulmonary ventilation, lung volume, or pulmonary diffusing capacity are measured as expired volume in 1 or in 3 s, in forced vital capacity, or peak flow rate. Surprisingly few data exist, however, on the extent to which variation in these parameters reflects genetic variability, and in no instance do differences among individuals or groups appear to represent the action of single genes. Use of these variables, if inferences are to be rigorous, requires genetic studies in addition to the adaptational ones. The former could embrace conventional family inquiries, the study of twins, or "family sets" (for the latter, see Chakraborty *et al.* 1976). Not all of these are likely to be equally practicable; twins are after all uncommon events. However, family studies should be relatively easily integrated into the overall data collection. Indeed, if villages are the units of study, and if they are exhaustively studied, family data will automatically accrue. It would seem naive, however, to believe that these data will disclose simple, single-gene effects of a common sort; past studies have failed to do so. It is more likely that variation will appear to be multigenic, and that demonstratability of systematic changes will depend, therefore, on the heritability of the variable measured.

Whatever the ultimate array of variables chosen for measurement, some presumably will be continuously distributed, others ordered, and still others will doubtlessly be of the so-called nominal variety, i.e., will denote the presence or absence of a particular attribute. All are conceivably influenced by other variables, some more so than others, of course. Recognition and measurement of as many of these influencing variables as practicable is essential. This not only will markedly improve the detection of differences of interest through the removal of extraneous sources of variation, but it can provide insights into the dynamics of adaptive change. For example, in the present context, the permanency of one's residence in a given village at a particular altitude can vary; this variation can lead either to a confusion of acclimatic with adaptive changes (if oxygen tension has also varied) or, possibly, if the villages are the ultimate unit of inquiry, confound differential migration effects with adaptive ones. These problems, to the extent that they materialize, should be construed not as failures in the design but as opportunities that await exploitation. Patently, however, one's strategy cannot be so loose as to be unresponsive to any single hypothesis, nor should it be so inflexible as to be unresponsive to interesting, indeed important, unforeseen contingencies. The measure of a really good study is its capacity to be structured but flexible.

Sampling and Sample Sizes

If the population of the locale of interest were sufficiently large or possibly disperse, studies of the sort that are implicit in our previous remarks could entail the definition of a sampling frame and the choice of some one of the many different sampling techniques that exist (see, e.g., Kish 1965). We shall not examine the merits of these alternatives here, for we surmise that adaptational studies are more likely to involve locales or villages in which participation is sought of everyone without regard to age and sex. Rarely will all individuals participate, so it becomes important to ascertain compliance rates. Estimation of the latter obviously requires knowledge of the population—the number of its members, their sexes, ages, and perhaps biologic affinities. It is highly desirable, therefore, to have a complete, current census of the population in the study area to identify and, it is hoped, evaluate biases that may stem from dissimilar participation rates among the groups of interest.

What size of sample should one seek, however, and how is the number to be allocated over villages or niches if such allocation seems appropriate? These are not simple issues with which to grapple. Whatever the sample number may be it will be adequate to exclude some hypotheses and inadequate to exclude others. Adequacy is not some property inherent in a number but depends on the alternatives between which one hopes to discriminate. If the variable of interest is binomially distributed, then adequacy is a function of the frequency of the event of interest in some "standard" population, and the difference between that standard and a value that we should like to identify as different if, in fact, it obtains. If the variable is continuously distributed, then one must consider the difference between the means of the "standard" population and the one of investigative interest, and the coefficient of variability. In either event, we must establish some minimum difference of moment, or alternatively, specify the precision (the complement of the error we will tolerate) we seek in any estimate that may be generated from the data. Neither is an easy task, for rarely do theory or prior experience dictate the difference or precision that should be sought; often one's only recourse is to intuition or approximation.

Two lines of evidence or theory can be used to guide sample size calculations in the current situation. First, with respect to quantitative genetic characteristics, there is a fairly voluminous literature on the expected response to a program of artificial selection (see Falconer 1960). Its applicability here clearly depends on two assumptions, namely, (1) that artificial selection is a reasonable paradigm for natural selection, and (2) that to a first approximation at least deadaptation mirrors adaptation. Whereas the general processes may be equivalent, and these assumptions warranted, we want to reemphasize that the rates may not be the same. Otherwise put, as variability is constricted through a process of selection, the amount of diversity in a population, if stochastic elements can be ignored, will be dependent solely on the initial degree of variation and the extent to which selection diminishes variability in each generation—the origin of new variation through mutation affects this dampening process only trivially, if present estimates of gene mutation are correct. However, once restricted, the growth in variability in an isolated population when selection is relaxed will depend much on the mutation rate and the size of the population. For simple phenotypes (those which are essentially synonymous with genotypes) one measure of variation in a population is its heterozygosity, where the latter is defined as:

$$h = 1 - \sum_{i=1}^{k} p_i^2 \tag{10.1}$$

and p_i is the frequency of the ith allele at a particular locus ($i = 1, 2, \ldots, k$). Now the expected value of h at infinite time is (see Li and Nei 1975)

$$E(h_\infty) = M/(1 + M) \tag{10.2}$$

where M is equal to $4Nv$, that is, the product of the mutation rate v and the effective population number N. The variance of heterozygosity, h_∞, expected under the hypothesis of neutral mutations is (see Li and Nei 1975):

$$\text{VAR}(h_\infty) = \frac{2M}{(1 + M)^2(2 + M)(3 + M)} \tag{10.3}$$

We use these notions as follows: At time t_0, we assume the heterozygosity of the "parent" population, that is, the one at low oxygen tension, to be zero. This, of course, is tantamount to asserting that as new mutations arise they are removed by selection, and the population remains homozygous at the locuses of interest. Suppose that at time t_0 a group of individuals from the "parent" population established a colony at some more favorable oxygen tension, one at which all new mutations would be effectively neutral or nearly so. At time t_n, this population will have heterozygosity h_n, which will presumably not be zero because of the persistence of some of the neutral mutations. Our task is to assess the significance of h_n when contrasted with zero.

Li and Nei (1975) have shown that:

$$E(h_n) = E(h_\infty) + [E(h_0) - E(h_\infty)]e^{-[(M+1)/2N]n}$$
$$= E(h_\infty)[1 - e^{-[(M+1)/2N]n}] \tag{10.4}$$

since here $E(h_0) = 0$. Note that the expected heterozygosity will be a function of the effective size of population established by the colonists, the mutation rate, and time. The variance of h_n is known (see Li and Nei 1975). Given some assumptions about N and the mutation rate per locus, one can estimate the difference in heterozygosity that may obtain. Unfortunately, conventional sampling theory (see Kish 1965, Weiner and Lourie 1969) suggests that a difference of the anticipated magnitude is marginally demonstrable at best.

This further emphasizes, we believe, the need for a multidimensional approach to enhance the discriminatory power of the study. Although no simple sampling theory is known to us that predicts the added value of multiple measurements upon the same population, the results of tests of significance of a variety of attributes in a study of the similarity or dissimilarity of several populations can be combined. One of the more conventional devices through which this can be done is the χ^2 transformation of the probabilities of the individual tests (Fisher 1950). This strategem, however, encounters difficulties if the tests are not independent.

The effect of selection, and hence of its relaxation, upon quantitative genetic traits clearly depends on the heritability of the latter, that is, on the genetic contribution to the variation that is observed among individuals. If selection rather than its relaxation is involved, theory states that the response to selection R is equal to the selection differential S times the heritability of the trait h^2, that is:

$$R = h^2 S \qquad\qquad (10.5)$$

The selection differential in this context is the difference with respect to the trait in question of the mean of the population of offspring born in a particular generation and the mean of those offspring who will become parents. For example, if the trait were stature, where heritability is about 50 percent, if the selection differential were 2 percent per generation, and if selection proceeded over 10 generations, a mean initially 170 cm, say, would be expected to be 187 cm if selection favored the tall or about 153 cm if it favored the short. Would a difference of this dimension be statistically demonstrable with samples of 700 adults when the means of both the reference as well as the selected population must be estimated? If we assume that the effect of selection is largely confined to the mean, that is, that the variances are little changed (and we can avoid such concerns as the Fisher–Behrens problem in the test of the difference of two means), the answer must be yes. This assumes of course a coefficient of variability no less than 5 percent, a defensible value in the case of stature. As we have repeatedly emphasized, however, the restoration of variability may, indeed, is likely to proceed at a different pace. Suppose new variation accumulates at only 10 percent of the rate of loss which we have previously conjectured. This is tantamount to asserting that relaxation of selection would, in 10 generations and if selection were initially toward a smaller more compact body mass, merely move the mean of 153, to 155. Again, if we assume that the coefficient of variability remains essentially unchanged, samples of 145 would be needed to demonstrate a shift of this size.

At this juncture it seems important to call attention to the fact that an investigator's interest may not be focused on tests of significance, and indeed should not be so focused. The weight of a variety of lines of evidence may make such tests not only uninteresting, but ludicrous. Mutation, for example, is a universal biologic phenomenon; to test that man's mutation rate is zero based upon a few thousand or even ten thousands of observations is, in effect, meaningless because the null hypothesis will doubtlessly be accepted. The capacity to discriminate otherwise with numbers of these magnitudes is essentially zero. The aim of an investigation, therefore, may be not some conventional but irrelevant test of significance; instead, it may be an unbiased estimate of the extent of some phenomenon of universal occurrence, e.g., mutation. With this objective, un-biasedness clearly becomes the catchword; even so, the sampling error of an estimate, unbiased or otherwise, cannot be ignored. We therefore accept the reality of the phenomenon and are concerned merely with the precision with which we can estimate its impact. An argument of this kind may lead to uneven allocation of sample sizes; the "changed population" becomes the only unknown, in effect.

Analysis of Quantitative Measurements

We have more or less tacitly assumed that a study of adaptation to the hypoxia of altitude will embrace a number of villages, all relatively small, perhaps, and each at a somewhat different altitude (that is, partial oxygen pressure). Within a village, a variety of reasons is likely to dictate the examination of individuals of both sexes and all ages. Irrespective of their age and sex, these persons may have lived in their current village of residence for differing periods of time; that is, they may have come to the village from another village at a higher, a lower, or the same altitude at some time in the past; also, they may enjoy different nutritions or other perquisites as a consequence of their socioeconomic status, leadership role within the community, and the like. In short, these individuals may differ with respect to a number of concomitant sources of variation that may produce spurious differences between villages or obscure ones properly ascribable to varying oxygen tension.

Clearly, these sources of variation must be "controlled," either by matching their distributions exactly over villages so that their effects are randomized, that is to say, become a part of the "error variance," or by covariance analysis. The former route seems to us the less desirable for two reasons primarily. First, "matching" will necessarily lead to a loss, possibly substantial, of data gained at considerable effort and expense. Second, randomization of the effects of these variables will inflate the "error variance" (as contrasted with that which would obtain if these effects were removed) and so lead to a loss in precision in tests of significance. The "error variance" is, of course, the standard of variability against which the significance of the effect of a particular factor is to be judged.

We advocate, therefore, a covariance analysis, and a model that may appear as follows:

Let $y_l^{(p)}$ be an observation on the lth individual in the (p)th population $(p = 1, 2, \ldots, s)$; it can be single valued or a vector of values. We shall assume the former in the succeeding paragraphs but this involves no loss in generality of the basic argument. The expectation of this observation we take to be:

$$E(y_l^{(p)}) = \mu^{(p)} + \sum_c \beta_c^{(p)} C_{cl}^{(p)} + \text{error} \tag{10.6}$$

where $\mu^{(p)}$ is a constant (the mean) associated with the (p)th population, $\beta_c^{(p)}$ is the regression coefficient associated with the cth concomitant variable in the (p)th population, and $C_{cl}^{(p)}$ is the observation on the lth person in the (p)th population with respect to that concomitant variable.

The model set out above and formalized in Equation (10.6) allows us to identify the "best" possible model, among many conceivable linear models, from which to estimate parameters of interest. We do this through examining such questions as:

1. Does the cth concomitant variable have a significant effect upon the observation y in population p?
2. Does the cth concomitant variable fail to exert a significant effect upon y in any population?
3. Is the effect of the cth concomitant, if different from zero, the same in all populations?

If the populations have structure vis-à-vis one another, that is, if either they can be aggregated on the basis of the sex of the individuals examined or the villages can be aggregated into "niches" based upon the similarity in ecologic factors, of which we will presume oxygen to be the most important, still further questions can be asked. Suppose, for example, that populations 1 through k are male and $k + 1$ through s are female; suppose further that within each sex group half the populations fall within one niche and half within another. We can now ask:

4. Is the effect of the cth concomitant the same in all populations within a given sex–niche group?
5. Is the effect of this concomitant the same within a given niche for the two sexes?

Clearly other questions occur, such as:

6. Is the effect of this concomitant the same within the various niches but possibly different between the sexes?

It may be asked why there should be this preoccupation with concomitant sources of variation. Is it not sufficient merely to remove their effects and get on to the task at hand, namely, to identify systematic differences in the variable under scrutiny with changing oxygen pressure? Such concern can be justified on the grounds that the factor of interest, partial oxygen tension, may exert its effect upon a particular variable either directly or indirectly or both. An indirect effect can occur through an impingement upon the "normal" dependencies associated with that variable and other variables. Careful analysis of the similarity or dissimilarity of the β_c will permit exploration of this latter contingency.

Presumably it will ultimately be determined that there is or is not an effect exerted upon the dependent variable y through the concomitants that have been measured. The observed value of the variable under examination can now be replaced by its "residual," that is, by the deviation between the measurement and its expected value, the latter based upon Equation (10.6) or some similar model. Let us designate the residual on the lth individual on the (p)th population as $d^{(p)}$; this new variable can be represented as follows:

$$d_l^{(p)} = y_l^{(p)} - \mu^{(p)} - \sum_c \beta_c^* C_{cl} \qquad (10.7)$$

where β_c^* implies the "best"[1] estimate of the effects of c. Note that through the use of $\mu^{(p)}$ we will have removed differences in the means of the concomitants in the various populations, but sex and village effects exerted directly on the ys remain.

Two options are now open. First, if each of the j villages has an oxygen tension that differs substantially from every other village, one may choose to fit a model such as:

$$d_l^{(i)} = y_l^{(i)} - \mu^{(i)} - \beta^{(i)} P_l^{(i)} \qquad (10.8)$$

where l now extends over all individuals of a given sex ($i = 1, 2$). This model asserts, of course, that the deviation is a function of the sex of the lth individual and of the partial oxygen pressure $P_l^{(i)}$ associated with his or her village of examination. The significance of the regression coefficient can be tested for each sex and the comparability of the two examined. Second, if the villages are not strikingly different in partial oxygen pressure, one may be inclined to view the villages as assignable to "ecologic niches" and to examine differences among the latter as if "niches" were a way of classification. Thus, villages become replications, as it were, in a two-way, sex and niche, analysis of variance. This approach would be likely to entail only one departure from a simple analysis of variance if the number of replications (that is, villages) within a niche were the same for all niches, and that is provision for the probable nonorthogonality of the "main effects" and "interactions" as a result of the numbers of observations in the various sex–niche cells being unequal and disproportionate. Allowance for this nonorthogonality should in this instance be conceptually and computationally relatively simple (see, e.g., Scheffe 1959).

If the number of villages within a sex–niche group is not the same for all such groups, the analysis is less straightforward, and several different strategies come to mind. First, one may arbitrarily "balance" the analysis by randomly excluding villages within a sex–niche until all of the latter have the same number of villages. Like most losses of data this is not an especially appealing course of action.

---

[1] "Best" in this context implies the effect as measured with the maximum permissible pooling of observations.

It would, however, regularize the analysis. Second, it is possible to adjust the observations within a village within a given sex–niche group and then to ignore in effect, the potential differences between village means. This is tantamount to collapsing the data over replications. Clearly some information is sacrificed; the gain is a two-way analysis that almost certainly will be nonorthogonal, but manageable. This approach would subsume differences between villages within a sex–niche group in the "error variance." Finally, one may follow the tack just described but instead of using the mean square deviation within a sex–niche group averaged over such groups as the "error variance" use the mean square deviation within villages within a sex–niche again averaged over the sex–niches. This strategy takes cognizance of differences between villages within a sex–niche group, if such exist. It could, therefore, be viewed as a somewhat more conservative analysis than the second, for if no differences in fact exist, the two analyses should be essentially equivalent.

Imbedded in all of the analyses described in the foregoing paragraphs are, of course, certain statistical assumptions. These include the normality of the distribution of the "experimental" variate (or multivariate normality if the "variate" is, in fact, a vector), homoscedasticity of the variances (or variance–covariance matrices) associated with the groups (sex, niche, village) under scrutiny, etc. Patently, the validity of these assumptions should be examined. Appropriate methods will be found in a variety of references (see, e.g., Rao 1965); we shall not enlarge upon these methods here.

Analysis of the Qualitative Measurements

Two somewhat different kinds of noncontinuous data can be anticipated; on the one hand, the observation on a given individual may be the designation of the presence or absence of an attribute where the role of genetic factors may be obscure, e.g., a clockwise frontal vector loop revealed on electrocardiography; on the other hand, it may be the genotype of an individual in respect of some biochemical or immunologic system. In the latter instance, as previously suggested, an enzymatic system may or may not be known to play a direct role in oxygen transport or the tissue utilization of oxygen in the present instance.

Conceptually, the data might be envisaged as multiple classified observations of a presence–absence nature or, alternatively, as an investigation of the dependence of an observed response, possibly binomial but not necessarily so restricted, on one or more independent variables. At a minimum, these classifications or independent variables would include sex, niche (or some other suitable recognition of differences in oxygen tension), and village; they might also include age, residential permanence, ethnicity, and the like. Patently, the number of such variables that can be examined has limits, but the independent variables themselves can be continuous, ordered, or nominal.

If the data are viewed as multiple classified observations, then we have, in effect, a contingency table analysis of some sort; whereas if they are viewed as the reponses of a dependent variable to some series of independent variables, we have a quantal-response problem. We shall examine these alternatives briefly and separately. The first could result in a factorial analysis wherein the contingency tables that arise are complete, that is, observations occur in every cell in the table, or incomplete. Both, however, would resemble analysis of variance models that do not consider trend effects. A fairly extensive literature and reasonably rigorous techniques have been formulated for their analysis (e.g., Roy and Kastenbaum 1956; for an application of their approach see Neel and Schull 1956, Darroch 1962, Kullback and Ireland 1968). Haberman (1974) has argued cogently that these are all essentially ramifications of a more fundamental log-linear model. As unstructured tests of intergroup differences, the procedures cited above may all be more or less equivalent. Here, however, one would like to believe there is some fundamental relationship between genetic variability and oxygen tension, that is, that there should be a response proportional, simply or otherwise, to an increase in oxygen tension. Obviously, this point of view makes quantal-response models more attractive. Among the more common such are the logit and the multinomial logit (for applications of the former as well as another, the exponit model, see Schull and Neel 1965); Haberman (1974) has shown all of these to be log-linear models in the sense in which he defines such models and so to possess all of the properties of the latter, which are unexpectedly general.

Inferences and Speculations

To speculate, Webster (1974) asserts, is to review something idly or casually and often inconclusively. Presumably few if any scientific investigations are initiated to generate a series of speculations in the sense just stated. Instead, the investigator seeks a number of propositions that he considers to be true and from which other truths follow. These latter truths we term inferences. Unfortunately, many so-called inferences upon close scrutiny prove to be speculations. Either the study has been poorly conceived, or the inferential process has not been rigorous. In short, more is claimed than is justified. Few studies of processes as complex as adaptation are faultless, however. Consider, for example, the following: We have stated that the ecologic niches of Arica, to which reference has been made, differ markedly one from another in partial oxygen pressure, temperature, and humidity. Indeed, the latter two are inextricably confounded with the former. This can pose formidable logical difficulties, for to infer that a particular difference between altiplano and coast is the result of the difference in partial oxygen pressure may be wrong. It may be a temperature effect, ascribable to differences in moisture, or both. Interpretation of the causal origin of an effect must be guarded, therefore, or clearly identified as speculative. A well-designed study certainly warrants an equal degree of care in the identification and interpretation of the "truths" it reveals.

Conclusions

It would be actively misleading and possibly presumptuous to assume that these final remarks will in some fashion summarize those which have gone before. Instead, they will attempt to speak to a philosophy appropriate, it is hoped, to a particular class of problems. We set out, therefore, a series of questions or issues that the individual investigator should confront. First, can one deduce a clear, unambiguous, and non-self-serving statement of the rationale and objectives of the research? Second, how large must the changes or differences be for such samples to detect? Third, how, or possibly why, were the variables that were measured chosen? Fourth, how are the data to be analyzed? Are they to be viewed as the bases for a series of tests of significance, or is the primary concern one of unbiased estimation? If the latter, what is the sampling precision that is sought? Fifth, what form is the data analysis to take? Is it to be a series of ostensibly independent, but actually correlated analyses? If so, how are these analyses to be subsequently integrated? Finally, has consideration been given to the manner in which these data, if they do not lead to unambiguous inferences in themselves, may be integrated with prior or subsequent studies?

Acknowledgments

We gratefully acknowledge the support of the National Institutes of Health through the grant, HL-15614, the Multinational Genetics Program, Organization of American States, Chile, and the program RLA 75-047 from PNUD/UNESCO.

References

Baker, P.T. (1969). Human Adaptation to High Altitudes. *Science* 163:1149–1154.

Bird, J.B. (1943). Excavations in Northern Chile. *Anthropol. Pap. Am. Mus. Nat. Hist.* 38: 171–316.

Buskirk, E. (1971). Work and Fatigue in High Altitude. *In* E. Simonson, Ed., *Physiology of Work Capacity and Fatigue.* Springfield, Ill.: Charles C. Thomas.

Chakraborty, R., W.J. Schull, E. Harburg, and M.A. Schork. (1976). Heredity, Stress and Blood Pressure, a Family Set Method. V. Heritability Estimates. *J. Chron. Dis.* 30:683–700.

Darroch, J.N. (1962). Interactions in Multifactor Contingency Tables. *J. Roy. Stat. Soc.,* Ser. B 24:251–263.

Falconer, D.S. (1960). *Introduction to Quantitative Genetics.* Edinburgh: Oliver and Boyd.

Fisher, R.A. (1950). *Statistical Methods for Research Workers.* 11th ed. New York: Hafner.

Frisancho, R. (1975). Functional Adaptation to High Altitude Hypoxia. *Science* 187:313–319.

Haberman, S.J. (1974). *The Analysis of Frequency Data.* Chicago: Univ. Chicago Press.

Harrison, G.A., C.F. Kucheman, M.A.S. Moore, A.J. Boyce, T. Baju, A.E. Mourant, M.J. Godber, B.G. Glasgon, A.C. Kopeć, D. Tills, and E.J. Clegg. (1969). The Effects of Altitudinal Variation in Ethiopian Populations. *Phil. Trans Roy. Soc. London Ser.* B 256:147–182.

Hurtado, A. (1964). Acclimatization to High Altitudes. *In* W.H. Weihe, Ed., *Physiological Effects of High Altitude*. Oxford: Pergamon Press.

Hurtado, A. (1971). The Influence of High Altitude on Physiology. *In* A. Porter and J. Knight, Eds., *High Altitude Physiology: Cardiac and Respiratory Aspects*. Edinburgh: Churchill Livingstone.

Kish, L. (1965). *Survey Sampling*. New York: John Wiley and Sons.

Kullback, S., and C.T. Ireland. (1968). Contingency Tables with Given Marginals. *Biometrika* 55:179–188.

Li, W.-H., and M. Nei. (1975). Drift Variances of Heterozygosity and Genetic Distances in Transient States. *Genet. Res.* 25:229–247.

MacNeish, R.S., R. Berger, and R. Protsch. (1970). Megafauna and Man from Ayacucho, Highland Peru. *Science* 168:975–977.

Neel, J.V., and W.T. Schull. (1956). *The Effect of Exposure to the Atomic Bombs on Pregnancy Termination in Hiroshima and Nagasaki*. Washington, D.C.: NAS–NRC.

Rao, C.R. (1965). *Linear Statistical Inference and Its Application*. New York: John Wiley and Sons.

Roy, S.N., and M.A. Kastenbaum. (1956). On the Hypothesis of "No Interaction." *Ann. Math. Stat.* 28:749–757.

Scheffe, H. (1959). *The Analysis of Variance*. New York: John Wiley and Sons.

Schull, W.J., and J.V. Neel. (1965). *The Effect of Inbreeding on Japanese Children*. New York: Harper and Row.

Schull, W.J., and J.V. Neel. (1965). *The Effect of Inbreeding on Japanese Children*. New York: Program: A Study of Adaptation to the Hypoxia of Altitude. *In* J.S. Weiner, Ed., *Physiological Variation and its Genetic Basis*. London: Society for the Study of Human Biology.

Stewart, T.D. (1973). *The Peoples of America*. New York: Charles Scribner and Sons.

Webster's New World Dictionary. (1974). New York: World Publishing Co.

Weiner, J.S., and J.A. Lourie. (1969). *Human Biology, a Guide to Field Methods*. Oxford: Blackwell Scientific Publications.

Chapter 11

Patterning of Skeletal Pathologies and Epidemiology

T. Dale Stewart

Introduction

In 1892, four centuries after Columbus' discovery of America, it would have been impossible to make more than a few reliable statements about the skeletal pathologies[1] of the First Americans. In 1926, only 34 years later, much more was known about the American Indians, but still very little about the Aleuts and Eskimos of Alaska. Beginning in 1926, Aleš Hrdlička, who already had assembled the largest collection of Indian skeletons, undertook the first of 10 collecting seasons in Alaska. Thanks mostly to Hrdlička's collections and his studies thereon, the paleopathology of the First Americans is on a sounder footing than that of any other major human group.

Although data on the pathologies of the First Americans now are voluminous, space limitation makes it necessary for me to select for presentation here those for which the patterns are demonstrable, particularly as regards the Aleuts–Eskimos versus the Indians, the two components of the First Americans generally regarded as the most different. Their differences extend beyond the physical and cultural to the geographic, thereby reflecting the relatively late arrival of the one located near the entrance to the hemisphere, and the much earlier arrival of the other

---

[1] In medical school half a century ago I was taught that the term "pathology" meant the branch of medical science concerned with morbid conditions, their causes, nature, etc., and therefore that morbid conditions by themselves should not be referred to as "pathologies." The latter, by this definition, means more than one of this particular branch of medical science, something which was regarded as being virtually inconceivable. Although my inclination is to abide by my teachers' admonition, I am departing therefrom here because common usage of the term in its "forbidden" sense, and the convenience of expression this implies, has led to its inclusion in some recent American dictionaries.

occupying the rest of the hemisphere. The order in which the pathologies are considered is not in itself significant.

In the past, "epidemic diseases" meant primarily such dreaded scourges as smallpox, typhus, yellow fever, cholera, and plague. This class of diseases also is labeled "acute" because it is characterized by a tendency to spread quickly and rapidly to reach high frequencies of occurrence; in other words, to become "epidemic." Today most epidemic diseases are under control and epidemiologists are able now to devote much of their attention to chronic diseases and such afflictions as appear to represent interactions between people and their environments.

Anthropologists view epidemics more abstractly as one of the factors that in the past often must have played a role in human evolution and more recently have acted as important governors of population size. Beyond this, physical anthropologists may see reflections of epidemics in certain concentrations of skeletons but are at a loss to confirm their suspicions because the swiftness with which early epidemics struck down their victims left insufficient time for bone involvement. For these reasons and because knowledge of the epidemiology of the First Americans has not increased much over that known in the last century, I shall subordinate acute diseases to the skeletal pathologies that had long duration in life. General comments on both subjects are postponed until the final section.

Skeletal Pathologies

Syphilis

The catalog cards of Hrdlička's collection of Aleuts and Alaskan Eskimos include only 17 labeled syphilitic or possibly syphilitic. None of these appears to be ancient. Even when the skull lesions are in the active stage they are seldom accompanied by long bones swollen and pitted with periostitis and osteitis. Support for these statements appears in Hrdlička's writings (1930 pp. 320–321, 1944 p. 404, 1945 p. 371).

Present knowledge of the frequency of syphilitic bones in American Indian populations is very little better than that offered here for the Aleuts–Eskimos. The foremost authorities on venereal diseases who have examined the seemingly most likely examples of bone syphilis recovered by archaeologists from Indian sites have disagreed in their diagnoses. For example, in a letter to Hooton dated 1926 Herbert U. Williams said about the Pecos pathologic bones:

> I do not believe that we are warranted in saying of any single dried bone specimen that it is certainly syphilis. However, your three cases . . . together with the various thickenings of long bones from other cases, make powerful evidence for the existence of syphilis among the people from whom the bones were derived. (Hooton 1930, p. 311)

Commenting on this opinion and on another by James Ewing denying that the three cases in question were syphilitic, Hooton said:

> . . . it is evident that the existence of syphilis among the American Indians in pre-Columbian times cannot be proved until pathologists become more certain of their diagnoses from dry bones of the lesions caused by this disease. It is unfortunate that qualified experts should have disagreed in the three cases under discussion, since all of them are definitely and indisputably prehistoric—a statement which cannot be made concerning most remains of American Indians thought to be syphilitic. (Hooton 1930, pp. 311–312)

In the only attempt so far to look objectively at the situation, Stewart and Quade (1969) recorded every frontal lesion observed in a wide sampling of pre- and protohistoric Indian skeletal populations. They found wide variations in the frequencies of these lesions. This does not mean, of course, that frontal lesions always bespeak syphilis; a high percentage of them may be caused merely by trauma. Besides, relatively few were accompanied by long-bone lesions. The point is, however, that syphilis is a chronic disease that seems to have appeared among the First Americans long after their arrival in America, but whether or not before the appearance of the Second Americans is still debatable.

Tuberculosis

Another major chronic disease that affects the skeleton is tuberculosis. The signs of its presence are quite different from those of syphilis. Whereas in the latter new bone formation accompanies the bone destruction, in the former the lesions are solely of the destructive type. A difference of this sort might be expected to make the tubercular lesions easily recognizable. However, according to Morse (1961), there are a number of other diseases that produce similar effects.

Probably because of the diagnostic problem, no reliable figures are available on the presence and distribution of tubercular lesions in the skeletal collections of any of the First Americans. The neglect of this matter would be unfortunate indeed were it not for the fact that Allison *et al.* (1973) have established convincingly the presence of tuberculosis in a prehistoric Peruvian mummy. This discovery adds to the likelihood that the First Americans brought the disease with them from Asia.

Neural Arch Defect

I have reported (Stewart 1931a, 1953) an unusually high incidence of an inconspicuous vertebral abnormality in the Aleuts–Eskimos. This consists of an interruption of bony continuity of the neural arch, most often in the pars interarticularis, on one or, more usually, on both sides of the fifth lumbar vertebra and/or of one or more others. On analysis four striking patterns emerged: (1) for the Alaskan region as a whole, a progressive increase in the frequency of the defect with age (sexes combined) from 0 at birth to 33.9 percent after 40 years of age; (2) for each of three subdivisions of the region (north of the Yukon, Yukon south through Bristol Bay, Aleutians and Kodiak Island) a different frequency of the defect (all ages and both sexes combined), 40.3, 15.3, and 24.7 percent, respectively; (3) more defects in males than in females (35.4 vs. 24.0 percent); and (4) an increase with age of cases with two or three defective arches (5.6 percent).

At first I assumed from available information that the defect was present at birth, but obviously a defect that increases so much in frequency from birth onward is acquired; it can be said to be congenital only in the sense that an inherited predisposition of some sort may have led to it. The orthopaedic surgeon Leon L. Wiltse (1962, Wiltse *et al.* 1975) has shown that this is indeed so. His findings on American Whites indicate that the defect tends to run in family lines and that with such family lines combined the incidence therein rises above the 5.8 percent for the rest of the white population so as actually to approach the figures for the Aleuts–Eskimos given above. According to Wiltse *et al.* (1975), in Japan Ohta also has reported higher incidences of the condition in certain family lines.

In contrast to the Aleuts–Eskimos, the Indians are poorly documented as regards their proneness to neural arch defect. In my search of the literature in 1931 I found only five reported cases in 79 Indian skeletons of both sexes (6.3 percent) from widely scattered parts of the United States (New Jersey, Arkansas, Louisiana, and California) and only seven cases in 115 skeletons (6.1 percent) from South America (Argentina). Since then, Snow (1948) has reported a much higher frequency for the Indian Knoll collection (Table 11.1). To this record I can now add my previously unreported findings on the Arikara of South Dakota as represented in the National Collection (Table 11.1).

Osteoarthritis

The defective arch described in the preceding section often induces bony lipping around the vertebral centrum with which it is associated. Lipping often appears also on adjacent, otherwise normal vertebrae, the amount depending on the natural mobility of the segment, individual age, and sex, in other words, on the wear and tear during life resulting from customary weight bearing and hard usage of joints.

Years ago I undertook, in connection with my study of the arch defect, to demonstrate graphically the differing patterns of vertebral osteoarthritis between the Eskimos and Pueblo Indians represented in the National Collection. Except for an abstract (Stewart 1947), the only published part of this work is the illustrations (Stewart 1966), two of which are reproduced here as Figure 11.1.

In this figure each vertebral centrum is represented by a horizontal black bar with the differing lengths of the upper and lower borders indicating their differing proneness to lipping. To establish the dimensions of the bars I scored the lipping on each centrum, above and below separately, on a scale of four and averaged all of the scores for each border. No attempt has been made to treat the two sides separately, so the illustrations appear unnaturally symmetrical. The

Table 11.1. *Neural Arch Defect: Prevalence in Two Indian Populations*

Segment(s) affected	Males		Females	
	No.	%	No.	%
Kentucky (171 males, 139 females)				
L4	—	—	1	**0.7**
L4-5	—	—	2	**1.4**
L5 (left only)	1	**0.5**	1	**0.7**
L5 (or L6)	29	**17.0**	26	**18.7**
Totals	30	**17.5**	30	**21.6**
South Dakota (94 males, 69 females)				
L3, 5	—	—	1	**1.6**
L4	3	**3.2**	2	**3.1**
L4-5	—	—	2	**3.1**
L5 (right only)	4	**4.2**	2	**3.1**
L5 (left only)	—	—	3[a]	**4.7**
L5 (or L6)	21[b]	**21.3**	8	**12.5**
L?	1	**1.1**	—	—
Totals	29	**30.8**	18	**28.1**

[a] One is questionable because of damage to the right side.
[b] All show separation through the pars interarticularis on both sides, except one which on the right shows separation between the spinous process and the inferior articular facet.

numbers between the bars are the actual counts, by side, of arthritic posterior facets seen. The pattern differences shown in the figure are mainly: (1) greater vertebral lipping in the part of the Pueblo column above the lumbar region, and (2) greater posterior-facet involvement throughout the Eskimo column.

A nearly comparable study of the arthritic involvement of the joints of the appendicular skeleton, again in Eskimos and Pueblo Indians, was undertaken for the purpose of a doctoral dissertation by Jurmain (1975). He looked at a large number of specimens in the Terry dissecting-room collection to decide on the most significant degenerative changes in each of four joints (knee, hip, shoulder, and elbow) that could be scaled progressively and systematically. In these joints marginal lipping is accompanied by breakdown of the articular surfaces, ending in cartilagenous erosion and bone polishing (eburnation). Taking into account all such changes, Jurmain arrived at criteria for assigning those of each joint to one of three categories: None–slight, moderate, and severe.

Figure 11.2 shows how the frequencies of the second and third categories for each of the joints compare by sex and side. In all cases the frequencies for the Eskimos are higher than those for the Pueblos. However, whereas the knee and shoulder show the most change in both groups, the joint subject to the least change is the elbow in the Pueblos and the hip in the Eskimos. Surprisingly, the right side is not consistently involved in either group, but not surprisingly the males have more arthritis than the females.

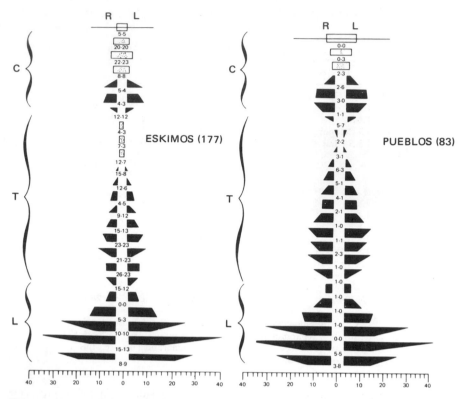

Figure 11.1 Patterns of vertebral osteoarthritis: Eskimos versus Pueblo Indians. For explanation see text. (From Stewart 1966, Figures 7 and 8 combined.)

As is shown in the next section, the Pueblos probably do not provide a typical Indian base from which to judge the relative amount of osteoarthritis as between Eskimos and Indians. In this connection it should be noted that Blumberg *et al.* (1961) undertook to evaluate the prevalence of osteoarthritis in living Alaskan Eskimos by radiographic means and to compare this prevalence with that in Whites (USA). Without going into details, they concluded that the Eskimos have somewhat less osteoarthritis than Whites (as judged by hand and wrist x rays only). This led them to speculate that:

> ... what appears to be a low prevalence of osteoarthritis in the Eskimos may have arisen as a result of selection over many generations against individuals prone to develop osteoarthritis, since such individuals would be less likely to survive the rigours of Arctic life. This may imply that some trait related to the development of osteoarthritis is inherited, which is consistent with some previous studies. Any further conjectures must await more complete investigations on this interesting population. (Blumberg *et al.* 1961, p. 339)

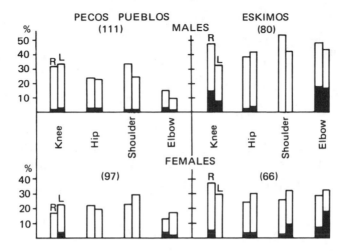

Figure 11.2. Relative arthritic involvement of four appendicular joints: Eskimos versus Pueblo Indians. □, Moderate; ■, severe. For other explanation see text. (From Jurmain 1975, parts of Figures 3, 9, 15, and 21 combined.)

Osteoporosis

Unlike osteoarthritis, which is localized in the joints and results in the formation of some new bone, osteoporosis (not to be confused with so-called "osteoporosis symmetrica," which is restricted to the skull) affects all the bones in their entirety and results in a loss of bone substance that weakens the structure and reduces the overall weight. To some extent osteoporosis is a normal age change. Yet there is more to it than this because the extent of the change often becomes pathological, especially as regards vertebral involvement in older females.

The varying expressions of true osteoporosis in the skeletons of some of the First Americans was the subject of a doctoral dissertation by Ericksen (1973). She selected her populations for the contrasting environments under which they lived: Eskimos from northern Alaska, Pueblo Indians from the southwestern United States, and Arikara Indians from the Great Plains of the northern United States, all available in the National Collection.

First, she separated the skeletons of each group into males and females. Then for each sex she set up three adult age groups: young (18 to 25), middle aged (30 to under 50), and old (50 and above). The gap between the first two was deliberate and aimed at making more evident any early developing age changes. From these samples she selected three bones (femur, humerus, L4 vertebra) for anthropometric, radiographic, and microscopic study. For microscopic sectioning she took circular plugs from set locations on each bone by means of a $\frac{5}{8}$ in. diamond core drill bit.

Figure 11.3 combines two of Ericksen's graphs showing rather marked group differences. The top graph shows the femur cortical index, which expresses the

T.D. Stewart

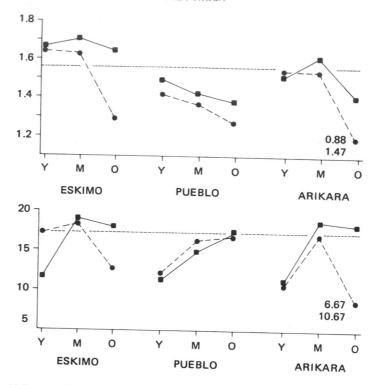

Figure 11.3. Age changes in two skeletal indicators of osteoporosis: comparisons between Eskimos, Pueblo Indians, and Arikara Indians. ■, Males; ●, females. Y, young; M, middle aged; O, old. Upper: Femur cortical index (except in five cases, 20 or more individuals in all three age categories). Lower: osteon count (except in three cases, ten or more individuals in all three age categories). Note that for both indicators only two old Arikara females were available. For further explanation see text. (From Ericksen 1973, Figures 11 and 22 combined.)

mean of medial and lateral cortical thickness, measured on the radiographs, as a percentage of the maximum femoral length. As is evident, the Pueblo males and females enter the adult stage with a lower index than the males and females of the other two groups and lose steadily but not as much as the latter; in other words, throughout adult life the Pueblos have less cortical thickness relative to bone size than the other groups and this thickness changes less.

The graph at the bottom of Figure 11.3 shows the osteon counts. Both the Eskimos and Arikara Indians experienced a rise in number of osteons between youth and middle age, after which the numbers dropped, most noticeably in the females. In contrast, the Pueblos experienced counts that rose steadily but never reached the levels of the other groups, either in middle age or in old age.

Although these two examples are fairly similar, they are not altogether typical of all the findings in this study. As a matter of fact, the showings of the three groups, one against another, are mostly so variable that Ericksen could formulate no concise answer to the question: Do the three populations differ from each other in the way aging changes take place? Her failure in this regard may result from the scarcity of older individuals in these populations. In any case, the meaning of the aging changes between the Eskimos and Indians is still unclear.

Hyperostosis Spongiosa

When Hrdlička first saw the signs of this disorder in the skulls he collected during his visits to Peru in 1910 and 1913 he was impressed by their symmetrical locations on the two sides of the vault and their coral-like, porous appearance. Hence his name "osteoporosis symmetrica" (Hrdlička 1914). This name has fallen into disuse since the one used here was introduced first by Müller (1934) and then by Hamperl and Weiss (1955). The current name emphasizes the elevation of the areas of involved bone as well as the sponginess of this bone.

The spongy or porous structure of the bone also can be likened to a sieve. This idea must have been in Welcker's mind when he gave the name " cribra orbitalia " to lesions of identical appearance on the roofs of the orbits (Toldt 1886). Opinions are divided, however, as to whether the vault and orbital lesions are manifestations of the same disorder. They do not always appear together. Moreover, if this disorder is an anemia caused by iron deficiency, as now generally believed (El-Najjar and Robertson 1976), the hemopoietic properties of the bones of the two areas are totally dissimilar (D.J. Ortner, personal communication 1976).

To the best of my knowledge only two investigations have dealt exclusively with cribra orbitalia—one by Welcker (1887), the other by Nathan and Haas (1966). The 6500 skulls from all over the world that Welcker examined includes those of 72 eastern Eskimos without signs of involvement, 222 Indians other than Peruvian with 8 (3.6 percent) showing signs of involvement, and 272 Peruvians with 22 (8.1 percent) showing signs of involvement.

Nathan and Hass' worldwide sample of 718 skulls includes 90 Alaskan Eskimos of which 18 (20.2 percent) showed involvement, 120 Indians from the North American Southwest of which 38 (31.6 percent) showed involvement, 60 Mexican Indians of which 13 (21.6 percent) showed involvement, and 105 South American Indians (Argentina, Bolivia, Chile) of which 11 (10.4 percent) showed involvement.

The great difference between the results of these two investigations could be due mainly to Welcker's failure to recognize any but "active" lesions (gross porotic areas). Nathan and Haas, on the other hand, recognized three types of lesions: porotic (scattered, isolated fine lesions), cribrotic (conglomerate of larger but still isolated apertures), and trabecular (bone trabeculae formed from confluent apertures). The differing proportions of immature skulls in the samples also could be a factor influencing the results.

Most other observers of hyperostosis have given primary attention to the vault lesions, but often without indicating developmental stages of the disorder. How much these dissimilar approaches account for the reported group differences is anyone's guess. For instance, as regards 262 adult and 16 immature skulls from the north coast of Peru, Hrdlička (1914) simply noted that 8 (3.0 percent) and 3 (18.8 percent), respectively, had the lesions; as regards the Pecos Pueblos, Hooton (1930) singled out from 581 skeletons the skulls of 19 adults (3.2 percent) and 9 subadults (1.5 percent) with "active manifestations of the disorder"; and as regards a collection of 21 children's skulls recovered from the Sacred Cenote at Chichén Itzá in Yucatan, Hooton (1930) reported 14 (66.7 percent) with the lesions, presumably also in the active stage.

Hrdlička's work in Alaska in the 1920s and 1930s left him with little to say on the subject. Nowhere in his writings thereon does he mention the vault lesions of what he continued to call "osteoporosis." However, he did report three cases of cribra orbitalia from Kodiak Island (Hrdlička 1944). This implies agreement with Welcker's earlier indication of a scarcity of any manifestation of the disorder among the Eskimos.

Now for a shocker. Juan Munizaga of Chile claims (1965) that many of the sign of vault hyperostosis have gone unrecognized; that the disorder was prevalent throughout the First Americans. Working with Hrdlička's Peruvian collections, Munizaga set up standards of normality for surface appearance and vault thickness and identified four age-related types of vault lesions. By this means he classified the pathologic signs in a total of 2145 Peruvian skulls from two coastal areas. As developed in graph form, his surprising findings, arranged in seven age groups, are reproduced here as Figure 11.4. The regularity of the progressive age changes shown is indeed impressive. However, because the percent-

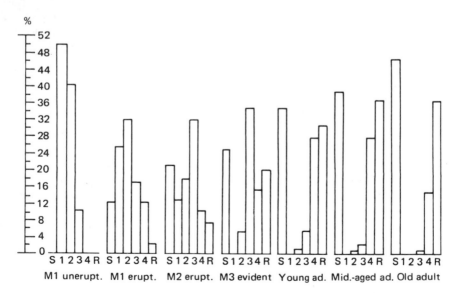

Figure 11.4. Hyperostosis spongiosa in Peruvian Indians: pattern of increasing cranial vault involvement through seven age categories. S, Normal; 1, 2, 3, and 4, types of cribriform surface; R, "natiforme" (sagittal grooving caused by midparietal thickening). (From Munizaga 1965, Lamina IV.)

ages are so much higher than those previously reported for the vault lesions, the natural inclination of anyone inspecting the graph is to wonder whether something besides hyperostosis has been included. His finding of equally high percentages in small series of Alaskan Eskimos and Florida Indians reinforces the skepticism. Yet, taking also into account the findings of Nathan and Haas, I find it difficult to avoid the conclusion that there is more to hyperostosis of the skull than is generally recognized.

Scaphocephaly

While on the subject of the skull vault it is appropriate to consider abnormal closure of the sagittal suture, the commonest type of craniostenosis. Most of this suture normally closes at between 26 and 30 years (Todd and Lyon 1925, p. 39). In rare cases, however, all of the suture closes before birth or shortly thereafter. When this happens no further sideways growth from the midline is possible, but growth leading to elongation of the vault continues to take place at the coronal and lambdoid sutures. Because this abnormal growth pattern appears while the brain is still enlarging, the earlier the closure occurs, the more the skull takes on the abnormally elongated shape known as "scaphocephaly," at the same time temporarily giving rise to some, but usually not serious, elevation of the endo-cranial pressure. In the most extreme case I have seen (a Maryland Indian) the cranial index is 56.6. The disorder appears to occur a little more frequently in males than in females.

Although numerous examples of scaphocephaly have been observed in skulls of American Indians—for example, in Kansas by Eiseley and Asling (1944), in California by Hohenthal and Brooks (1960), in Mexico by Comas (1965), and in Arizona by Bennett (1967)—none has been reported, so far as I can discover, in the skulls of Aleuts–Eskimos.

The cause or causes of this disorder are still obscure. In a study of cleft-palate patients Moss (1957) found a positive correlation between premature closure of the metopic suture and malformations of the skull base. He reasoned, therefore, that premature suture closure in general is mediated by an alteration within the brain case. However, because he arrived at this assumption by studying skulls in which premature suture closure had already taken place, one wonders how he could tell which abnormality came first. Bennett (1967) notes that Nance and Engel (1967) lean to a polygenic origin.

Ear Exostosis

Reports on the occurrence in the First Americans of a bony outgrowth (osteoma) from the wall of the external auditory meatus have been accumulating for a century. In some cases the exostosis is large enough to occlude the lumen of the canal and cause deafness.

Hrdlička (1935) was the most recent person to summarize and add significant-ly to the subject. His observations show only two cases in 1000 Eskimo skulls

(0.2 percent), but a great range of frequencies for Indian skulls, as for example: 12 in 500 Pueblo skulls (2.4 percent), 35 in 395 Florida skulls (6.9 percent), 46 in 435 California skulls (10.6 percent), 522 in 3651 Peruvian skulls (14.3 percent), and 30 in 109 Arkansas skulls (27.2 percent). So far as the evidence goes, the highest frequencies of ear extosis anywhere in the world occur among some of the Indians of the United States. These frequencies increase to nearly 50 percent when males alone are considered. Also, the left ear tends to be more affected than the right, but often both ears are equally involved. Hrdlička (1935) could not find an explanation that satisfactorily accounted for all the variation in frequency.

Ear Cholesteatoma

The external auditory canal can be occluded not only by a bony tumor, but also by a soft-tissue tumor. The cranial evidence of the existence of such a tumor in life consists of a greatly dilated or ballooned canal. Accompanying the dilation usually is a thinning and perforation of the tympanic plate. My records on this affliction in Aleuts–Eskimos and Indians are probably the only ones in existence: 15 cases in skulls of the former and only one in skulls of the latter (from Peru). In none of these cases does the enlargement exceed the size of an acorn.

Dr. J.B. Gregg, the otolaryngologist well known to physical anthropologists for his studies of ear diseases in prehistoric Indians, believes that this type of tumor is a cholesteatoma (personal communication, January 13, 1976). He describes it as a mass of epithelial debris resulting from an alteration in the squamous epithelium of the canal. When this mass is not removed it builds up pressure within the canal and produces the ballooning observed. Cases have been reported in the clinical literature by Greene (1933), McGuckin (1961), and Bunting (1968).

If the Alaskan natives do indeed have an unusual incidence of cholesteatoma, possibly it may have something to do with the cold climate in which they live. In this connection it is worth noting that Maynard et al. (1972) reported that:

> In Alaska, otitis media has been the second most frequent cause of morbidity in Aleut, Indian and Eskimo children. During a study of Eskimo infants in western Alaska, 38% were found to have had one or more episodes of otorrhea during the first year of life, with 19% of these having their initial episode before the age of 4 months. Audiometric testing also revealed that 31% of the children had hearing deficiencies in one or both ears. (Maynard et al. 1972, p. 597)

Mandibular Torus

Another skeletal position that a tumor elects to occupy is the upper lingual side of the mandible in the premolar region. The tumor in this instance is bony and takes the form of one or more knobs of varying size arranged in a row. In extreme bilateral cases not enough space remains between the tumors of the two sides to permit the introduction of a finger tip. This intrusion into the space between the mandibular rami probably interfered with tongue motion. Although often called a hyperostosis, I am calling it a torus to avoid confusion with hyperostosis spongiosa, already discussed. I shall not attempt to deal with the similar tori on the palate and maxilla, because they are not so numerous and/or so well documented.

Fürst and Hansen (1915) reported finding the torus in 182 of 215 Greenland Eskimo mandibles of both sexes (84.8 percent). Reporting on the mandibles of Eskimos from the northern part of Alaska, Hrdlička (1930) gave figures of 79.6 percent for 350 males and of 60.1 percent for 360 females. By including the Eskimos of southern Alaska in his 1940 report, he lowered these figures to 57.1 percent and 32.0 percent, respectively. The same report gives 63.6 and 58.6 percent, respectively, for the Aleuts–Pre-Aleuts and 64.7 and 48.9 percent, respectively, for the Koniags–Pre-Koniags.

By contrast, Hrdlička (1940) found the American Indians to have much lower and more variable percentages. For instance (sexes combined): Peruvian Indians, 16 of 465 (3.4 percent); Florida Indians, 22 of 497 (3.4 percent); California Indians, 47 of 410 (11.5 percent); and Pueblo Indians, 128 of 526 (24.3 percent).

Explanations of the torus have ranged from a need for reinforcement of the underlying bone and a response to some form of irritation to a racial trait of long standing. So far none of the explanations has been agreed upon.

Dental Caries

Under certain conditions the enamel covering of the teeth is subject to decay, which may progress to abscess formation at the apices of the roots. Unfortunately, the reported group frequencies derived from skeletal remains are seldom comparable, mainly because so many of the teeth tend to be lost postmortem. Also, in cases of antemortem tooth loss it is impossible to tell whether caries has been the cause of the loss. For these reasons I shall limit the data cited (Table 11.2) to

Table 11.2. *Crown Caries of Mandibular Molars (Both Sides): Prevalence in Adult Eskimos and Indians*

Molar	Sex	Eskimos[a]		Peruvian Indians[b]	
		No. of teeth	Percent with caries	No. of teeth	Percent with caries
M_1	Male	622	0.8	163	1.2
	Female	639	1.4	163	0.6
M_2	Male	535	2.4	138	16.7
	Female	560	2.0	104	12.5
M_3	Male	394	5.6	98	7.1
	Female	380	2.9	87	6.9

[a] Goldstein (1932, Table 2).
[b] Stewart (1931b, Table 2).

two sets of figures for mandibular molars (the teeth least frequently lost post-mortem and most frequently carious) merely to make the point that the Eskimos have fewer caries than a representative group of Indians (those from coastal Peru). It is my impression that in this respect the Aleuts closely resemble the Eskimos, whereas the Peruvians may be near the upper extreme of Indian variation. The comparison in Table 11.2 would be more impressive if the Peruvian's high per-centages of cervical caries were shown. The latter type of caries, which tends to ring the neck of the tooth below the enamel line, does not occur in the Eskimos.

Realistically, so many etiologic factors are known to combine in producing variations in prevalence of caries that little, if anything, can be deduced from the latter regarding population relationships.

Epidemiology of Acute Disease

It is customary to look upon Columbus' voyage to America in 1492 as the first significant contact between the First and Second Americans. Whatever one may think of the Viking contact—the only earlier one with any claim to validity—it left nothing of an epidemiologic nature that has been detected. As far as the evidence goes, the Viking contact took place above the 50° north latitude line and was very brief. Considering the cold climate of that part of the world, it is unlikely that any Vikings reaching the American shores would have been carrying the germs of an acute disease.

The circumstances of the Spanish voyages beginning in 1492 were very different. The part of the world in which they took place was in the low latitudes, where the warm climate favors the perpetuation of acute diseases. Also, the first settlements served as staging points for further explorations to the west, which in turn repeated the process. Once an acute disease reached the first settlements, therefore, the ensuing epidemics there ensured its transmission along the further lines of exploration and settlement.

Historical records indicate that this is actually what happened. Only about a year after the outbreak of a smallpox epidemic in Santo Domingo, Mexico received it. In both places the Indians were decimated. As recently as the late eighteenth and early nineteenth centuries the continuing spread of smallpox, especially that among the Indian tribes of the United States, was still a tragic affair (cf. Stearn and Stearn 1945). Possibly the Arikara skeletons referred to in the foregoing sections represent some of the victims of these late epidemics.

Smallpox receives most credit for the abrupt and immense reduction in numbers of the First Americans after 1492. Probably other less recognizable acute diseases, such as typhus and measles, also contributed importantly to the death toll. In any case, it seems clear that the First Americans had not been exposed previously to the acute diseases that the Europeans brought with them to America. Their response was not unnatural; just extreme. Early in the present century both the First and Second Americans responded in a like manner to a disease new to both of them: influenza.

In 1960 I advanced the idea that the acute diseases known to us today might have been left behind by the First Americans during their passage from Asia through the Far North (Stewart 1960). To account for this I likened the Far North to a "cold filter." Although the cold does seem to play a role in inhibiting the spread of disease, I am not so sure now that it alone accounts for the failure of the Old World diseases to reach America prior to 1492 (cf. Stewart 1974). Were the presently known acute diseases in existence in the Old World at the time of the first peopling of America? Migrating people cannot leave behind something that has not yet come into existence.

Instead of pursuing this aspect of the problem, I shall look at its other aspect. Did the First Americans have any such acute diseases for which the Second Americans had no immunity? The answer seems to be "No." Does this negative answer mean that the First Americans had not been isolated from the Old World long enough to develop acute diseases of their own? Contrary to my view on this matter in 1960, I am inclined now to think that we do not know enough about what is involved in disease origins to estimate the time required for their development.

Conclusions

The data presented in the foregoing sections are summarized in Table 11.3. Perhaps the most noteworthy point brought out in this table is the distinctiveness of the Aleuts–Eskimos as regards the presence or absence of at least half of the 12 pathologies considered. In three of these (neural arch defect, ear cholesteatoma,

Table 11.3. *Summary of Pathologies Discussed*

Pathologies	Aleuts–Eskimos	Indians
Skeletal populations		
Syphilis	Not evident in ancient populations	Not evident in ancient populations
Tuberculosis	Poorly documented	Present in Ancient Peru
Neural arch defect	Highest known frequency north of Yukon	Moderately high frequency in area of United States
Osteoarthritis	More than in Pueblos	Less than in Aleuts–Eskimos
Osteoporosis	Few clear differences	Few clear differences
Hyperostosis spongiosa	Rare (?)	Fairly common (?)
Scaphocephaly	No cases reported	Numerous cases reported
Ear exostosis	Rare	Low to moderate frequencies
Ear cholesteatoma	Several cases known	Only one case known (Peru)
Mandibular torus	Highest known frequencies	Low to moderate frequencies
Dental caries	Rare	Low to high frequencies
Living populations		
Smallpox, typhus, measles, etc.	No immunity	No immunity

and mandibular torus) the Aleuts–Eskimos have the highest frequencies known for any First Americans; in the other three (scaphocephaly, ear exostosis, and dental caries) they rank among the lowest.

On each side of this dichotomy there is one pathologic entity that may represent primarily an environmental response. At the present state of knowledge I am only expressing a hunch when I say that, among those with high frequencies of occurrence, ear cholesteatoma is likely to be the one most influenced by the environment. Among those with low frequencies of occurrence certainly dental caries is the one most influenced by the environment. The diet of the Aleuts–Eskimos, which is rich in vitamins and low in carbohydrates, yields sound teeth. This is proved by the fact that the abandonment by the Aleuts–Eskimos of this diet in favor of white man's food promptly resulted in carious teeth (Collins 1932).

As for the two remaining pathologic entities on each side of the dichotomy (neural arch defect and mandibular torus on the high-frequency side and scaphocephaly and ear exostosis on the low-frequency side), I know only that neural arch defect runs in family lines. This has led me to wonder whether the Aleuts–Eskimos have become through inbreeding one grand family line so far as this type of defect is concerned. The same probably applies to mandibular torus. The process of inheritance, however, can operate in the other direction through failure to acquire or retain a trait. This may explain the rarity of ear exostosis and scaphocephaly among the Aleuts–Eskimos.

As regards the greater variability in frequencies of skeletal pathologies among the Indian groups, I suggest that this is consistent with the longer exposure of these groups to more varying environments. If so, the still emerging picture of this aspect of the First Americans offers a means—once the time interval is ascertained —of learning how and at what rate a recent branch of mankind has acquired its disorders.

Finally, the absence from all of the foregoing of comparable data from Asia on the closest relatives of the First Americans is lamentable. If I had known of the existence of such data, naturally I would have included them. If they exist unbeknown to me, I hope someone will bring them to wider attention. Moreover, if they do not exist but possibilities exist of undertaking the studies to provide them, I hope the present account will serve as a stimulus.

References

Allison, M.J., D. Mendoza, and A. Pezzia. (1973). Documentation of a Case of Tuberculosis in Pre-Columbian America. *Am. Rev. Resp. Dis.* 107:985–991.

Bennett, K.A. (1967). Craniostenosis: A Review of the Etiology and a Report of a New Case. *Am. J. Phys. Anthropol.* 27:1–10.

Blumberg, B.S., K.J. Bloch, R.L. Black, and C. Dotter. (1961). A Study of the Prevalence of Arthritis in Alaskan Eskimos. *Arth. Rheumat.* 4:325–341.

Bunting, W.P. (1968). Ear Canal Cholesteatoma and Bone Absorption. *Trans. Am. Acad. Ophthalmol. Otolaryngol.* 72:161–172.

Collins, H.B. Jr. (1932). Caries and Crowding in Teeth of the Living Alaskan Eskimo. *Am. J. Phys. Anthropol.* 16:451–462.

Comas, J. (1965). Crânes Mexicains Scaphocéphales. *L'Anthropologie* 69:273–301.

Eiseley, L.C., and W. Asling. (1944). An Extreme Case of Scaphocephaly from a Mound Burial Near Troy, Kansas. *Trans. Kansas Acad. Sci.* 47:241–255.

El-Najjar, M.Y., and A. Robertson, Jr. (1976). Spongy Bones in Prehistoric America. *Science* 193:141–143.

Ericksen, M.F. (1973). *Age-Related Bone Remodeling in Three Aboriginal American Populations.* Ph.D. Thesis, George Washington University, Washington, D.C., *xii* + 228 pp.

Fürst, C.M., and Fr.C.C. Hansen. (1915). *Crania Groenlandica.* Copenhagen: A.F. Host and Son.

Goldstein, M.S. (1932). Caries and Attrition in the Molar Teeth of the Eskimo Mandible. *Am. J. Phys. Anthropol.* 16:421–430.

Greene, L.D. (1933). Cholesteatoma-like Accumulation in the External Auditory Meatus. *Arch. Otolaryngol.* 18:161–167.

Hamperl, H., and P. Weiss. (1955). Über die spongiöse Hyperostose an Schädeln aus Alt-Peru. *Virchow Arch.* 327:629–642.

Hohental, W.D., and S.T. Brooks. (1960). An Archaeological Scaphocephal from California. *Am. J. Phys. Anthropol.* 18:59–65.

Hooton, E.A. (1930). *The Indians of Pecos Pueblo: A Study of Their Skeletal Remains.* New Haven: Yale Univ. Press.

Hrdlička, A. (1914). Anthropological Work in Peru in 1913. With Notes on the Pathology of the Ancient Peruvians. *Smithsonian Misc. Coll.* 61(18):*vi* + 69 pp.

Hrdlička, A. (1930). Anthropological Surveys in Alaska. *46th Ann. Rept. Bur. Am. Ethol. 1928–1929,* pp. 19–374.

Hrdlička, A. (1935). Ear Exostoses. *Smithsonian Misc. Coll.* 93(6):*lv* + 100 pp.

Hrdlička, A. (1940). Mandibular and Maxillary Hyperostoses. *Am. J. Phys. Anthropol.* 27:1–67.

Hrdlička, A. (1944). *The Anthropology of Kodiak Island.* Philadelphia: The Wistar Institute, *xx* + 486 pp.

Hrdlička, A. (1945). *The Aleutian and Commander Islands.* Philadelphia: The Wistar Institute, *xx* + 630 pp.

Jurmain, R.D. (1975). *Distribution of Degenerative Joint Disease in Skeletal Populations.* Ph.D. Thesis, Harvard University, Cambridge, Mass., *xi* + 506 pp.

McGuckin, F. (1961). Concerning the Pathogenesis of Destructive Ear Disease. *J. Laryngol. Otol.* 75:949–961.

Maynard, J.E., J.K. Fleshman, and C.F. Tschopp. (1972). Otitis Media in Alaskan Eskimo Children. *J. Am. Med. Assoc.* 219:597–599.

Morse, D. (1961). Prehistoric Tuberculosis in America. *Am. Rev. Resp. Dis.* 83:489–504.

Moss, M.L. (1957). Premature Synostosis of the Frontal Suture in the Cleft Palate Skull. *Plast. Reconstr. Surg.* 20:199–205.

Müller, H. (1934). Het Voorkomen van de zgn. "Osteoporosis Symmetrica Cranii" op Java en Haar Verband met de Rhachitis. *Geneesk. Tidschr. Nederl.-Indië* 74:1·84–1093.

Munizaga, J.R. (1965). Espongio Hiperostosis (Hamperl y Weiss) u Osteoporosis Simetrica (Hrdlička). Diagnostico-Epidemiologia-Antigüedad. *Antropol. Rev. Centro Est. Antropol. Univ. Chile* Año 3, 3:31–63.

Nance, W.E., and E. Engel. (1967). Autosomal Deletion Mapping in Man. *Science* 155:692–694.

Nathan, H., and N. Haas. (1966). "Cribra Orbitalia," A Bone Condition of the Orbit of Unknown Nature: Anatomical Study with Etiological Considerations. *Israel J. Med. Sci.* 2:171–191.

Snow, C.E. (1948). Indian Knoll Skeletons of Site Oh2, Ohio County, Kentucky. *Univ. Kentucky Rept. Anthropol.* 4(3, Pt. 2):370–532.

Stearn, E.W., and A.E. Stearn. (1945). *The Effect of Smallpox on the Destiny of the Amerindian.* Boston: B. Humphries.

Stewart, T.D. (1931a). Incidence of Separate Neural Arch in the Lumbar Vertebrae of Eskimos. *Am. J. Phys. Anthropol.* 16:51–62.

Stewart, T.D. (1931b). Dental Caries in Peruvian Skulls. *Am. J. Phys. Anthropol.* 15:315–326.

Stewart, T.D. (1947). Racial Patterns in Vertebral Osteoarthritis. *Am. J. Phys. Anthropol.* 5:230–231. (Abs.)

Stewart, T.D. (1953). The Age Incidence of Neural-Arch Defects in Alaskan Natives. Considered from the Standpoint of Etiology. *J. Bone Joint Surg.* 35-A(4):937–950.

Stewart, T.D. (1960). A Physical Anthropologist's View of the Peopling of the New World. *Southwest J. Anthropol.* 16:295–273.

Stewart, T.D. (1966). Some Problems in Human Palaeopathology. *In* Saul Jarcho, Ed., *Human Palaeopathology.* New Haven: Yale Univ. Press, pp. 43–55.

Stewart, T.D. (1974). Perspectives on Some Problems of Early Man Common to America and Australia. *In* A.P. Elkin and N.W.G. Macintosh, Eds., *Sir Grafton Elliot Smith, the Man and His Work.* Sydney: Univ. of Sydney Press, pp. 113–135.

Stewart, T.D., and L.G. Quade. (1969). Lesions of the Frontal Bone in American Indians. *Am. J. Phys. Anthropol.* 30:89–110.

Todd, T.W., and D.W. Lyon, Jr. (1925). Cranial Suture Closure, Its Progress and Age Relationship. II. Ectocranial Closure in Adult Males of the White Stock. *Am. J. Phys. Anthropol.* 8:23–45.

Toldt, C. (1886). Über Welcker's Cribra Orbitalia. *Mitt. Anthropol. Ges. Wien* 16:20–24.

Welcker, H. (1887). Cribra Orbitalia, ein ethnologisch-diagnostisches Merkmal am Schädel mehrerer Menschenrassen. *Arch. Anthropol.* 17:1–18.

Wiltse, L.L. (1962). The Etiology of Spondylolisthesis. *J. Bone Joint Surg.* 44-A:539–560.

Wiltse, L.L., E.H. Widell, Jr., and D.W. Jackson. (1975). Fatigue Fracture: The Basic Lesion in Isthmic Spondylolisthesis. *J. Bone Joint Surg.* 57-A:17–22.

Chapter 12

Anthropometric Variation of Native American Children and Adults

Francis E. Johnston and Lawrence M. Schell

Introduction

The study of human morphology and the analysis of the patterns of its variability have been focal concerns of physical anthropology since the beginnings of the discipline. Typologic classifications gave way to investigations that were oriented around the concept of evolutionary process, and especially to the role of genetic adaptation in shaping morphologic variability.

In recent years, several developments have occurred that have caused a renewed interest in studies of human morphology. Among the most important has been the formalization of ecologic anthropology as a subdisciplinary approach. Human ecologists have begun to model the sets of interactions that they study in ways which permit quantitative estimates of environmental and biocultural parameters. As they have done so, they have come to appreciate once again the importance of the phenotype as the interface between the organism and its environment (Little and Morren 1976).

A second development has been the gradual recognition and more gradual acceptance of the fact that the study of biochemical genetic polymorphisms has contributed, and will continue to contribute, relatively little to our understanding of adaptation and natural selection. The rapid rise of the neutralist view of monogenic variation and the serious challenges it has raised to traditional neo-Darwinian theory (Salzano 1975) must be taken seriously. Although natural selection may very well be operating at the many loci now known to be polymorphic in humans, techniques at our disposal have been unsuccessful at establishing the existence of selection except in the case of falciparum malaria and certain loci determining red cell characteristics (Livingstone 1967, Crawford and Workman 1973). The failure of thousands of studies to establish the existence and intensity of specific selective responses has resulted in many anthropologists' turning back to the organism itself as the locus of adaptation, as seen in the individual, either through genetic mechanisms not yet isolated or through the body's innate plasticity.

The final development related to a renewed concern with morphology has been the accumulation of evidence pointing to the incredible sensitivity of the body to surrounding environmental pressures. This sensitivity is seen in the interaction between individuals and their environments and the responses that result. Morphologic adaptation has been hypothesized for much longer than we have had a theory of evolution (Smith 1810), but only in more recent years has the extent and the subtlety of interaction been realized. In particular, the sensitivity of the organism during the growing years suggests that adaptive responses may be greatest when tissue synthesis and differentiation is greatest. The "heritability" of quantitative traits can no longer be defined as a fixed quantity but must be seen as a variable that is dependent on the strength of specific environments. Where the environment is harsh, genotypic variability will be relatively insignificant in causing phenotypic variability, but where the environment is less harsh, the genotypes will be less constrained and the heritability of a trait will "rise" (Mueller 1975).

Anthropologic Studies of Growth and Development

Anthropologists have studied growth and development for decades; in fact, much of the initial systemization of theory and methodology came from the work of anthropologists (Boas 1940, Tanner 1959). During the first third of the twentieth century, the contributions of such individuals as Boas, Wissler, Matiegka, Skerlj, Todd, and Krogman set the stage for the scientific study of growth processes and for the utilization of the findings throughout the biologic, behavioral, and medical sciences.

Several rationales for growth studies have emerged over the years, all of which have developed out of empirical and theoretical bases, and all of which serve to provide the justification for the inclusion of investigations of growth and development as an essential component of physical anthropology and human biology.

First, growth may be studied as the response to a set of environmental pressures. Viewed in this way, an individual's pattern of development becomes a record of response to that environment. The adaptive process results from the expression of the genotype within an ecosystem, and this process may be measured in studying growth and development (Frisancho and Baker 1970, Thomas 1973, Wolanski 1970).

Second, because children of different populations differ in their patterns of growth, these differences are part of the range of human biologic variability. At one time, physical anthropologists dealt almost exclusively with the heads (or skulls) of adult males. Gradually, they allowed their interest to drift downward to include the entire body (but still only men were measured). The next stage came with the realization that the information gained from the measurement of females was at least as valuable, and qualitatively different, from that gained by the measurement of males. Finally, children and youth were let into the circle of anthropologically relevant subjects and, in more recent years, there have been several surveys of geographic variation in the size and patterns of growth of individuals from various populations (Meredith 1970, Eveleth and Tanner 1976, Schell and Johnston 1978).

The third rationale for growth studies is the realization that the developmental patterns of children are excellent indicators of the quality of the environment of the entire population. Deviations of these patterns from the expected serve as sensitive indications of nutritional status and of the general level of health of the entire population (Jelliffe 1966, World Health Organization 1968, Tobias 1972).

Finally, there is the study of the process of development itself. Such studies are directed toward such questions as the nature of the growth process, the extent of interactions among its various components, and the operation of internal regulators (Marshall 1974, Bock *et al.* 1973, Bogin 1977).

Morphologic Studies of Native Americans

Studies of the morphologic variability of children, youth, and adults of Native American populations have been carried out by physical anthropologists since the nineteenth century. In particular, growth studies of American Indians and Eskimos were among the first to be conducted systematically in any group. Since the early research of Boas (1901, 1940) and Hrdlička (1908), additional studies have been conducted by anthropologists, nutritionists, and public health workers. Likewise, there is an impressive list of studies of anthropometric variability among adult populations indigenous to the New World.

Unfortunately, many of these data are unusable for any scientific purpose. All too often, ages were not obtained, or age grouping was done in such a way as to render meaningless any interpretations. Frequently the sample sizes were so small as to preclude any analysis and so failed to inspire any confidence in the means themselves. The absence of any statement of quality control of measuring techniques in many instances, particularly when the means or variances are aberrant, excludes a major proportion of studies. Finally, the lack of any standardized measurement list, and the haphazard selection of measurements, with no apparent thought given to hypothesis or problem, has resulted in a hodge-podge of measurements.

Even though a "wealth" of data exists, therefore, the wealth is misleading because of the many inadequacies such as those listed above. To include such data in any survey may not only mask significant trends and patterns, it may also lead to false conclusions.

Morphologic Variation of Native Americans

Adults

In surveying the morphologic variability of Native Americans, we have decided not to attempt a comprehensive survey of the entire literature, or of the entire range of measurements that have been taken. There have been surveys of the literature for groups of Native Americans which present a wide range of means, and a large list of bibliographic references (e.g., Casey and Downey 1969, Comas 1966, 1971). Instead, we have selected samples from the literature to indicate the range of variability that exists, based on several criteria:

1. The selected samples must cover much of the area from northern North America into tropical South America.
2. The samples must be large enough in size, adequately characterized, and based upon acceptable measuring techniques to give them sufficient reliability.
3. Both males and females must be measured.

Even though the selected samples are "large enough," they are in some instances rather small. Where this is the case, it is because the population itself is small, as for example with the isolate groups of the South American rain forest.

No analysis of adults by age could be done because not enough investigators had adequate sample sizes specified by age. Therefore, as much as possible, we selected samples that spanned the adult ages from 20 through 60 years.

The following measurements were utilized:

height
weight
sitting height
biacromial breadth
bicristal breadth
triceps skinfold
subscapular skinfold.

Unfortunately, not enough investigators took biacromial, bicristal, or sitting height to permit any analysis, and we present only the means for information.

In addition to the measurements we also derived, from the means, the value of weight-for-height, as $(height)/(weight)^2$, and the sitting height as a percentage of stature. Finally, skinfold thickness was analyzed in relationship to body weight.

Table 12.1 presents the 21 adult samples selected along with the sample sizes for each sex. It can be seen that both Eskimo and Indian populations are represented, and that samples have been taken from North, Central, and South America.

Tables 12.2 and 12.3 present the means, by sex, for each of the measurements, where one is available. Also included are the means of United States adults; these data come from the Health Examination Survey (Stoudt et al. 1965, 1970) and represent the entire population as determined by a national probability sample.

Table 12.1. *Adult Samples Selected for Comparison*

Group	Reference	Sample size Male	Female
North America			
Eskimo	Jamison and Zegura (1970)	43	37
Northern Eskimo	Mann *et al.* (1962)	82	53
Southern Eskimo	Mann *et al.* (1962)	107	76
Blackfoot	ICNND (1964b)	98	133
Assiniboine–Gros Ventres	ICNND (1964a)	77	100
Ahousat	Birkbeck *et al.* (1971)	36	45
Anaham	Birkbeck *et al.* (1971)	36	55
Alaskan Athabascan	Mann *et al.* (1962)	24	25
Navajo	Darby *et al.* (1956)	272	245
Seminole	Pollitzer *et al.* (1970)	94	143
Mesoamerica			
Otomie	Faulhaber (1955)	101	100
Nahua	Faulhaber (1955)	100	100
Mayan	Steggerda (1932)	104	94
South America			
Quechua	Frisancho and Baker (1970)	50	50
Cayapo	Da Rocha and Salzano (1972)	130	156
Caingang	Da Rocha (1971)	340	244
Trio	Glanville and Geerdink (1970)	114	140
Wajana	Glanville and Geerdink (1970)	54	70
Cashinahua	Johnston *et al.* (1971)	38	51
Yanomamo	Spielman *et al.* (1972)	316	316
Xingu	Eveleth *et al.* (1974)	356	366

As expected, the 21 samples reveal a wide range of variability among themselves in any single measurement. Native Americans do not present us with any homogenous stereotype; when we examine the actual data, we find an impressive range of variability. The height means span a range, in males, of almost 22 cm and, in females, of 19 cm. Weight means for either sex cover a range of some 25 kg. Even though there are fewer means available, the same is true for the other measurements as well.

Table 12.4 presents the two indices of adiposity and relative sitting height, derived from the means and again indicating variation from sample to sample.

Compared to the adult population of the United States, Native Americans tend to be smaller in all dimensions, but especially in linear measurements. The only exception among our samples are the Blackfoot Indians, who are taller. However, despite the greater linear dimensions, United States adults are, on the average, lighter in weight than a number of Indian and Eskimo samples. This is especially true among females, where six samples have greater mean weights; of these six, five are from northern North America and one, the Seminole, from the southeastern USA. Only two male samples display higher mean weights than the United States population: the Ahousat Indians of British Columbia and Seminole Indians of Florida.

Table 12.2. Mean Values of Adult Males

Group	Height (cm)	Weight (kg)	Sitting height (cm)	Biacromial (cm)	Bicristal (cm)	Triceps (mm)	Subscapular (mm)
Blackfoot	174.8	75.8				8.6	11.4
Assiniboine–Gros							
Ventres	171.8	75.4				7.8	11.5
Ahousat	170.4	79.6				8.8	
Anaham	170.3	70.7				7.9	
Navajo	169.5	65.0					
Alaskan Athabascan	167.2	65.0					
Eskimo	166.3	67.2	87.8	38.5		9.5	12.2
Southern Eskimo	163.0	64.5					
Northern Eskimo	168.9	71.0					
Seminole	169.5	76.5					
Otomie	157.0	52.6	83.7	37.3	27.4		
Nahua	154.8	49.5	83.1	37.0	27.1		
Maya	155.4	54.2	85.9	38.0	28.1		
Quechua	160.0	55.9	84.5			4.0	8.5
Cayapo	165.4	61.4	84.9				
Caingang	161.0	56.3	82.4				
Trio	157.7	58.2	79.8	37.4		5.9	8.9
Wajana	156.6	61.3	79.9	36.7		5.7	9.2
Cashinahua	154.7	61.3				5.4	8.5
Yanomamo	153.2		81.3				
Xingu	160.0		81.7	37.3	26.8		
United States							
(national)	173.2	76.4	90.4	39.6		13.0	15.0

There seem to be clear geographic differences among Native American populations. Those of northern North America are the largest, those of Mexico the smallest, and those of South America intermediate. To test this, we performed analyses of variance on the means of the samples grouped as follows:

1. Northern: Blackfoot, Assiniboine–Gros Ventres, Ahousat, Anaham, and Alaskan Athabascan
2. Eskimo: the three Eskimo samples
3. Mesoamerican: Otomie, Nahua, Maya
4. South America: Cayapo, Caingang, Cashinahua, Trio, Wajana

The Seminoles were not utilized because they were the only sample from the southeastern USA. In like manner, we did not use the Navajo as the only southwestern group or the Quechua as the sole high-altitude sample. The Yanomamo and the sample from the Brazilian Xingu were not used because they lacked measurements of body weight.

For the remaining 16 groups, ANOVAS were carried out for height, for weight, and for relative weight (weight-for-height). These ANOVAS are presented in Table 12.5. Height shows clear differences in all three main effects

Table 12.3. *Mean Values of Adult Females*

Group	Height (cm)	Weight (kg)	Sitting height (cm)	Biacromial (cm)	Bicristal (cm)	Triceps (mm)	Subscapular (mm)
Blackfoot	161.1	69.5				15.4	15.8
Assiniboine–Gros Ventres	159.4	69.0				17.0	17.1
Ahousat	158.3	65.6				15.9	
Anaham	156.5	65.8				22.6	
Navajo	156.0	58.1					
Alaskan Athabascan	154.3	60.4					
Eskimo	155.8	66.3	83.7	35.6		23.3	25.2
Southern Eskimo	150.7	55.6					
Northern Eskimo	157.0	65.5					
Seminole	157.2	69.1					
Otomie	144.5	46.9	77.8	34.5	28.3		
Nahua	143.5	44.4	76.6	33.2	26.9		
Maya	141.8	47.5	75.0	34.3	28.8		
Quechua	148.0	54.0	80.0			9.5	10.7
Cayapo	153.9	51.6	79.9				
Caingang	149.1	50.0	76.6				
Trio	147.5	48.7	75.5	33.3		9.3	12.9
Wajana	146.2	51.4	75.0	33.5		9.1	12.5
Cashinahua	145.1	52.4				11.4	14.2
Yanomamo	142.3		75.8				
Xingu	149.4		76.4	32.9	26.8		
United States (national)	160.0	64.5	84.6	35.5		22.0	18.0

tested: sex, geographic areas, and samples. The ranking of heights, from highest to lowest, is as follows: northern, Eskimo, South America, Mesoamerica.

The analysis of variance of the weight means also reveals significant differences for all three effects. As with height, the means rank in the following sequence, from highest to lowest: northern, Eskimo, South America, Mesoamerica. However, the relative magnitude of the effects, as indicated by their F-ratios, differs. For height, the effect of sex was the greatest, with an F-ratio of 1384.26, while the area effect was less, at 427.17. For weight, the effect of area was greater than that of sex. In both instances, the between-sample, within-area effect is the smallest of the three tested.

Further information may be obtained by examining the ANOVA for the relative weight, expressed here as weight/(height)^2. As with height and weight, significant differences exist among the means for all three effects. As with weight, the differences in relative weight are greatest between the four geographical areas, exceeding the F-ratios of both sex and sample. Since relative weight adjusts the mean body weight for the mean height, this analysis indicates that the differences among the samples in weight are independent of the differences in height and significantly related to geographical area.

Table 12.4. *Anthropometric Indices among Adults*[a]

Group	(Weight)/(height)^2 Male	Female	(Sitting height)/(Stature) Male	Female
Blackfoot	24.8	26.8		
Assiniboine–Gros Ventres	25.5	27.2		
Ahousat	27.4	26.2		
Anaham	24.4	26.9		
Navajo	22.6	23.9		
Alaskan Athabascan	23.3	25.4		
Eskimo	24.3	27.3	52.8	53.7
Southern Eskimo	24.3	24.5		
Northern Eskimo	24.9	26.6		
Seminole	26.6	28.0		
Otomie	21.3	22.5	53.3	53.8
Nahua	20.7	21.6	53.7	53.4
Maya	22.4	23.6	55.3	54.1
Quechua	21.8	24.7	52.8	54.1
Cayapo	22.4	21.8	51.3	51.9
Caingang	21.7	22.5	51.2	51.4
Trio	23.4	22.4	50.6	51.2
Wajana	25.0	24.0	51.0	51.3
Cashinahua	25.6	24.9		
Yanomamo			53.1	53.3
Xingu			51.1	51.1
United States (national)	25.5	25.2	52.2	52.9

[a] Derived from published means.

This analysis suggests that body weight is particularly sensitive to geographic area in these 16 Native American samples. Our data do not allow us to assign any reason to this. However, we may point to the increased adaptation to low temperature afforded by a high weight-for-height. Roberts has demonstrated that various ethnic groups show a significant regression of body size upon environmental temperature and this analysis would seem supportive of Roberts' studies (Roberts 1973). The extent to which this is a genetic phenomenon, produced by natural selection, is still, to us, open to question. The Navajo are a population of Athabascan speakers, genetically related to Indians of the north rather than to other southwestern groups (except, of course, the Apache). However, the relative weight and the body weight of the Navajo are less than the means for the northern samples analyzed here.

Another problem with a straightforward genetic interpretation of these data is in the differences between the Mesoamerican and the South American samples. There seems to be no climatic reason for Indians of Mesoamerica to be smaller than those of the South American tropical rain forest. However, the deficits in energy intake among traditional Mesoamerican populations is well documented and may have contributed to the small body size of the samples we have analyzed. Whatever the case, it is clear that temperature and nutrition must interact in a way that allows body size to become an adaptation to the climate.

Table 12.5. *Analyses of Variance of Adult Means*

ANOVA					Means	
Source	df	MS	F	Population	Male	Female
Height						
Total	31			Northern	170.9	157.9
Sex	1	1135.26	1384.26[a]	Eskimo	166.1	154.5
Samples	15			Mesoamerican	155.7	143.2
Areas	3	350.28	427.17[a]	South American	159.1	148.4
Samples in areas	12	18.09	22.06[a]			
Error	15	0.82				
Weight						
Total	31			Northern	73.3	66.1
Sex	1	401.86	89.30[a]	Eskimo	67.6	62.5
Samples	15			Mesoamerican	52.0	46.7
Areas	3	669.35	148.74[a]	South American	59.7	50.8
Samples in areas	12	22.28	4.95[b]			
Error	15	4.50				
(Weight)/(height)^2						
Total	31			Northern	25.1	26.5
Sex	1	5.28	5.68[b]	Eskimo	24.5	26.1
Samples	15			Mesoamerican	21.5	22.6
Areas	3	22.85	24.57[a]	South American	23.6	23.1
Samples in areas	12	2.49	2.68[b]			
Error	15	0.93				

[a] $p < 0.01$.
[b] $p < 0.05$.

There are not enough samples with means for biacromial, bicristal, sitting height, or skinfold thickness to allow any analysis of the data. We may only make a few observations. First is that, in general, skinfold thickness means tend to vary as do weight means. Compared to the United States probability samples, the mean thicknesses of the Native American males are all below the United States means. The females are much closer to the United States females and, in fact, the average thickness of the triceps skinfold in the Anaham and the Wainwright Eskimos, and the subscapular in the Wainwright Eskimos, exceed the United States means.

Although we have selected a relatively small number of samples for presentation here, they are representative of other samples from the same geographic areas. For example, the three Mesoamerican samples do not differ significantly from those presented in the comprehensive reviews of Comas and Faulhaber (1965), Comas (1966), and Jaen *et al.* (1976). The same is true for the other areas, although there are no equivalent reviews of the literature available. The one exception is with the Pima Indians of the desert southwest of the United States. The Pima have very high relative weights, a high prevalence of obesity and diabetes (Savage *et al.* 1976), and, although displaying the general tendency toward a high relative weight found among Native Americans (especially of North America), are at the extreme.

Children and Youth

The most comprehensive reviews of the patterns of physical growth of Native Americans are those by Eveleth and Tanner (1976) and by Schell and Johnston (1978). Both are reviews of the literature; that by Schell and Johnston deals only with North America but provides a more comprehensive survey of the literature than the review by Eveleth and Tanner. Eveleth and Tanner, in contrast, present data on more dimensions and, in addition, they are reviewing growth variation on a worldwide basis.

Both of these reviews are consistent in their findings. Native American children and youth are smaller than their age and sex peers of most other populations, and especially populations of European and African ancestry also living in the Western Hemisphere. Even though smaller, Native American individuals display during their growing years the same high relative weights that adults display, compared to European reference standards (including data from the United States). In some samples, these differences relative to other Americans may be related to the poor environmental conditions in which most Native Americans live, but even when the environments do not differ so markedly the patterns persist.

The genesis of these different developmental patterns reaches into the prenatal years. The mean birthweights of native North American infants tend to be higher than those of other ethnic groups (Meredith 1970, Adams and Niswander 1973). During childhood Native Americans are smaller and the differences seem to become more marked in height during the years of adolescence (Eveleth and Tanner 1976, Johnston et al. 1978). In other words, the adult differences are seen, to a lessened extent, during infancy. These differences among infants persist throughout childhood and then become intensified during the adolescent years.

Discussion

In terms of anthropometry, Native Americans present us with a generalized picture of smallness, relative to Europeans, which includes linear measurements as well as measures of body mass and composition. Yet the smallness is more pronounced in linear dimensionality such that both Indians and Eskimos have relatively greater weight-for-height.

Significant areal variation is seen within Native Americans. Larger body sizes and larger weights-for-height characterize populations of northern North America, both Eskimo and Indian. Mesoamerican groups are the smallest and those of the South American rain forest, although intermediate, are still closer to Mesoamericans. This gradient would seem to be of significance as an adaptation to the colder temperatures of the Arctic and Subarctic regions, just as the smaller relative weights may also be adaptive in Mesoamerica and South America. Nutritional factors cannot be overlooked, however, and none of these differences can be attributed to genetic factors with certainty.

Whether or not these differences are attributable to the environment or not, they seem to predispose North American Indians to obesity, especially when their diets become similar to the diets of other North Americans. In one study, Johnston *et al.* (1978) have demonstrated even greater weights-for-height among urban Chippewa Indians than among their relatives from a reservation in the northern United States. Data from this study are presented in Table 12.6. The table presents the prevalence of obesity in urban and reservation children and youth. Obesity has been defined as a relative weight equal to or in excess of 115, where relative weight is:

$$\left(\frac{\text{actual weight}}{\text{predicted weight}} \right) \times 100$$

Predicted weight was obtained from the national probability sample of the United States Health Examination Survey (HES) (Hammill *et al.* 1970, 1973). For HES children and youth, median weight was regressed upon height for each sex–age group and the equations used to predict weight in the Native American sample.

As the table shows, there is, in general, a high percentage of obesity among Native Americans, and especially so in the urban sample. Comparable samples for United States low-income Black youths show percentages of obesity of between 20 and 25 percent (Stunkard *et al.* 1972, Johnston and Mack 1978). The frequency of obesity in the urban Chippewa is significantly higher than that in the reservation Chippewa, regardless of sex. We may hypothesize that the changed environmental conditions associated with urbanization are associated with the increased prevalence of obesity.

The increased fatness of Native Americans of the United States is found distributed on the body in a pattern which differs from that of Americans of European and African ancestry. Figures 12.1 through 12.4 show the distribution of skinfolds in the urban Chippewas from infancy through 18 years of age. The skinfold thicknesses are plotted relative to the medians for probability samples of United States children and youth from the Health Examination Survey and the Health and Nutrition Examination Survey (Johnston *et al.* 1972, 1974, Abraham *et al.* 1975). The American Indians show increased subscapular skinfold thicknesses at all ages, the differences being statistically significant. However, the thicknesses of the triceps skinfolds are significantly less, at all ages, in females. In

Table 12.6. *Prevalence of Obesity in Urban and Reservation Chippewa*

	Male			Female		
	Total	Obese[a]	Percent	Total	Obese[a]	Percent
Urban (6–17 years)	226	60	26.5	280	111	39.6
Reservation (6–12 years)	280	33	11.8	255	78	30.6

[a] Obesity = relative weight ≥ 115.

F.E. Johnston and L.M. Schell

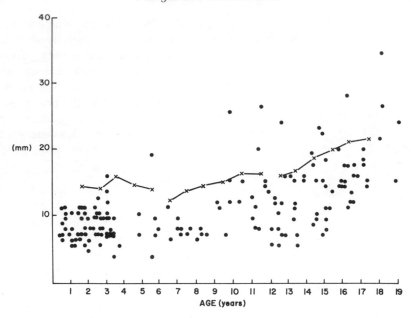

Figure 12.1. Distribution of triceps skinfolds in urban American Indian females. Solid lines indicate United States national medians.

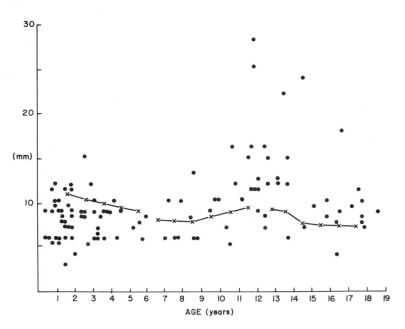

Figure 12.2. Distribution of triceps skinfolds in urban American Indian males. Solid lines indicate United States national medians.

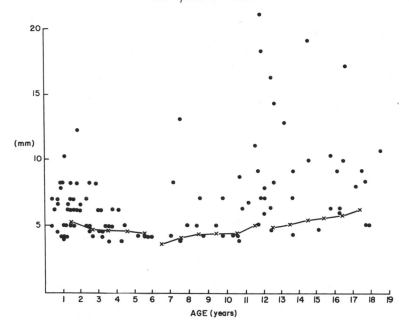

Figure 12.3. Distribution of subscapular skinfolds in urban American Indian females. Solid lines indicate United States national medians.

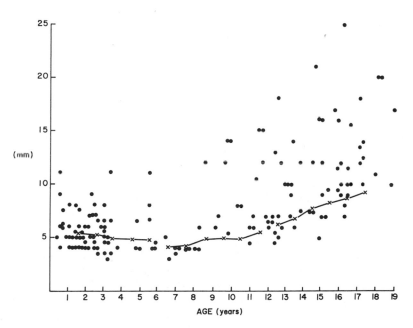

Figure 12.4. Distribution of subscapular skinfolds in urban American Indian males. Solid lines indicate United States national medians.

males, the triceps skinfolds are significantly less below 6 years of age but significantly greater from 6 through 17 years. Although there seems to be little inherited variation in the absolute thicknesses of the skinfolds (Brook *et al.* 1975), the relative thicknesses are more persistent and show a higher heritability (Garn 1955, Mueller and Reid 1977). It may be that the differences in thickness patterns between American Indians and the United States population generally have an inherited basis. However, this remains to be verified.

Conclusions

Anthropometric variation is of demonstrated adaptive significance. Unfortunately, in the past 20 years, anthropometry fell into disuse as a reaction to the excesses and errors of those who, earlier, had analyzed data in a typologic context. With the renewed interest in human morphology, there is the need for accurate, well-defined surveys in which measurements are taken that are now known to be of utility in analyzing relationships to the environments of the populations and that may be of genetic significance as well. Despite the many studies of Native Americans, more data are needed. The early studies are of but limited usefulness and we must await additional research before we are able to analyze more satisfactorily the patterns of variation that exist among Native American populations and that may be related to particular ecologic and genetic mechanisms.

References

Abraham, S., F.W. Lowenstein, and D.E. O'Connell. (1975). Preliminary Findings of the First Health and Nutrition Examination Survey, United States, 1971–1974. Anthropometric and Clinical Findings. Dept. Health, Education and Welfare Publ. (HRA) 75-1229. Washington, D.C.: U.S. Govt. Printing Office.

Adams, M.S., and J.D. Niswander. (1973). Birth Weight of North American Indians: A Correction and Amplification. *Hum. Biol.* 45: 351–358.

Birkbeck, J.A., M. Lee, G.S. Meyers, and B.M. Alfred. (1971). Nutritional Status of British Columbia Indians. 11. Anthropometric Measurements, Physical Examinations at Ahousat and Anaham. *Can. J. Publ. Health* 62: 403–414.

Boas, F. (1901). A.J. Stone's Measurements of Natives of the Northwest Territories. *Bull. Am. Mus. Nat. Hist.* 14: 53–68.

Boas, F. (1940). *Race, Language, and Culture.* New York: Free Press.

Bock, R.D., H. Wainer, A. Peterson, D. Thissen, J. Murray, and A. Roche. (1973). A Parameterization for Individual Human Growth Curves. *Hum. Biol.* 45: 63–80.

Bogin, B.A. (1977). *Periodic Rhythm in the Rates of Growth in Height and Weight of Children and its Relation to Season of the Year.* Ph.D. Dissertation, Temple University, Philadelphia, Pa.

Brook, C.G.D., R.M.C. Huntley, and J. Slack. (1975). Influence of Heredity and Environment in Determination of Skinfold Thickness in Children. *Brit. Med. J.* 2: 719–721.

Casey, A.E., and E.L. Downey. (1969). *Compilation of Common Physical Measurements of Adult Males of Various Races*. Birmingham, Ala.: Amite and Knocknagree Historical Fund.

Comas, J. (1966). *Caracteristicas Fisicas de la Familia Linguistica Maya*. Mexico D.F.: Universidad Nacional Autonoma de Mexico.

Comas, J. (1971). Anthropometric Studies in Latin American Indian Populations. *In* F.M. Salzano, Ed., *The Ongoing Evolution of Latin American Populations*. Springfield, Ill.: Charles C Thomas, pp. 333–404.

Comas, J., and J. Faulhaber. (1965). *Somatometria de los Indios Triques de Oaxaca, Mexico*. Mexico D.F.: Universidad Nacional Autonomia de Mexico.

Crawford, M.H., and P.L. Workman. (Eds.). (1973). *Methods and Theories of Anthropological Genetics*. Albuquerque: Univ. New Mexico Press.

Darby, W.H., C.G. Salsbury, W.J. McGanity, H.F. Johnson, E.B. Bridgforth, and H.R. Sandstead. (1956). A Study of the Dietary Background and Nutriture of the Navajo Indian. *J. Nutrit*. 60 (Suppl. 2):3–85.

Da Rocha, F.J. (1971). *Antropometria em Indigenas Brasilieros*. Porto Alegre: Ministerio de Edecacaoe Cultura.

Da Rocha, F.J., and F.M. Salzano. (1972). Anthropometric Studies in Brazilian Cayapo Indians. *Am. J. Phys. Anthropol*. 36:95–102.

Eveleth, P.B., and J.M. Tanner. (1976). *Worldwide Variation in Human Growth*. Cambridge: Cambridge Univ. Press.

Eveleth, P.B., F.M. Salzano, and P.E. De Lima. (1974). Child Growth and Adult Physique in Brazilian Xingu Indians. *Am. J. Phys. Anthropol*. 41:95–102.

Faulhaber, J. (1955). *Antropologia Fisica de Veracruz*. Veracruz: Gobierno de Veracruz.

Frisancho, A.R., and P.T. Baker. (1970). Altitude and Growth: A Study of the Patterns of Physical Growth of a High Altitude Peruvian Quechua Population. *Am. J. Phys. Anthropol*. 32:279–292.

Garn, S.M. (1955). Relative Fat Patterning: An Individual Characteristic. *Hum. Biol*. 27:77–89.

Glanville, E.V., and R.A. Geerdink. (1970). Skinfold Thickness, Body Measurement and Age Changes in Trio and Wajana Indians of Surinam. *Am. J. Phys. Anthropol*. 32:455–461.

Hamill, P.V.V., F.E. Johnston, and W. Grams. (1970). *Height and Weight of Children. United States*. Natl. Ctr. Health Stat., Ser. 11, No. 104. Washington, D.C.: U.S. Govt. Printing Office.

Hamill, P.V.V., F.E. Johnston, and S. Lemeshow. (1973). *Height and Weight of Youths 12–17 Years. United States*. Dept. Health, Education and Welfare Publ. No. (HSM) 73-1606. Washington, D.C.: U.S. Govt. Printing Office.

Hrdlička, A. (1908). *Physiological and Medical Observations among the Indians of the Southwestern United States and Northern Mexico*. Bull. 34, Bur. Am. Ethnol. Washington, D.C.: U.S. Govt. Printing Office.

Interdepartmental Committee on Nutrition for National Defense (ICNND). (1964a). *Fort Belknap Indian Reservation Nutrition Survey, August–September, 1961*. Washington, D.C.: U.S. Govt. Printing Office.

Interdepartmental Committee on Nutrition for National Defense (ICNND). (1964b). *Blackfeet Indian Reservation Nutrition Survey, August–September, 1961*. Washington, D.C.: U.S. Govt. Printing Office.

Jaen, M.T., C. Serrano, and J. Comas. (1976). *Data Antropometrica de Algunas Poblaciones Indigenas Mexicanas*. Mexico D.F.: Universidad Nacional Autonoma de Mexico.

Jamison, P.L., and S. Zegura. (1970). An Anthropometric Study of the Eskimos of Wainwright, Alaska. *Arctic Anthropol.* 7:125–143.

Jelliffe, D.B. (1966). *The Assessment of the Nutritional Status of the Community.* Geneva: World Health Organization.

Johnston, F.E., and R.W. Mack. (1978). Obesity in Urban, Black Adolescents of High and Low Relative Weight at One Year of Age. *Am. J. Dis. Child.,* 132:862–864.

Johnston, F.E., P.S. Gindhart, R.L. Jantz, K.M. Kensinger, and G.F. Walker. (1971). The Anthropometric Determination of Body Composition among the Peruvian Cashinahua. *Am. J. Phys. Anthropol.* 34:409–416.

Johnston, F.E., P.V.V. Hamill, and S. Lemeshow. (1972). *Skinfold Thickness of Children 6–11 Years. United States.* Dept. Health, Education and Welfare Publ. No. (HSM) 73-1602. Washington, D.C.: U.S. Govt. Printing Office.

Johnston, F.E., P.V.V. Hamill, and S. Lemeshow. (1974). *Skinfold Thickness of Youths 12–17 Years. United States.* Dept. Health, Education and Welfare. Publ. No. (HRA) 74-1614. Washington, D.C.: U.S. Govt. Printing Office.

Johnston, F.E., J.I. McKigney, S. Hopwood, and J. Smelker. (1978). Physical Growth and Development of Urban Native Americans: A Study in Urbanization and Its Implication for Nutritional Status. *Am. J. Clin. Nutrit.* 31:1017–1027.

Little, M.A., and G.E.B. Morren, Jr. (1976). *Ecology, Energetics, and Human Variability.* Dubuque, Io.: W.C. Brown.

Livingstone, F.B. (1967). *Abnormal Hemoglobins in Human Populations.* Chicago: Aldine.

Mann, G.V., E.M. Scott, L.M. Hursh, C. Heller, J.B. Youmans, C.F. Consolazio, E.B. Bridgforth, A.L. Russell, and M. Silverman. (1962). The Health and Nutritional Status of Alaskan Eskimos: A Survey for the ICNND-1958. *Am. J. Clin. Nutrit.* 11:31–76.

Marshall, W.A. (1974). Interrelationships of Skeletal Maturation, Sexual Development, and Somatic Growth in Man. *Ann. Hum. Biol.* 1:29–40.

Meredith, H.V. (1970). Body Weight at Birth of Viable Human Infants. *Hum. Biol.* 42:217–264.

Mueller, W.H. (1975). *Parent–Child and Sibling Correlations and Heritability of Body Measurements in a Rural Colombian Population.* Ph.D. Dissertation, University of Texas, Austin.

Mueller, W.H., and R.M. Reid. (1977). A Multivariate Analysis of Relative Fat Patterning in Adults. *Am. J. Phys. Anthropol.* 47:151–152.

Pollitzer, W.S., D. Rucknagel, R. Tashiah, D.C. Shreffler, W.C. Leyshon, K. Namboodiri, and R.C. Elston. (1970). The Seminole Indians of Florida: Morphology and Serology. *Am. J. Phys. Anthropol.* 32:65–82.

Roberts, D.F. (1973). *Climate and Human Variability.* Module in Anthropology, No. 34. Reading, Mass.: Addison-Wesley.

Salzano, F.M. (Ed.). (1975). *The Role of Natural Selection in Human Evolution.* New York: American Elsevier.

Savage, P.J., R.F. Hamman, G. Bartha, S.E. Dippe, M. Miller, and P.H. Bennett. (1976). Serum Cholesterol Levels in American (Pima) Indian Children and Adolescents. *Pediatrics* 58:274–282.

Schell, L.M., and F.E. Johnston. (1978). Physical Growth and Development of American Indian and Eskimo Children and Youth. *In Handbook of North American Indians.* (In Press.)

Smith, S.H. (1810). *An Essay on the Causes of the Variety of Complexion and Figure in the Human Species.* New York: J. Simpson.

Spielman, R.S., F.J. Da Rocha, L.R. Weitkamp, R.H. Ward, J.V. Neel, and N.A. Chagnon. (1972). The Genetic Structure of a Tribal Population, the Yanomama Indians. VII. Anthropometric Differences among Yanomama Villages. *Am. J. Phys. Anthropol.* 37: 345–356.

Steggerda, M. (1932). *Anthropometry of Adult Maya Indians: A Study of Their Physical and Physiological Characteristics.* Publ. No. 434, Carnegie Inst., Washington, D.C.

Stoudt, H.W., A. Damon, R. McFarland, and J. Roberts. (1965). *Weight, Height, and Selected Body Dimensions of Adults. United States, 1960–1962.* Natl. Ctr. Health, Stat., Ser. 11, No. 8. Washington, D.C.: U.S. Govt. Printing Office.

Stoudt, H.W., A. Damon, R. McFarland, and J. Roberts. (1970). *Skinfolds, Body Girth, Biacromial Diameter, and Selected Anthropometric Indices of Adults. United States, 1960–1962.* Natl. Ctr. Health Stat., Ser. 11, No. 35. Washington, D.C.: U.S. Govt. Printing Office.

Stunkard, A., E. d'Aquili, S. Fox, and R.D.L. Filion. (1972). Influence of Social Class on Obesity and Thinness in Children. *J. Am. Med. Assoc.* 221: 579–584.

Tanner, J.M. (1959). Boas' Contribution to Knowledge of Human Growth and Form. *In* W. Goldschmidt, Ed., *The Anthropology of Franz Boas.* Washington, D.C.: Am. Anthropol. Assoc., Mem. No. 89, pp. 76–111.

Thomas, R.B. (1973). Human Adaptation to a High Andean Energy Flow System. *Occasional Papers in Anthropology* No. 7. University Park, Pa.: Pennsylvania State Univ. Press.

Tobias, P.V. (1972). Growth and Stature in Southern African Populations. *In.* D.J.M. Vorster, Ed., *Human Biology of Environmental Change.* London: Unwin, pp. 96–104.

Wolanski, N. (1970). Genetic and Ecological Factors in Human Growth. *Hum. Biol.* 42: 350–368.

World Health Organization. (1968). *Nutritional Status of Populations: A Manual on Anthropometric Appraisal of Trends.* WHO/NUTR/70.129. Geneva: World Health Organization.

Chapter 13

Endogamy and Exogamy in Two Arctic Communities: Aleut and East Greenlandic Eskimo

Joëlle Robert-Lamblin

Introduction

Of the American Arctic hunters, the groups that are most geographically separated from one another are the Aleuts and Eskimos of East Greenland. They are to be found at the extreme ends of the Eskimo migration route; that is to say, to the west, the large chain of islands that extends from the Alaska Peninsula towards Kamchatka, and to the east, the eastern coast of Greenland which is bordered by a broad ice pack (Figure 13.1). Their major point in common is the long isolation in which they have remained because of the difficulties of access to both these regions.

These two groups have been the subject of my investigations. I studied the Aleut, emphasizing Qigeron Aleuts, while visiting the Department of Biobehavioral Sciences, University of Connecticut, in 1971, and the Ammassalimiut Eskimos as a member of the Centre de Recherches Anthropologiques of the Musée de l'Homme in Paris, from 1965 onward.

For data concerning the geography, climate, and ecology that constitute the specific environment in which these populations have developed and live at present, I refer the reader more particularly to the publications of W. S. Laughlin (1967, 1972), P. Robbe (1971) and J. Robert-Lamblin (1971, 1975a).

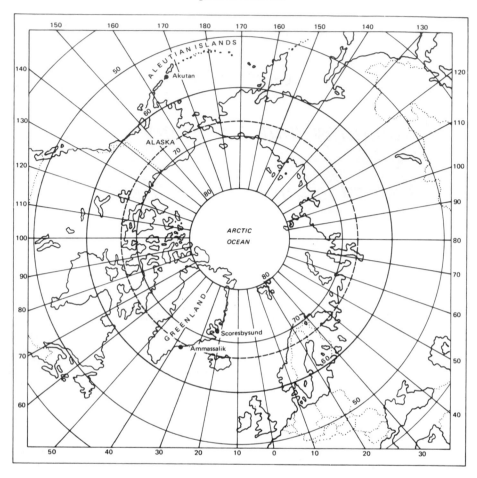

Figure 13.1. Geographic locations of the Aleuts and East Greenlanders.

In the present study of Aleuts and Ammassalimiut I first sought to trace the traditional marriage systems (in the sense of the way one chooses one's mate), each corresponding to a specific type of socioeconomic structure, and then the concommitant evolution of the matrimonial and socioeconomic structures since the entry of these populations into Western history.

The same methods were employed in considering both populations, including study of genealogies, historical demographic analysis, and the effect that contact with western civilization has had upon altering these formerly isolated populations. By using identical methods I have been able to draw parallels between the two groups.

Mating in the Traditional Society

Aleuts

The Tribes (Endogamous Units) at the Time of Their Discovery

When Bering and Chirikov discovered the Aleutians in 1741, the Aleuts were the only inhabitants of this volcanic archipelago extending from Port Moller to Attu. The absence of sea ice and the exceptional richness of marine fauna as a result of the location of the islands between the Pacific Ocean and the Bering Sea created conditions that had long sustained a population of substantial size along the coastline. It has been estimated that during that period (the mid-18th century) the Aleut population reached a level of about 16,000 persons, with a heavier concentration in the east than in the west.

Thus, from what we now know about the human density in this area (particularly high for a population of Arctic hunter-gatherers) and about the exclusively warlike contacts between the Aleuts and their Eskimo neighbors, we can assume that marriages between Aleuts and non-Aleuts were rare and, therefore, that their gene pool contained few, if any, genes from external sources. Nevertheless, the possibility of descendants of slaves being taken prisoner during wars, particularly against the Koniag Eskimos from Kodiak Island, should not be excluded.

The Aleut population, which constituted a large isolate, was in fact subdivided into subgroups or endogamy circles. The matrimonial and sociocultural structures, which are no longer perceptible today, can be traced from genealogies, travelers' narratives, anthropological data (Veniaminov 1840, Hrdlička 1945, Lantis 1970), and, to a certain extent, from oral tradition that has reached our time.

Aleut marriages were contracted within definite boundaries, namely those of the tribe. The various Aleut tribes constituted a type of small isolated groups, separated from one another by some "barriers" of unknown origin. It is possible that these "barriers" were of social rather than geographic nature. Indeed, there were no geographically impassable areas between the different tribes. As the Aleuts were excellent navigators, the distance between remote tribes could not have acted as an impossible impediment to exchanging wives or goods. Further proof of the Aleut capacity to cover long distances is illustrated by the warlike expeditions they undertook against each other or against Eskimo enemies.

From a morphological point of view, physical anthropologists distinguish two main divisions within the Aleut population: an eastern one and a western one, with a middle category of Aleuts whose parents belong to both of these divisions. The linguists, for their part, distinguish three main Aleut dialects: an eastern one, a central one, and a western one. As far as mating and socioeconomic structures are concerned, when gathering available historical data and field investigations, I have been able to determine at least seven subdivisions, or tribes, within the large Aleut territory (Figure 13.2).

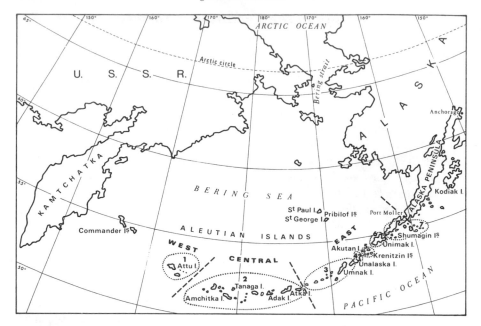

Figure 13.2. Linguistic divisions (west, central, east) and former tribal divisions of the Aleut population. Names of the tribes (according to my oldest informants in 1971): (1) Saskinach, (2) Neqoren, (3) Qawalengen, (4) Tewiurion, (5) Qigeron, (6) ? (this tribe has not existed since 1847), (7) Aleskin.

These tribes showed the following characteristics:

1. A well-defined territory (each tribe covered a group of villages or contiguous islands). The other Aleut tribes were "enemies" and the intertribal wars were almost unceasing.

2. A name for its members.

3. Dialectal particularities.

4. Its own traditions.

5. Last, but not least, these tribes formed endogamous units: the marriage partners were chosen within the same tribal group. However, some of the raids on enemy villages were aimed at abducting wives.

The Choice of Mate

As mentioned, marriages were performed, as a general rule, between Aleuts belonging to the same tribe; furthermore, they were often consanguineous, as the preferential marriage was between cross-cousins (whereas parallel cousins were regarded as brother and sister and their union would have been incestuous[1]).

---

[1] Incestuous relations that the community prohibited, however, could be in some cases regarded as confering some special strength. One of my aged informants told me that when he was a young man, his father ordered him to sleep one night with his older half-sister, so that he would become strong and get good luck while hunting.

The custom of entrusting a young boy's education to his maternal uncle was conducive to unions between cross-cousins; the "tutor uncle" would become later his nephew's father-in-law.

From the first visitors' narratives, polygyny seemed to have been frequent, but they disagree on the number of co-spouses: up to three (Sarychev), four (Coxe and Shelikhov), six (d'Orbigny), or seven or eight (Merck). Cases of polyandry have been reported, but they were rare. The couple's residence was generally patrilocal.

Each tribe thus constituted a homogeneous unit, with probably very few extratribal genes in its gene pool. This led to morphologic and genetic differentiation between Aleut subgroups, particularly perceptible between Western Aleuts and Eastern Aleuts, in spite of the similarity of their environment (climate, resources) and of their diet. Laughlin emphasizes differences in stature, weight, dentition, and metrical characters of the skull, face, and mandible, between Western and Eastern Aleuts (Laughlin 1951, 1966, oral communication 1971).

Ammassalimiut Eskimos

The Tribe at the Time of Discovery

Ammassalimiut Eskimos of the eastern coast of Greenland were discovered by the western world much later than the Aleut population. It was in 1884 that a Danish officer, Gustav Holm, reached them and established the first household census. There were 413 at that time in the Ammassalik area (Thalbitzer 1914 and 1923).

At this time there were no longer any Eskimo north of the region (66° North lat.). Half-way south, between Ammassalik and Cape Farewell, there was a small group of 135 Eskimos, last remnants of the Southeastern tribe. It was not long before they joined the Southwestern population of Greenland.[2]

Since the whales disappeared during the nineteenth century, and until their colonization, the Eskimos of the East coast appear to have lived in a precarious demographic balance, with successions of alternating growth and reduction in their numbers, for several sources disclose that they were subjected to severe famines, notably that of the winter of 1881–1882, when 70 persons died of starvation in the Ammassalik region.

At the time Holm discovered them, the principal food resource of these nomads was the seal, which also provided them with light, fuel, and clothing. They were scattered along the coastline and inside the fjords, living in large patriarchal houses that sheltered about 30 persons.

---

[2] However, we know that afterwards, in 1896, several members of that group joined the Ammassalimiut returning to their country of origin (see Figure 13.4) and that 12 among them (6 men and 6 women) intermarried with the Ammassalimiut.

The Choice of Mate

Until 1884 the Ammassalimiuts had lived without direct contact with Europeans, and their relations with the Southeastern tribe, from what we know, had been limited to bartering on some occasions (Gessain and Robert-Lamblin 1974).

Thus, due to its nearly total isolation, mating occurred within the Ammassalimiut tribe itself. However, it seems that there were also some kinds of subdivisions within this territory, and that some of the small scattered groups exchanged mates preferentially, while others had almost no matrimonial exchanges.

In contrast to Aleut customs, one ought not marry a near relative, up to and including one's first cousin. This meant, in practice, exclusion as an eligible mate all persons who grew up with oneself in the patriarchal house, because it grouped together one's grandparents, uncles and aunts, first cousins, brothers and sisters, half-brothers, and half-sisters.

In that society, where the division of labor made men and women perfectly complementary—the man as the provider, the woman to deal with the product of the hunt and to transform it into food, clothes, and combustibles—marriage was an economic necessity. It was performed around the age of 18–20 years, when the man had become a sufficiently skillful hunter and the woman was skilled in sewing. In the traditional society, single people were very few. As for widows (hunting accidents were a frequent cause of death among adult males), remarriage was the only solution to the problem of survival for themselves and their children.

The crucial problem of the surplus of females was then in part solved by the practice of bigamy. In 1884, 9 out of 89, or 10% of the married men were bigamous (see Table 13.3).

The same observation was made by Dumas concerning the Iglulik Eskimos, where the frequency of polygamy, higher than among the Netsilik and the Copper Eskimos, is related to the particular sex ratio of that population: "there are more polygynous unions among the Iglulik, which can be explained by the fact that they had a surplus of adult females [where] as the other groups [Netsilik and Copper Eskimo which practiced female infanticide] did not" (Damas 1975, p. 416).

The residence of the couple was sometimes matrilocal, sometimes patrilocal, according to the economic necessities of the distribution of the producers and consumers. This distribution occurred at the end of the summer, when the patriarchal family regrouped itself in the winter house.

Thus, at the time of its discovery, the Ammassalimiut tribe constituted a real isolate that could be characterized by:

1. Its own dialect (hardly understandable by West Greenlanders).

2. Its well-marked territory, of which the hunters and their families migrated following the game (ca. 500 km of coastline).

3. Its own way of life and customs.

4. It was an endogamous unit within which marriages were performed between a restricted number of eligible mates; this number was all the more reduced, as unions between first cousins were forbidden.

Evolution in Marriage and the Present Society

Aleuts

The Progressive Removal of the Tribal Boundaries

The changes that occurred under the influence of the Russian colonization led to the opening of the tribal units. These structural alterations had two causes: (1) the large numerical decrease of the Aleut population, (2) the influence of the Russian Orthodox missionaries. After contacts with Russian colonists, the Aleut population decreased dramatically: 50 years after their discovery, the Aleuts, decimated by diseases and massacres, numbered only 2500. The number of eligible mates within each tribe being considerably reduced, mates had to be sought among the other tribes.

Furthermore, the Christianization of the Aleut population during the first half of the nineteenth century included very strict opposition of the church to the traditional preferential marriages between cross-cousins. The Aleuts had to forsake any union between relatives, including all second cousins.

The Qigeron Tribe

The Aleut Qigeron tribe (No. 5 on Figure 13.2), which I have most closely investigated, numbered at the time of discovery (middle of the eighteenth century) around 1800 inhabitants, scattered in 25 villages on the seven islands of the Krenitzin group. At the time when the population density seems to have been the highest, the furthermost points of the Qigerons' habitat were no more distant than about 90 km.

After 25 years of contact between Russians and Aleuts, less than half of the tribe remained, and, 50 years later, only 20 percent of the Qigeron group had survived (Figure 13.3). Since 1890, there has been only one Qigeron village left in this territory, Akutan. Its population has fluctuated between 60 and 100 people.

The study of marriages recorded on the church registers at the end of the nineteenth century and the beginning of the twentieth century and the genealogies I have established lead to the following conclusions as far as the evolution of matrimonial structures in the Qigeron tribe is concerned: Up to the beginning of the twentieth century, one continued to look preferentially for a mate within one's own tribe; and, if there was no nonconsanguineous available mate, the chief of the village or the priest of the district arranged a marriage with someone from another Aleut tribe. In such cases, the spouses had never seen each other.

As shown on Table 13.1, of more than 100 marriages performed between 1871 and 1928, half were still endogamous, although the number of Qigerons decreased from 150 to 65 persons during the same period. From that table we can

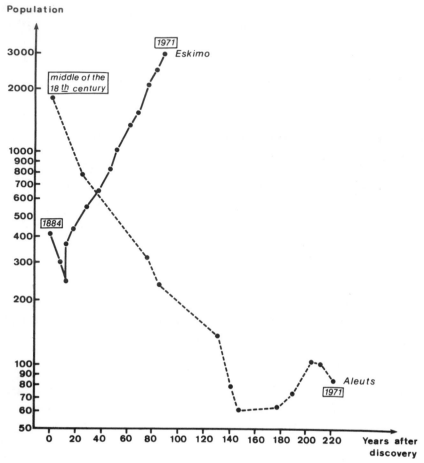

Figure 13.3. Demographic evolution of the Qigeron and the Ammassalimiut from the time of their discovery.

Table 13.1. Exchange of Mates Between the Qigeron Tribe and Other Aleut Tribes for Marriages Performed Between 1871 and 1928

Qigeron mate	Origin of the other mate	Aleut tribes[a]							Pribilof Islands[b]	Continental Whites	Unknown origin	Total
		1	2	3	4	5	6	7				
Unions where the man is Qigeron		1	1	4	7	25	—	5	3	—	2	48
Unions where the wife is Qigeron		2	—	5	16	25	—	1	3	1	2	55

[a] The numbering of tribes refers to Figure 13.2: No. 5 is the Qigeron tribe. No. 6, Unimak Island, was not inhabited anymore at that time.

[b] The Pribilof Islands are populated by Aleuts transferred from the eastern tribes by the Russian colonists in 1786.

also see that there was not necessarily a reciprocal exchange of marriage partners between tribes. For example, Unalaska (No. 4, Figure 13.2) attracted more Qigeron brides than it sent to Qigeron men, whereas the Alaska Peninsula (No. 7) provided wives to Qigerons but attracted very few women from that tribe.

The Opening of Aleutian Islands to the Outside World

The opening of the small Aleut tribal units to each other was a result of the changes in the mating system (end of the consanguineous marriages) as well as of the decrease in the number of eligible mates (decline of the population). Furthermore, the Aleut tribes came into contact with non-Aleut people; the admixture started under Russian domination, with the presence of administrators and traders, and continued under the American colonization, after 1867.

Concerning the history of the Qigeron in particular, one should mention that a Scottish trading agent named Hugh McGlashan arrived from England and settled in Akutan around 1887. Married to an Aleut from Attu, he had numerous descendants, a certain number of whom married Whites, and his impact on the technological and sociological evolution of the small village was considerable.

Another important event for the people of Akutan was the presence between 1912 and 1941 of a whaling station in the bay where the village is located. Employed as seasonal workers, the Akutans worked in this industry side by side with Norwegians and Americans.

However, it has been mostly since World War II that frequent contacts between Aleuts and Westerners have occurred. In 1942, the native population of the Aleutian Islands was wholly evacuated. The inhabitants of Attu, prisoners of the Japanese, were transferred to Hokkaido, while all other Aleuts were transported to camps in southeastern Alaska, where they were in daily contact with the western world.

Since their return to their territory in 1945, and especially in the last few years, with the improvement of the means of transportation, the Aleuts have many opportunities to travel to other parts of Alaska, for their education or for jobs.

During their travel, young Aleuts have now the possibility of finding mates from places outside the Aleutian or Pribilof Islands, among Whites, Indians, or Eskimos of Alaska. However, the Aleuts still possess a certain prejudice against Eskimos. Nowadays, young women attempt to marry white men.

Thus, the geographic distance for finding a mate, very small at the time when the first cousin was the preferred spouse, has increased by successive stages and today sometimes exceeds a thousand kilometers.

Ammassalimiut Eskimos

Sociologic Changes in Marriage

Among Ammassalimiuts, as among Aleuts, Christianization altered not only the system of religious beliefs but the social structures as well. In Ammassalik, the conversion of the Eskimos to the Lutheran religion started at the very end of the

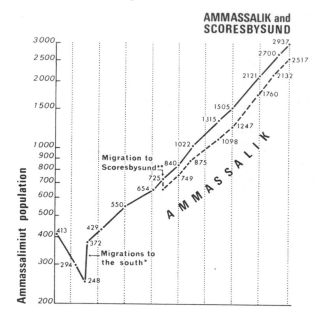

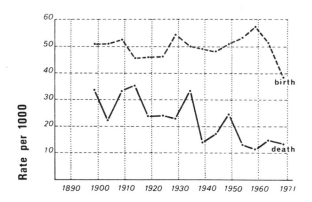

Figure 13.4. Demographic evolution of the Ammassalimiut population of East Greenland from the time of its discovery to the present day. Upper chart: The number of inhabitants based on household censuses. *After Holm's stay in the community, 118 persons emigrated southward in small groups over several years, and 120 persons returned in 1896 because a trading post was opened in Ammassalik. **In 1925, 10% of the population of Ammassalik left to establish the colony of Scoresbysund, 1000 km farther north. Lower chart: Variation in the birth and death rate during the twentieth century (by five-year periods), emphasizing the demographic explosion, particularly strong during the 1960s, and the effects of the birth control program started in 1968.

nineteenth century and was completed in 1921. As far as marriage is concerned, Christianization had as one of its main effects the removal of polygamy. Officially, there were no more bigamous men in the 1911 census (whereas there were still three of them in the 1901 census); the Lutheran minister who made the census regarded the former cospouses as "widows" or "single women." The rotation of wives, frequent before, stopped with the new religious marriage, which, when performed, could not be dissolved (divorce on the east coast of Greenland was allowed only after 1967).

Another consequence of the Danish colonization of the Ammassalimiuts— which was quite contradictory to the attitude of the Russian Orthodox missionaries in the Aleutian Islands—was to allow marriages between first cousins, as they are by the Danish law. A few marriages of this type were performed, but the ancestral taboos are not easily removed, even in a changing society. In fact, public opinion still disapproves of unions between first cousins, which are very rare.

In the Ammassalik area, because of the unusual demographic growth (the population increased sevenfold in 75 years) (Figures 13.3 and 13.4) and the breakup of the families into small nuclear units, young people do not know their genealogies as well as they did in the past. I have been told that two first cousins only discovered their kinship after they were married (Robert-Lamblin 1975b).

In Scoresbysund, 1000 km north of Ammassalik, where a few related families from Ammassalik settled in 1925, the number of nonconsanguineous eligible mates is even more reduced. In 1970, I only met one couple of first cousins in this small population of 420 persons (Robert 1970, 1971). There had been three other cases where two first cousins had a child born out of wedlock: but in the first case, the father committed suicide, in the second he emigrated to the west coast, and in the third he died while kayaking. One should note in this case that the society tacitly disapproves of the incestuous relations between first cousins, but not illegitimate births.

Since World War II, a new phenomenon has occurred and is increasing: Other types of wage-earning activities have appeared besides the traditional seal hunting— commercial cod fishing and salaried jobs within the different departments of the Danish administration. For these new socioeconomic classes, i.e., fishermen and wage earners, where young people are numerous, there is no need of complementarity male and female activities. With the money he earns, the man can buy from the store what his wife would have made in the past (namely clothing); and the wage-earning woman can provide for her own needs (Robert-Lamblin 1972).

At the present time, singleness is frequent among young Ammassalimiuts; children born out of wedlock (more than one third of all births in 1971) are welcomed in the families of unmarried mothers, if they do not want to bring up the children themselves. One might find something reminiscent of the earlier matrimonial customs (the rotation of wives mentioned above) in the present-day practice of late religious marriages, performed after the women had already had one or several children, sometimes from different fathers.

Exogamic Tendancies and the Opening of the Ammassalimiut Isolate

The demographic, social, and cultural changes that followed the beginning of the Danish colonization did not involve an alteration of the Ammassalimiut gene pool up until the Second World War.

Europeans and West Greenlanders in charge of the local administration were then very few in number, and the Danish government kept the number of exchanges between this area and the outside world under strict control.

From the settlement of a trading post in 1894 (when the colonization of eastern Greenland really started) up to 1941, among all Ammassalimiut children born in Ammassalik or Scoresbysund, less than 1 percent had a Danish father and less than 1 percent, likewise, had a parent from western Greenland (Gessain *et al.* 1975).

Thus, the Ammassalimiut Eskimos remained a very endogamous community until the Second World War created a crucial turning point in their history. For over 5 years, the east coast of Greenland was cut off from any contact both with the west coast and with Copenhagen and was supplied and defended by the United States. A large American base was established in the Ammassalik area, in Ikatek (located between the villages of Kummiut and Sermiligak), and from 1942 to 1947 the presence of about 800 American soldiers had considerable repercussions on the life of the surrounding villages.

When Denmark resumed communications with eastern Greenland in 1945, the general trend of evolution accelerated (Gessain 1969, 1975). Admixture increased in Scoresbysund area with the setting up of a meteorologic station in Kap Tobin in 1947, and with the building, in Ammassalik district, of an airstrip and a radar station in Kulusuk (which entailed the presence of 400 Danish workers in 1958).

As a result of the increasing number of westerners living in both areas[3], the total number of Ammassalimiut hybrid births occuring in the 10-year period 1962–1971 rose to 11 percent. Like Aleut women, Ammassalimiut women today attempt to marry white men or to live with them, either in Greenland or in Denmark.

Last, a number of West Greenlanders, who have come to settle temporarily or permanently on the East coast, play a certain part in the evolution of the Ammassalimiut gene pool. These intermarriages between East and West Greenlanders occur more frequently in Scoresbysund than in Ammassalik (Table 13.2).

However, as shown in Table 13.2, due to an important emigration of the hybrids, admixture is still little in the whole of the East Greenlandic population; it is more marked in Scoresbysund (Jacquard 1973) than in Ammassalik (Langaney *et al.*, 1974).

[3] In 1971, the Danish colony in Ammassalik represented about 250 persons for 2517 Ammassalimiuts, and in Scoresbysund 80 persons for 420 East Greenlanders. In addition, there is in the Ammassalik area an American base of about 12 persons and a meteorologic station comprised of some 20 Europeans.

Table 13.2. *Composition in Percentage of the East Greenland Population Living in 1971*

	Ammassalik (2517 persons) (%)	Scoresbysund (420 persons) (%)	Total East Greenlanders (2937 persons) (%)
Individuals whose ancestors are all Ammassalimiut	88.8	76.2	87.0
Individuals the father of whom is European or American[a]	5.5	4.0	5.3
Individuals the grandfather of whom is European or American	1.0	—	0.9
Individuals one parent of whom is a West Greenlander	1.3	10.7	2.6
Individuals one grandparent of whom is a West Greenlander	0.9	7.4	1.8
Individuals whose father is unidentified in the genealogies	2.3	—	2.0
West Greenlanders settled in East Greenland[b]	0.2	1.7	0.4

[a] A number of hybrids born in eastern Greenland have left the area, either adopted by Danish families or because of their parents' emigration.

[b] These West Greenlanders have mixed with East Greenlanders owing to intermarriages.

Numerous migrations of Ammassalimiuts towards West Greenland or Denmark that have developed in the last fifteen years (Perrot 1974, 1975) account for, among other factors, the opening of the Ammassalimiut isolate to the outside world. They relate mostly to women and to young people who continue their education or seek employment.

It is interesting to analyze, in conclusion, this phenomenon of the emigration of women. We have seen that the practice of bigamy played a social and economic role in a closed population such as that of Ammassalik before its discovery. Today, polygamy is no longer allowed, and with the isolate being opened to the outside world, the problem of an excess of adult females is solved mainly by the emigration of a number of them. Perrot emphasizes that more women than men emigrate,

Table 13.3. *Sex Structure of the Ammassalimiut Population (Number of Males for 100 Females)*

Age group	Years	
Ammassalik District	*1884*	*1971*
Under 15	83	90
15 and over	92	101
All ages	88	95
Scoresbysund District	*1925* (founders' group)	*1970*
Under 15	90	85
15 and over	78	85
All ages	84	85

principally to Denmark (Perrot 1975). The phenomenon is particularly pronounced in Scoresbysund, where the imbalance of the sex structure is 'strong. In 1968 I observed that one third of the women over twenty years of age had left Scoresbysund (Robert 1971). As can be seen in Table 13.3, the emigration of women successfully restored the balance of the sexes at Ammassalik, where I found 101 men to 100 women over the age of 15 in 1971. But as far as Scoresbysund is concerned, the female emigration, although very important, did not succeed in reestablishing a balanced sex ratio among adults, owing to a singular sex ratio at birth (92 boys to 100 girls were born since the founding of the colony in 1925) and to a high mortality rate among adult males.

Summary

    The study of earlier and recent marriage practices of two Arctic communities, Aleut and Ammassalimiut Eskimo, emphasizes their total endogamy before their contact with the western world. Due to geographical as well as social factors, the two groups therefore constituted true biological, cultural, and linguistic isolates.

    Since their discovery by western people, they have opened themselves to the outside world while undergoing transformations in their matrimonial structures, as well as in other fields. The search of white mates by young Aleut and Ammassalimiut women increases today the exogamic tendencies.

    The parallel established here between two small Arctic societies is based on the use of the same methods of investigation and data analysis, i.e., the methods of genealogies and historical demography.

References

Damas, D. (1975). Demographic Aspects of Central Eskimo Marriage Practices. *Am. Ethnol.* 3(2):409–418.

Gessain, R. (1969). *Ammassalik ou la Civilisation Obligatoire.* Paris: Flammarion, 251 pp.

Gessain, R., and J. Robert-Lamblin. (1974). Migrations des Ammassalimiut au XIXè Siècle, d'après les Archives des Frères Moraves. Cahiers du Centre de Recherches Anthropologiques 13. *Bull Mém. Soc. Anthropol. (Paris)* 2(13):153–159.

Gessain, R., and J. Robert-Lamblin (Eds.). (1975). Tradition et Changement au Groenland Oriental. Dix années d'Enquêtes du Centre de Recherches Anthropologiques. Musée de l'Homme. *Objets et Mondes,* été, XV(2):117–285.

Gessain, R., J. Robert-Lamblin, and P. Robbe. (1975). Dictionnaire Anthroponymique Codé des Ammassalimiut. Paris: Documents du Centre de Recherches Anthropologiques du Musée de l'Homme, 3.

Hrdlička, A. (1945). *The Aleutian and Commander Islands and their Inhabitants.* Philadelphia: The Wistar Institute of Anatomy and Biology.

Jacquard, A. (1973). Distances Généalogiques et Distances Génétiques. Applications aux Indiens Jicaques du Honduras et aux Eskimo du Scoresbysund. *Cah. Anthropol. Ecol. Hum.* 1:11–124.

Langaney, A., R. Gessain, and J. Robert. (1974). Migration and Genetic Kinship in Eastern Greenland. *Soc. Biol.* 21(3):272–278.

Lantis, M. (1970). *Ethnohistory in Southwestern Alaska and the Southern Yukon.* Lexington, KY: Univ. of Kentucky Press.

Laughlin, W.S. (1951). The Alaska Gateway Viewed from the Aleutian Islands. *Papers on the Physical Anthropology of the American Indians.* New York: Viking Fund, pp. 98–126.

Laughlin, W.S. (1966). Genetical and Anthropological Characteristics of Arctic Populations. *In* P. Baker and J. Weiner, Eds., *The Biology of Human Adaptability.* Oxford: Clarendon Press, pp. 469–495.

Laughlin, W.S. (1967). Human Migration and Permanent Occupation in the Bering Sea Area. *In* D. Hopkins, Ed., *The Bering Land Bridge.* Stanford, CA: Stanford Univ. Press, pp. 409–450.

Laughlin, W.S. (1972). Ecology and Population Structure in the Arctic. *In* G.A. Harrison and A.J. Boyce, Eds., *The Structure of Human Populations.* Oxford: Clarendon Press, pp. 379–392.

Perrot, M. (1974). Les Ammassalimiut au Danemark 1972. Etude Démographique et Psycho-sociologique de l'Émigration de la Côte Orientale du Groenland. Cahiers du Centre de Recherches Anthropologiques 13. *Bull. Mém. Soc. Anthropol. (Paris)* 2(13):5–102.

Perrot, M. (1975). L'Émigration du Groenland Oriental, un Nouveau Shème Culturel. *Objets et Mondes, été,* XV(2):169–176.

Robbe, P. (1971). Climat d'Angmagssalik. Cahiers du Centre de Recherches Anthropologiques 11–12. *Bull. Mém. Soc. Anthropol. (Paris)* 8(12):137–197.

Robert, J. (1970). Démographie et Acculturation. Une Nouvelle Phase dans l'Histoire des Ammassalimiut Émigrés au Scoresbysund: l'Introduction du Contrôle des Naissances (Côte Orientale du Groenland 1970). *J. Soc. Am.* 59:147–155.

Robert, J. (1971). Les Ammassalimiut Émigrés au Scoresbysund. Etude Démographique et Socio-Economique de leur Adaptation (Côte Orientale du Groenland 1968). Cahiers du Centre de Recherches Anthropologiques 11–12. *Bull. Mém. Soc. Anthropol. (Paris)* 8(12):5–136.

Robert-Lamblin, J. (1972). Ammassalik 1967–1972. Les Tendances Actuelles de l'Evolution Démographique et Economique (Côte Orientale du Groenland). *J. Soc. Am.* 61: 258–270.

Robert-Lamblin, J. (1975). Eskimo Ammassalimiut et Aléoutes Qigeron: Différences Ecologiques, Historiques, Démographiques. *L'Anthropologie* 79(3):519–536.

Robert-Lamblin, J. (1975). Mortalité et Expansion Démographique à Ammassalik au XXè Siècle. *Objets et Mondes,* Été, XV(2):223–135.

Thalbitzer, W. (Ed.). (1914 and 1923). The Ammassalik Eskimo. *Med om Grönland, 39,* I–II, 752 pp. and III 294 pp.

Veniaminov, I. (1840). *Notes on the Islands of the Unalaska Division.* 3 Vols. St. Petersburg.

Chapter 14

Life Expectancy and Population Adaptation: The Aleut Centenarian Approach

Albert B. Harper

Introduction

Human adaptability transcends the biologic world but is produced and maintained by genetic variability between groups of the species (Laughlin 1968a). The pathways to "attuned existence" (Dobzhansky 1955) are multiple and are expressed in a variety of ways in the subpopulations of the species. No population possesses an inside path or monopoly on genetic variation, hence, adaptability essential to a successful evolutionary future (Ginsburg and Laughlin 1966, Ward *et al.* 1975).

The ability of the human species to expand, occupy, and exploit almost all ecosystems demonstrates the success the species has had in adapting to new environmental situations as they occur. The enormous range of adaptive mechanisms available to man are conferred by his extremely labile biobehavioral repertoire (Baker 1969), an obvious example of which is the ability to learn from experience and, more importantly, to relay previous action to the uninitiated. No less important is the functional and morphologic plasticity that allows the individual to respond quickly and effectively to new challenges. The expansive literature on temperature and altitude acclimatization and adaptation has documented our species' ability to regulate new stresses by physiologic and/or morphologic change.

   Laughlin (1968a) has summarized the complex interaction between the biologic and behavioral responses that are integrated as human adaptability:

> Eskimos are so frequently cited as examples of cold adaptation that other
> and equally important aspects of their significance for human adaptation
> are overlooked. Cold adaptation in Eskimos includes their elevated
> basal metabolism and related physiological characteristics, as well as
> their food, clothing, shelter and behavior. Clothing tailored to the en-
> vironment is essential, but a traveling partner may also be. When one falls
> into the water, he can be helped out and rapidly rewarmed. The dry partner
> can share half his clothing with the wet person, and they can each tolerate
> wet and frozed clothing long enough to return home. The habit of walking
> and running slowly in order to prevent sweating is importantly adaptive,
> and certainly the habit of staying inside when it is especially stormy is
> adaptive wisdom of the most cunning sort. (Laughlin 1968a, pp. 15–16.)

   The importance of successful integration of adaptive mechanisms is also
found in a more general example relevant to the entire species. To the extent
that humans have evolved as hunters, we have placed a major emphasis on
developing an entire biobehavioral regimen of skills to secure nutrients. Successful hunting requires the integration of technology, ethology, anatomy, and sheer
physical and intellectual stamina. These, in turn, require extensive childhood
training and programming, with a heavy emphasis on tutoring of children by
former hunters (Laughlin 1968b). Hunting, too, is an adaptive response and sur-
passes the stress of altitude hypoxia, arctic cold, or tropical heat in configuring
human populations.
   Although no population has a monopoly on adaptive strategies or potentials,
it is clear that some human groups have flourished, while others have become
extinct. The evolutionary criteria for successful adaptation to an ecosystem is
simply survival of the population. The difficulty in comparing relative adaptive
success between populations, whether living under similar stresses with different
genomes or the reverse, lies in obtaining a measure that is applicable and general-
izable to the extent that it integrates the total human adaptive response simultaneously. The use of a single behavior or a single enzyme system as the only
axis on which comparisons between populations are based is innately limited
because only a very small portion of the total adaptive response is considered.
For example, the enzyme 2,3-DPG mutase can be shown to confirm some
advantage in terms of oxygen transport but represents only a part of the extensive functional, morphologic, and behavioral mechanisms also known to be
adaptive at high altitude (Baker 1966, 1969, 1978, Schull and Rothhammer 1976).
Each of these responses is adaptive; they confer an advantage to the individual
so that his probability for survival and opportunity to reproduce is enhanced.
However, as the interest is in measuring a population's level of adaptation, my
object is to sum all events (genes, morphology, behavior, culture) relevant to a
population's ability to interact with its current environment.

The difficulty in quantifying population adaptedness has not been neglected (Mazess 1975). Thoday (1953) suggested that the "fitness" of a population should be counted in terms of its probability of leaving descendants after some long period of evolutionary time. This measure is largely retrospective and is difficult to apply to living populations. Andrewartha and Birch (1954) considered the adaptedness of a population to be the innate capacity to increase in numbers measured as the maximum growth rate given an optimum environment. As most populations do not exist under maximized environmental conditions, including the absence of interspecific competition, this measure has found only limited application (Dobzhansky *et al.* 1964). Ayala's (1969) experiments on productivity and population size in *Drosophila* subspecies provide a more pragmatic approach but, again, are not experimentally applicable to human groups. Ayala was able to demonstrate that the ability of a population to transform available energy into adults of the same species is a useful measure of the level of adaptation. Moreover, he concluded that the rate at which a population adapts to an environment is proportional to the amount of genetic variability contained in the population.

Baker and Dutt (1972) are responsible for the provocative attempt to compare adaptive differences between Andean populations using demographic variables as measures of biologic adaptation. Baker and Dutt correctly saw the vast implications of the use of fertility and mortality data but unfortunately limited their approach because of the conceptual difficulty in associating the causal stress (altitude) with the measures at hand:

> The complexities of using demographic variables as adaptation measures for high altitude populations re-emphasizes the reasons why human biologists have generally rejected simple indices of fertility and mortality as adaptive measures for human populations. Even where the stress is clearly manifest and is only slightly modified by cultural form it is apparent that these indices cannot be easily interpreted. (Baker and Dutt 1972, p. 373)

The solution to this problem as I see it lies in two relatively simple assumptions: that human adaptation is more than a biologic response to some challenge and that the measure, if it is to be an integrated measure, will relate not only to a particular stress but to the entire spectrum of human existence in an ecosystem. Baker and Dutt must be credited with one of the few attempts by human biologists to enter the realm of human adaptability with the use of measures that transcend the usual genetic and morphologic approach to this topic.

Life Expectancy and Adaptation

Survivorship is most dependent on not dying, and death, regardless of its cause, is the ultimate stress with which all organisms must cope. A population whose members are better endowed to withstand the lifelong pressures of an early death must have made a more successful adaptation to all challenges than populations whose members undergo an untimely demise. It seems apparent that the application of life tables to the problem of quantifying the adaptive success of a population will produce results that are comparable between populations because we are using a common stress that all populations adapt to and a stress that also incorporates many of the environmental challenges facing the organism.

Human biologists, with some notable exceptions, have not made extensive use of the patterns of length of life, despite the fact that the actuarial application of life tables extends as far back as the work of Graunt (1662). Karl Pearson publishied a short note in the first issue of *Biometrika*, 1901–1902, "On the Change in Expectation of Life in Man During a Period of Circa 2000 Years." After examining the life expectancy curve of Roman era Egyptian mummies and comparing it with late nineteenth century England, he concluded:

> In the course of those centuries man must have grown remarkably fitter to his environment or else he must have fitted his environment immeasurably better to himself. . . . We have here either a strong argument for the survival of the physically fitter man, or for the survival of the civilly fitter society. Either man is constitutionally fitter to survive today, or he is mentally fitter, i.e., better able to organize his civic surroundings. Both conclusions point perfectly definitely to an evolutionary progress. (p. 264)

Social Darwinism aside, it is apparent Pearson had come upon an accurate method for quantifying population adaptedness.

Fisher (1958) was also acutely aware of the implications of the life table analysis. He begins his presentation of *The Fundamental Theorem of Natural Selection* with a discussion of "The life table: In order to obtain a distinct idea of the application of Natural Selection to all stages in the life history of an organism, use may be made of the ideas developed in the actuarial study of human mortality" (2nd ed., p. 22). In light of the early work of Pearson and Fisher, it is remarkable that this idea has been dormant for so long.

Life Expectancy Models

Although survivorship analysis is a powerful tool in demonstrating differences in length of life, it is dependent on accurate cohort representation in all age classes. If, for example, infant mortality is underestimated, as is so commonly the case in skeletal series, the survivorship probabilities of all age cohorts will increase dramatically. Angel (1969) has, in fact, objected to the use of life tables in paleodemography because underenumeration of infants profoundly affects all calculations derived directly from survivorship probabilities. Moore *et al.* (1975) demonstrated the unique property of life expectancy that makes it vastly superior to survivorship curves. In a simulation exercise, they added a 10-fold increase in infant mortality rates and found that while survivorship was thereafter disturbed, life expectancy was changed only in the infant cohort.

> Any enumeration error will become part of the cumulative history which determines later survivorship values. . . . Changes in infant representation affect only infant life table values and survivorship values; life expectancy and probability of dying are mathematically unaffected. (p. 60)

Because life expectancy is mathematically unaffected by cumulative error, it is a much superior measure than simple mortality and survivorship curves.

Our analyses follow the methodology of Acsadi and Nemeskeri (1970), Weiss (1973), and Ubelaker (1974), who have popularized demographic models suitable for analyzing human populations, both in skeletal and in aboriginal condition. The theory and assumptions are straightforward and derived primarily from Lotka's (1956) model of the stable population, which requires that the population be infinitely large, have no net migration, and have fixed rates of mortality and fertility within each age group. Given these conditions, which are so often closely met by real human populations, the population is described by its age distribution, which is determined solely by the rates of mortality and fertility. It follows that, once the birth and death rates are fixed, the age distribution will remain the same as the previous generation. Furthermore, Weiss (1975) has demonstrated that severe changes in the vital rates, such as a disaster that eliminates 25 percent of the population, have only minor effects on the stable age distribution and that temporary perturbations in fertility and mortality are overcome in a matter of a few years.

We strongly feel that the actual computation of individual life tables is a necessary procedure to describe the life experience of a population. In this way important demographic events relevant only to small populations, i.e., the loss of 15 men in a drowning accident or the death of a chief, are recorded. These chance events play a major role in the history of these populations and must be analyzed in context.

The alternative to our methods is the use of model tables that consider a "typical" age pattern of mortality to be a better generalization of the life experience. These models are an average pattern necessarily intended for use in situations where no direct reliable data are in hand. Furthermore, model tables are specific to the populations from which they are derived. The most authoritative model tables are those of Coale and Demeny (1966), which are based on data from industrially advanced European and Asiatic societies. As Coale and Demeny note, their life tables are simply models of different regions and are to be used only for similar regions. "The separate families of model life tables provide estimates that in our judgement are quite reliable when utilized judiciously for populations within the areas upon which each family of model tables is based" (p. 27). Clearly, Aleuts, Eskimos, and American Indians do not find their place in this work.

Advantages of Life Expectancy

Many advantages for the life expectancy approach to adaptation may be postulated; however, those that are especially relevant to this discussion include:

1. Only a single axiom is involved: A longer life is superior to a shorter life. This applies equally to an individual's existence and to the specific life expectancy of any age cohort in a population.

2. The calculation of life expectancy is straightforward and the theory can be generalized to include data derived from a single census or mortality records (Stolnitz 1956). Life expectancy is a much better statistic than survivorship probability because it is an age cohort-specific value. Whereas relative survivorship must have all cohorts accurately represented, life expectancy tables can be entered at any age. This feature eliminates the extreme bias of underrepresented children in skeletal series (Moore *et al.* 1975).

3. It integrates the entire repertoire of a population's adaptive options to their environment. This includes all factors that contribute to the mortality structure of a population.

4. It is a comparable statistic between all populations, even between those living under different environmental regimes. Intuitively, we know that modern industrialized populations are better adapted to accommodate a changing environment than Pleistocene man. This is, of course, a moot comparison; however, comparisons within the same population over time or between related populations seem especially relevant. Furthermore, differences in life expectancy are not trivial and relate to the division between success and extinction.

5. One of the more powerful features of life expectancy analysis is the fact that the secular changes in life expectancy and adaptedness can be followed over time. Comparisons between living populations and their skeletal antecedents can now be made on the basis of adaptive response to a changing environment.

6. It is a highly sensitive measure because life expectancy responds quickly to new environmental challenges, such as the case examples of epidemics or migration into a new environment to which the population has not yet adapted.

The Aleut Approach

The early Aleuts arrived at the terminus of the Bering Land Bridge at least 8700 years ago. The remains of the earliest Aleutian occupation on Anangula Island on Nikolski Bay display a complex lithic industry complete with manufacturing sites *in situ*. Although the antiquity and acidic soils of Anangula have erased all but the stone component for analyses, the location of Anangula dictates that these people were marine hunters. Their occupational skill is attested to by the cultural and temporal continuity of this population. The Aleuts adapted to and survived on Nikolski Bay from 8700 B.P. to the present. This base population, and perhaps several others like it, also expanded and eventually occupied the entire Aleutian archipelago. Because of their success in maintaining population continuity and because of their necessary technical skills as marine hunters, we have long felt that Aleuts were a well adapted population living at the interface of a marine coastal ecosystem.

Despite our preconceived acceptance of this idea, it was difficult to quantify the degree of Aleut adaptive success. One lead into this problem was suggested by Burch (1972) in his study of the caribou hunters of Northern Alaska. Burch demonstrated that those populations who depended on caribou as a primary resource had, in addition to other deficiencies, an extremely low population density. Those with more diversified economies, especially coastal populations, were capable of supporting larger numbers per habitat space. When Aleuts are added to this matrix of natural resources and population density, we find that Aleuts are vastly superior to even the best Eskimo populations (Table 14.1). Laughlin's (1949, 1972, 1975) estimates of 16,000 Aleuts render a density of over 1.2 Aleuts per habitat square kilometer. This is 10 times as great as the most densely populated Eskimo areas, the Siilivikmiut and the Tikiraqmiut, who are primarily

Table 14.1. *Population Densities of Selected Arctic Groups*[a]

Group	Resource base	Estimated population size	Estimated area (km^2)	Density (people/km^2)
Aleuts	Sea otters, sea urchins, pinnipeds, etc.	16,000	12,800	1.250
Siilivikmiut	Whitefish, sheefish, ling cod, small game, some caribou (for skins primarily)	950	5,000	0.190
Tikiraqmiut	Whales, seals, walrus, polar bear, some fish (grayling, char), some caribou (mostly for skins)	1,000	5,600	0.178
Kivalinaqmiut	More or less equal dependence on char, seals, and caribou	250	3,700	0.067
Kuuvakmiut	More or less equal dependence on fish (whitefish, sheefish, salmon, ling cod) and caribou	800	18,750	0.042
"Mountain people"	Primary dependence on caribou (also some mountain sheep, bears, grayling, char, and whitefish); in times of starvation, had recourse to neighboring groups with greater resources, especially fish	500	35,000	0.014
Patliqmiut	Primary dependence on caribou, with seasonal emphasis on char and seals; unable to transport sea mammal products to wintering area	400	30,000	0.013
Asiaqmiut	Primary dependence on caribou (also some musk oxen and lake trout); in times of starvation, no recourse to groups having more reliable resource base	700	70,000	0.010

[a] Modified from Burch (1972, p. 350)

marine hunters with some dependence on caribou. Those populations with a primary dependence on the wildly fluctuating caribou herds are some two orders of magnitude less than the Aleuts.

Although this emphasizes the remarkable difference in Aleut population biology, maintenance of a high density of people may be maladaptive, especially in a changing environment. Eventually an expanding population must reach a population density that crosses the carrying capacity of the environment, determined both by the natural resource productivity and by human ingenuity in extracting nutrients from limited supplies. Because of the marine ecosystems they exploit, and because of the upwelling systems in the passes between the islands, the Aleuts occupy one of the world's richest ecosystems (Kelley et al. 1971).

There is evidence that suggests the Aleut population had not exploited all available niches and occupation sites to their maximum potential. Frøhlich's (1976) Aleut site survey of the central and western Aleutians emphasizes the vast number of settlements, many of which are seasonal or satellite encampments. Although most good boat landing sites are occupied, the sites for the most part are small, fewer than 2500 m². This was partly predicted by Turner's (1967) estimates of maximum population size based on possible living sites determined from map readings. Using the criteria of a combination of bays, reefs, streams, and lakes, Turner counted 402 possible site locations and, assuming an average of 80 to 100 inhabitants per site, a maximum of 32,000 to 40,000 Aleuts was derived.

Hayden (1975) has emphasized the shortcomings of the carrying capacity argument, noting the difficulties in quantifying resource availability. Indeed, most populations, including humans, usually maintain population sizes far below the environmental potential (Wynne-Edwards 1962), 50 percent being a reasonable value. The biobehavioral control mechanisms of human population growth are at best speculative; however, it is possible through the balance of births and deaths to maintain a low population profile and still expand greatly in number, provided sufficient time is involved.

Our perspective on the potential for population expansion is unduly modified by the current Third World nations, whose uncontrolled expansion severely threatens to exceed 6 billion humans in less than another generation (Keyfitz and Flieger 1971). This combination of high growth rates (3–4 percent) and high population density for available resources must ultimately result in severe wastage and suffering.

In a striking contrast, the Aleut model represents a more probable pattern for human increase over the most of our history. An average growth rate of 0.0003, or an increase of 0.3/1000/year, is sufficient to account for a 16-fold increase in Aleuts in 9000 years. Even when these estimates are adjusted to major events in the ecosystem, e.g., tectonic uplift which created the vast Aleutian strandflats, the maximum rate of increase in any millennium cannot be more than 0.0007 per year. The Aleuts were able to expand at a very, very slow rate and still achieve one of the highest population densities in the New World (Kroeber 1939).

Aleut Centenarians

Just as we had intuitively known that Aleuts had necessarily become well adapted to their marine ecosystem, we also knew that Aleuts lived a long life. In 1963, Laughlin first called attention to the rather remarkable length of life among Aleuts dying between the years 1824 and 1836. In an analysis of age at death between Sadlermiut Eskimos of Southampton Island and Fox Island Aleuts, he noted that ". . . the maximum age at death among the Sadlermiut was between 50 and 55 years. In marked contrast, the maximum age at death among the Aleuts was between 90 and 100" (Laughlin 1963, p. 641). Indeed, Veniaminov's parish records, graciously translated by Dr. Lydia Black (personal communication 1976), reveal that of 427 deaths occurring during this period, 3.8 percent (16 of 427) were between 80 and 89, and, more remarkably, 1.6 percent (7 of 427) were over age 90. Comparable values for Massachusetts in 1900 place 0.64 percent of the population between age 80 and 89, and only 0.06 percent over ninety. Comparison of centenarians alone reveals a two order of magnitude difference between Aleuts of 1820 and citizens of Massachusetts of 1900, 0.94 percent versus 0.0015 percent, respectively.

This pattern is not an isolated example. In 1871, Pinart met an aged Aleut who remembered the arrival of the first Russians and was thus at least into the first decade beyond centenarianism, although Pinart believed him to be 120 (Robert-Lamblin 1976). In contrast, Steffansson (1958) was unable to find evidence of great length of life in the northern Arctic. He offered the example of the Eskimo lady, alive in 1914, who Jenness thought to be at least 75 as the most aged Eskimo, but adds, ". . . some persons looked as if they might be past 60, a very few as if past 70, and one or two as if past 80" (p. 19).

A short life history is more common than generally believed because members of small and preliterate societies may appear to be the epitome of health and vigor at age 30 but, in fact, join their ancestors before age 50. The observation by Neel and Salzano (1964) is applicable to so many populations that it serves as a basic generalization of human existence: ". . . despite this picture of abundant health in the young male, old men—and by that I mean men in their 50's—are uncommon. At 30 they look as if they should reach 100 at 50 they are for the most part gone" (p. 94). The Aleut pattern is also the same—healthy at 30—except they often become centenarians. Even when moved from their basic habitat, Aleuts tend to retain the feature of a long life as in Rychkov's and Sheremetyeva's (1972) study on the Commander Aleuts, but not to the same extent.

Aleut Life Expectancy

As Aleutian Island Aleuts live longer than other Arctic Mongoloid populations, including displaced Aleuts, I wish to pay special attention to temporal changes in life expectancy under the conditions in which optimal adaptation was achieved. The data base for the Aleutian Inland population is excellent. Pre-contact and contact Aleuts are well represented by large skeletal series preserved in the calcium-rich middens and in mummy caves. Our knowledge of the post-Russian contact period is exceedingly exact because of the Russian Orthodox Church's attention to death. The parish records abound with detailed mortality records extending back to the 1820s when Bishop Venaminoff arrived in Una-laska.

Aleut history may be conveniently divided into three epochs, the precontact period (8700 B.P. to 1741 A.D.), the Russian period (1741 to 1867), and the American period (1867 to present). Although the divisions follow political boundaries, they reflect some salient cultural and sociologic phenomena that bear directly on life expectancy.

Precontact Aleuts

The precontact Aleut population consists of Paleo-Aleuts and Neo-Aleuts, both representing major temporal subgroups of the same population. The long, narrow-headed Paleo-Aleuts are the earlier Aleut population, extending back as far as 4000 B.P., whereas the broad-headed Neo-Aleut population is more recent in time and is also represented by modern Aleuts. The dynamics of the evolution from Paleo-Aleuts to Neo-Aleuts has been discussed extensively by Hrdlička (1945), Laughlin (1963), and Harper (1975). The relevant point here is that the Neo-Aleut population evolved in the eastern Aleutian Islands slightly before the first Russian contact and diffused westward, perhaps with the aid of the Russian fur hunters.

Despite the limited sample sizes, especially in the Paleo-Aleut groups (Tables 14.2 and 14.3), the precontact series is a very important data source for inter-preting life expectancy before the epidemics brought on by contact with the western world. The noninteger sample sizes result from limitations in sex assign-ment, especially in the younger ages. Because of differential preservation and the technical inability to distinguish between young Paleo-Aleuts and young Neo-Aleuts, the 0 to 9 class was excluded from analysis. Although this does not effect subsequent life expectancy, we must enter all comparisons at age 10 for mortality data, and age 15 for census data.

Having reached age 10, the Paleo-Aleut males could expect to live an addi-tional 35.77 years (Table 14.4). This is a rather remarkable achievement because it is the longest life expectancy at this age for all Aleutian temporal units, except-ing modern Aleuts. Female Paleo-Aleut life expectancy is also among the longest, with an average of 37.71 years beyond age 10 (Table 14.5).

As we shall see, the Paleo-Aleuts represent one of the longest lived Arctic Mongoloid populations known. Their successful adaptation is no doubt con-tingent upon the maintenance of a complete population profile, with all age groups represented, including the elderly. However, limitations in determining the age at death of a skeleton beyond age 60 by traditional methods force the age 60 to 69 category to be an open-end terminal class. Were we able to assign

Table 14.2. *Raw Data for Population Pyramid, Males*

	0–9	10–19	20–29	30–39	40–49	50–59	60–69	70–79	80+	Total
				Age (years)						
Precontact period										
Paleo–Aleuts	—	2.5	6	7	9	12	9	—	—	45.5
Neo–Aleut	—	15.5	18	23	26	25.33	16.67	—	—	124.5
Russian period (1820–1867)										
1820–1829	23	12	16	18	20	14	6	6	5	120
1830–1839	57	25	45	29	38	38	20	11	7	270
1840–1849	104	38	32	19	28	28	13	6	—	268
1850–1859	91	19	24	16	9	14	6	4	—	183
1860–1867	37	6	5	9	4	3	4	1	—	69
American period (post 1867) in Aleutian Chain										
1897[a]	155	117	83	112	59	31	12	—	—	569
1948[a]	21	19	8	10	10	4	3	—	—	75
American period (post 1867) in Pribilof Islands										
1872–1879	77	8	8	10	8	6	1	—	—	118
1880–1889	101	12	26	23	23	10	6	—	—	201
1897[a]	52	29	21	13	14	7	1	1	—	138
1907[a]	44	37	23	15	9	7	—	—	—	135

[a] Census count; all others, number of deaths.

Table 14.3. *Raw Data for Population Pyramid, Females*

	0–9	10–19	20–29	30–39	40–49	50–59	60–69	70–79	80+	Total
				Age (years)						
Precontact period										
Paleo–Aleuts	—	4.5	2	1	10	11.33	8.67	—	—	37.5
Neo–Aleuts	—	20.5	27	20	27	28.67	21.33	—	—	144.5
Russian period (1820–1867)										
1820–1829	25	9	6	14	6	6	13	8	6	93
1830–1839	70	22	29	20	38	43	17	16	13	268
1840–1849	83	23	45	29	33	34	15	10	3	275
1850–1859	66	17	15	12	8	8	9	3	1	139
1860–1869	28	3	3	13	3	3	2	2	—	57
American period (post 1867) in Aleutian Chain										
1897[a]	135	101	133	104	69	21	15	2	—	580
1948[a]	17	9	14	7	4	4	1	1	—	57
American period (post 1867) in Pribilof Islands										
1872–1879	77	8	8	10	8	6	1	—	—	118
1880–1889	101	12	26	23	23	10	6	—	—	201
1897[a]	44	39	37	26	17	6	5	—	—	174
1907[a]	32	34	21	20	13	6	1	—	—	128

[a] Census count; all others, number of deaths.

Table 14.4. Life Expectancy of Aleutian Aleut Males

	Age (years)								
	0–9	10–19	20–29	30–39	40–49	50–59	60–69	70–79	80+
Precontact period in Aleutian Chain									
Paleo–									
Aleuts	N.D.	35.77	27.56	21.22	15.00	9.29	5.00	N.D.	N.D.
Neo–Aleuts	N.D.	31.24	24.97	18.92	13.63	8.97	5.00	N.D.	N.D.
Russian period in Aleutian Chain									
1820–1829	35.50	32.73	26.65	21.67	17.55	15.65	14.41	9.55	5.00
1830–1839	34.52	32.42	26.06	22.69	17.19	13.29	11.58	8.89	5.00
1840–1849	24.81	27.38	24.13	20.64	14.60	10.32	8.16	5.00	0.00
1850–1859	20.57	25.98	21.44	19.49	16.52	10.83	9.00	5.00	0.00
1860–1869	20.22	27.81	23.08	17.38	16.67	12.50	7.00	5.00	0.00
American period in Aleutian Chain									
1897[a]	31.39	29.56	22.95	17.05	12.29	8.87	5.00	0.00	0.00
1948[a]	32.30	25.17	26.09	22.87	15.84	10.29	5.00	0.00	0.00
1973[a]	45.56	41.44	37.05	32.28	27.01	21.06	14.21	6.95	5.00

[a] Data for age class midpoint.

Table 14.5. Life Expectancy of Aleutian Aleut Females

	Age (years)								
	0–9	10–19	20–29	30–39	40–49	50–59	60–69	70–79	80+
Precontact period in Aleutian Chain									
Paleo–									
Aleuts	N.D.	37.71	32.17	23.93	14.56	9.34	5.00	N.D.	N.D.
Neo–Aleuts	N.D.	30.56	24.78	20.29	14.26	9.27	5.00	N.D.	N.D.
Russian period in Aleutian Chain									
1820–1829	37.15	38.97	34.15	27.45	25.51	19.24	12.41	9.29	5.00
1830–1839	35.78	36.67	30.62	25.68	18.94	14.89	14.13	9.48	5.00
1840–1849	29.95	30.73	24.23	21.21	16.16	12.10	10.71	7.31	5.00
1850–1859	22.12	27.60	24.46	21.59	18.45	13.45	8.85	7.50	5.00
1860–1869	22.72	29.83	22.69	15.00	18.00	13.57	10.00	5.00	0.00
American period in Aleutian Chain									
1897[a]	37.94	31.15	23.55	15.49	10.51	13.10	6.33	5.00	0.00
1948[a]	29.90	26.76	24.00	17.81	16.34	13.90	10.99	5.00	0.00
1973[a]	38.82	35.90	32.93	29.85	26.57	22.92	18.68	13.48	5.00

[a] Data for age class midpoint.

older ages reliably, life expectancy at this age would be much greater, and life expectancy at earlier ages would increase somewhat, perhaps as much as 10 percent. We have recently turned to osteon counting as a method for determining age at death in skeletons (Kerley 1965). This technique is especially accurate in the older age ranges and should provide considerable power in rectifying this problem. In the interim, the life expectancy values for the precontact Aleuts are quite conservative.

The more recent precontact and contact groups, Neo-Aleuts, also have a relatively long life expectancy at age 10, although much reduced from Paleo-Aleut values. In males, the reduction is some 4 years, from 35.77 to 31.24 years, whereas in females, life expectancy drops over 7 years. This reduction is, in large part, a response to a new environmental challenge, the arrival of and subsequent large-scale massacre by the Russians. In adaptive terms, the Aleut population was faced with what must be considered to be a new predator and was unable to effectively meet this challenge. It is difficult to determine the total number of Aleuts lost in the early conflict, although some estimates run into the thousands (Bancroft 1886). Life expectancy is reduced in all age classes, suggesting a rather uniform destruction of the population.

Russian Period Aleuts

The establishment of the Russian Church near the end of the eighteenth century terminated the period of early contact strife. Our first data on this period begin in 1822 and are available for most years until 1867. The early church chronicles depict a stable Aleut environment with many customs, including dietary practices, remaining consonant with the old patterns. Life expectancy during the decade 1820 to 1829 increases for both males and females (Tables 14.4 and 14.5) and, in fact, reaches a high point for women, age 10, during this time. Because of Veniaminoff's copious records, we find an accurate description of the entire population profile. Life expectancy beyond age 60 is relatively great for both sexes and, as we previously learned, a large percentage of the population survived to become octogenarians.

The limitations with the skeletal data do not apply with these mortality records, so we are perhaps witnessing the characteristic Aleut pattern of population maintenance. The survival of the elderly is a critical feature of maintaining small populations. In this society they perform valuable services to the population along several lines. The elder Aleut has a number of intellectual and social roles that serve to increase life expectancy in this population. They perform a supervisory role in reducing infant mortality because the elderly are skilled midwives and resident physicians. They are the librarians of a nonliterary culture, serving to transmit knowledge from generation to generation. Additionally, they tutor the young in hunting skills, including the essentials of ethology and navigation, as well as physical training to endure hours of extended travel in small kayaks (Laughlin 1968b, 1972).

The perpetuating feature of this system is the fact that the elderly can also provide a high proportion of their own dietary needs from collecting on the extensive Aleutian reefs. The rich Aleutian ecosystem can provide several simultaneous niches for each cohort to exploit. Old persons, pregnant women, and children can fish in several different habitats that are readily accessible and also collect marine invertebrates when the tide is out (Table 14.6).

The early Aleut adaptive profile is therefore a cyclical system, highly dependent on the natural habitat that can maintain those disadvantaged persons who

Table 14.6. *Aleut Use of Resource Areas[a]*

Cohort[b]	Open sea (hunting and fishing)	Bay (fishing)	Beach (fishing)	Lake and stream (fishing)	Reef (collecting)	Cliff (collecting)	Inland (collecting and hunting)
Able Males	+	+	+	+	+	+	+
Able Females	−	+	+	+	+	+	+
Children	−	+	+	+	+	(+)	+
Old Males	−	+	+	+	+	−	−
Old Females	−	−	+	+	+	−	−
Infirm	−	−	(+)	(+)	(+)	−	−
Pregnant	−	(+)	(+)	−	(+)	−	(+)

[a] Modified from Laughlin (1972).
[b] All cohorts have access to more than one resource area. In some cases, access is qualified by use or by degree of disability; these are indicated by (+).

would otherwise become a nutritional burden. These people, especially the elderly, feed back directly in increasing infant and childhood survivorship, as well as by preparing the children for adult life as marine hunters.

Although Aleuts were well adapted to a marine hunter existence, the decade of 1830 brought a new challenge in the form of smallpox. Life expectancy is decreased only slightly from the previous decade, dropping 2.3 years in 10-year-old women, and 0.3 years in men. This is because the epidemic, which claimed 202 people, occurred in 1838 and the results are not fully appreciated until the subsequent decade. Most severely affected were the children, age 0 to 9, whose life expectancy dropped nearly 10 years in boys and 6 years in girls, and the elderly men. No man over 80 survived the 1838 epidemic and this age class has remained vacant until 1973.

In the 1840 decade, life expectancy at age 10 was nearly equally reduced for males and females, 5 and 6 years, respectively. Although Weiss' (1975) models would predict a rapid return to previous population profiles, a second epidemic of measles occurred during 1848 and 1849. As anticipated, life expectancy is again reduced, but most severely in children for the 1850s. During the decade, life expectancy at age 10 was only 25.98 years for males and 27.60 years for females. In the hundred years since Russian contact, the Aleut adaptive level had dropped nearly 10 years of life, or approximately 30 percent. Data for 1860 to 1867 are extremely limited, so the slight increase of 2 years for both sexes is only suggestive of the beginning of a return to previous levels.

American Period Aleuts

The ascension of the life expectancy curves continues with the first temporal intercept of the American period. The Hooper (1897) census of the entire Aleutian chain suggests that life expectancy at age 15 had risen to 29.56 years for men and 31.15 years for women, each representing a gain of about 2 years from the 1860 decade. Gone from the profile were men over age 70; however, childhood mortality was reduced from the last Russian decade.

Unfortunately, the data of the more recent period are sparse. The next intercept (1948) occurs over 50 years later and after several intervening bouts of influenza and almost constant tuberculosis ravaged the Aleuts. Laughlin's (1949) biomedical survey is only a sample of the population but reveals another decline in life expectancy. At age 15, female values dropped from 31 years, in 1897, to 26.8 years. Similarly, male life expectancy at this age declined over 4 years from 29.56 to 25.17. As this sample was taken very shortly after the people were relocated back to their native villages, the possibility of an altered demographic profile is high. The decline in overall adaptation is certainly consistent with the half century of disease.

By 1973 the population had dramatically revised the previous losses. The availability of medical care and an increased supply of store foods are probably implicated in the major increases in life expectancy. An Aleut male, age 15 in 1973, could expect to survive an additional 41.4 years, some 16 years longer than his 1948 ancestor. The women also display a substantial gain of nearly 10 additional years of life expectancy over the 1948 low. Whereas in 1948 the elderly were absent from the population profile, several of the 1973 samples were over 80. Many of the elderly also participate in the traditional Aleut adaptive mode by serving as tutors for child training in hunting and fishing and also collect invertebrates on the strandflats.

Over the long temporal history of human occupation in the Aleutian Islands, the Aleut pattern of life expectancy constantly moves in relationship to particular stresses of the environment. Of theoretical relevance is the possibility that life expectancy is an appropriate integration of the total fitness of the population. Wright's (1932) concept of evolution on an adaptive topologic surface of peaks and valleys that represent the fitness of various demic genomes relative to the environment is especially appropriate. A population cannot move from one plateau to another without first "deadapting" to a lower level. Much of the genetic contribution to the adaptation of recent Aleuts must have occurred many generations or millennia ago and is, therefore, inaccessible to study. However, the opportunity to study recent changes in adaptation, including "deadaptation" in Schull's words, is manifestly present.

The life expectancy profile indicates that the highest values occur in the Paleo-Aleut skeletal series (Figure 14.1) and that declines occur with circum-Russian Neo-Aleuts. The early Russian period represents a second adaptive plateau but declines in response to the smallpox and measles epidemics. After a small gain in 1897, additional bouts of epidemics, forced evacuation or relocation, and food shortages precipitate another adaptive valley. Modern Aleuts appear to have moved from this low and are aimed toward a third major adaptive peak.

The fluctuation in adaptive status of this population accurately reflects the severity of the stress to which adaptation occurred. The various descents into the valleys are always followed by a rise toward a new plateau and suggest that the

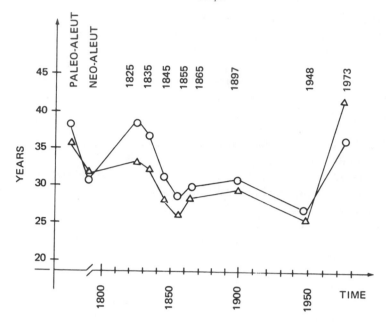

Figure 14.1. Trends in Aleut life expectancy at age class 10 to 19 years. △, Male; ○, female.

Aleut population has always maintained sufficient genetic variability to move in any adaptive direction.

The Adaptive Superiority of Aleuts

In spite of the recurrent transition in level of adaptation of the Aleutian Island Aleuts, they enjoy an adaptive superiority, as measured by life expectancy, over all comparable Arctic Mongoloids. This applies to migrant Aleut populations who share the basic Aleut genome but live under different environmental circumstances, as well as to Arctic Eskimos who have diverged genetically and ecologically from the original Bering Sea Mongoloid population.

Pribilof Aleuts

The most appropriate comparison with the Aleutian Aleuts are the Aleuts of the Pribilof Islands, some 200 miles north of the Aleutian chain. These two isolated islands are the principal breeding grounds of the fur seal but were never inhabited by Aleuts, who knew of their existence prior to the Russian colonizations (Laughlin 1976). The first human occupation began in the 1820s, when

Fox Island Aleuts were moved to the Pribilofs to harvest fur seal. Although many of the Pribilof Aleut surnames are Eastern Aleut in origin, the Pribilof population must necessarily have become genetically isolated from the Aleut population by stochastic processes and limited migration imposed by United States Government restrictions. They are a representative sample of an Aleut genome that has been partly diverted from the main Aleut stream and live under vastly different ecologic circumstances.

Our data do not begin until the 1870s but, thanks to United States Government interests in assuring a continual fur seal crop, detailed census records are available in the reports of the Pribilof agents. As found in Table 14.7, life expectancy for Pribilof Aleuts has always been rather low in comparison to Aleutian Aleut standards. During the first 50 years of American supervision, life expectancy at age 15 averaged 19.2 and 14.8 percent (male and female, respectively) less than cotemporal Aleuts of the same sex. At other age cohorts, the deficiency of Pribilof Aleut life expectancy is considerably greater. Infant mortality is especially high and census profiles display an unusual lack of Pribilof Aleuts over age 70.

Table 14.7. *Life Expectancy of Pribilof Aleuts*

	Age (years)								
	0–9	10–19	20–29	30–39	40–49	50–59	60–69	70–79	80+
Males									
1870–1879	15.34	24.76	19.55	14.20	10.33	6.43	5.00	0.00	0.00
1880–1889	20.47	26.10	18.98	14.84	10.64	8.75	5.00	0.00	0.00
1897[a]	21.92	25.10	18.75	15.00	10.71	8.33	5.00	0.00	0.00
1907[a]	25.68	19.59	18.48	15.67	12.78	5.00	0.00	0.00	0.00
1935[a]	20.77	23.28	19.09	15.67	11.00	10.00	5.00	0.00	0.00
1943[a]	31.04	23.24	19.94	15.63	17.33	12.61	11.01	5.00	0.00
Females									
1870 1879	15.34	24.76	19.55	14.20	10.33	6.43	5.00	0.00	0.00
1880–1889	20.47	26.10	18.98	14.84	10.64	8.75	5.00	0.00	0.00
1897[a]	34.55	28.33	19.59	15.77	11.47	13.33	5.00	0.00	0.00
1907[a]	33.96	26.96	25.32	16.34	12.44	11.12	8.75	5.00	0.00
1935[a]	25.72	20.55	20.92	17.04	15.00	15.00	8.33	5.00	0.00
1943[a]	24.33	26.07	23.10	18.75	17.00	19.00	12.00	10.00	5.00

[a] Data for age class midpoint.

The high mortality rate was a concern of the government agents as Mr. Tingle (1889) reported:

> The white population on both islands . . . [is] always remarkably healthy. It is a notable fact that not a single death from disease has occurred among them since the transfer of Alaska to the United States, whilst the percentage of mortality among the natives is much greater than can be found in any state or country of which we have statistics. (p. 357)

He also included a roll of natives, the oldest of whom was 66 and was described as "A confirmed invalid, very old (sixty-six years), oldest man on St. Paul" (p. 362). Mr. Tingle was unaware that a man of 66, had he lived in the Aleutian Islands, would not be described as "very old." Although the nutritional resources of the Pribilof Islands were excellent for millions of fur seal, and although Whites were not subject to high mortality, the Pribilof Islands were not a suitable environment for a community of Aleuts.

Among several environmental deficiencies, some are important in terms of Aleut adaptation (Laughlin 1976):

1. The absence of freshwater streams that empty lakes entails an absence of salmon and other fish that are trapped in beach seines or weirs. These fish are an extremely important resource in Aleutian communities and all segments of the population participate in their collection and storage for winter.

2. The common Aleutian complex coastline which would have provided sheltered coves and bays with many ecologic niches and diversity of resources is lacking. This eliminates utilization of this zone by the disadvantaged Pribilof cohorts.

3. The lack of extensive reef systems in the Pribilofs reduces both the availability of marine invertebrates and also reduces the protection from storm surges, thus eliminating the possibility of fishing from boats in protected waters while there are storms at sea.

4. An absence of off-shore islands also reduces complete cohort habitat utilization and lessens the amount of driftwood available for boats, hunting gear, homes, and utensils.

5. The presence of large numbers of fur seal further deprives the Pribilof Aleuts of using the coastline for long periods. The major invertebrates (sea urchin, limpets, mussels, chitons, whelks) cannot be reached while the fur seal population is in residence on the beaches.

The ecologic deficiencies are major parts of the high mortality and short life expectancy of the Pribilof Aleuts. Despite the resident medical care, the situation was little better in 1935 and 1943 as life expectancy did not respond appreciably to attempts by the government to improve living conditions. The environmental harshness of the Pribilofs succeeded in lowering the adaptedness of a previously well-adapted population. Indeed, the Aleutian Aleuts knew of the Pribilof Islands before Pribilof discovered them in 1786 but chose not to establish permanent residence there.

Arctic Eskimos

Human existence in the Arctic places a major emphasis on the successful bio-behavioral adaptation not only to the cold, but also to the extreme ecologic rigor of the environment. Human populations have lived in the Arctic for millenia; however, the degree of adaptive success of Arctic Eskimos is significantly less than that of cotemporal Aleuts (Harper 1975, 1976, Laughlin 1975).

For comparative purposes, I have chosen Eskimo populations that are primarily marine hunters and that are temporally similar to the various Aleut intercepts. Although it is not possible to follow secular change within these populations, the variation over time in Eskimo life expectancy seems to be somewhat reduced in relation to the fluctuation in the Aleut population (Table 14.8). At all temporal intercepts, Aleuts are better adapted than the Eskimo comparison. The overall distribution is highly significant, $p < 0.01$ and $p < 0.02$, males and females, respectively, based on the Mann–Whitney "U" test.

Table 14.8. *Life Expectancy of Aleut versus Eskimo*

	Male	Female
Paleo–Aleut[a]	35.8	37.7
Sadlermiut Eskimo[a]	19.7	23.8
Fox Aleut, 1820[a]	32.7	39.0
Labrador Eskimo, 1820[a]	30.7	30.7
Aleut, 1948[b]	25.2	26.8
Angmagssalik Eskimo, 1954[b]	21.1	21.1
Aleut, 1973[b]	41.4	35.9
Wainwright Eskimo, 1968[b]	28.7	24.1

[a] At age 10.
[b] At age 15.

A number of mitigating factors may be postulated for the major differences between Aleut and Eskimo adaptation. Eskimo populations are afflicted with several skeletal disorders that occur in extremely high frequencies, such as separate neural arches (Stewart 1931), vertebral fractures (Merbs 1969), and severe osteoporosis (Mazess 1971, Mazess and Mather 1975). The maintenance of adequate bone mineral content may be associated with longevity (Smith and Reddan 1976) and the high frequency of vertebral anomalies may suggest a heavy genetic load.

Conclusions

Although it is tempting to speculate on the genetic effects of cosmic radiation at high geomagnetic latitude, the environmental harshness in the Arctic renders the maintenance of the elderly difficult and thereby deprives the population of an important mechanism for survival. The most critical Aleut advantage is the accessibility of the reefs and streams to the infirm and elderly Aleuts who would otherwise be eliminated from the population. Indeed, the persistence of Aleuts over 9000 years may be attributed to the functions that old persons play. The alternative to this system is extinction, as in the Sadlermiut Eskimos, who were last seen in 1903.

References

Acsadi, Gy., and J. Nemeskeri. (1970). *History of Human Life Span and Mortality*. Budapest: Akademiai Kiado.

Andrewartha, H.G., and L.C. Birch. (1954). *The Distribution and Abundance of Animals*. Chicago: Univ. Chicago Press.

Angel, L.A. (1969) The Basis of Paleodemography. *Am. J. Phys. Anthropol.* 30:427–438.

Ayala, F.J. (1969). An Evolutionary Dilemma: Fitness of Genotypes versus Fitness of Populations. *Can. J. Genet. Cytol.* 11:439–456.

Baker, P. T. (1966). Ecological and Physiological Adaptation in Indigenous South Americans. *In* P.T. Baker and J.S. Weiner, Eds., *The Biology of Human Adaptability*. Oxford: Clarendon Press, pp. 275–303.

Baker, P.T. (1969). Multidisciplinary Studies of Human Adaptability; Theoretical Justification and Methods. *In* J.S. Weiner, Ed., *A Guide to the Human Adaptability Proposals, IBP Handbook No. 1*. Adington: Burgess and Son, pp. 61–69.

Baker, P.T. (1978). The Adaptive Fitness of High Altitude Populations. *In* P.T. Baker, Ed., *The Biology of High Altitude Populations*. Cambridge: Cambridge Univ. Press, pp. 317–350.

Baker, P.T., and J. Dutt. (1972). Demographic Variables as Measures of Biological Adaptation: A Case Study of High Altitude Populations. *In* G.A. Harrison and A.J. Boyce, Eds., *The Structure of Human Populations*. Oxford: Clarendon Press, pp. 352–378.

Bancroft, H.H. (1886). *The History of Alaska—1730–1885*. Darein, Conn.: Hafner Publishing. (1970 reprint.)

Burch, E.A. (1972). The Caribou/Wild Reindeer as a Human Resource. *Am. Antiq.* 37: 339–368.

Coale, A.J., and P. Demeny. (1966). *Regional Model Life Tables and Stable Populations*. Princeton, N.J.: Princeton Univ. Press.

Dobzhansky, Th. (1955). A Review of Some Fundamental Concepts and Problems of Population Genetics. *Cold Spring Harbor Symp. Quant. Biol.* 20:1–15.

Dobzhansky, Th., R.C. Lewontin, and O. Pavlovsky. (1964). The Capacity for Increase in Chromosomally Polymorphic and Monomorphic Populations of *Drysophila pseudoobscura*. *Heredity*. 19:597–614.

Fisher, R.A. (1958). *The Genetic Theory of Natural Selection*. 2nd ed. New York: Dover Publications.

Frøhlich, B. (1976). *Aleut Site Survey, 1975*. Paper presented to the Society of American Archaeologists, St. Louis, Mo., May, 1976.

Ginsburg, B.E., and W.S. Laughlin. (1966). The Multiple Bases of Human Adaptability and Achievement: A Species Point of View. *Eugen. Quart.* 13:240–257.

Graunt, J. (1662). *Natural and Political Observations Made on the Bills of Mortality*. London. (American edition edited by W.F. Willcox, Johns Hopkins Univ. Press, Baltimore, Md., 1939).

Harper, A.B. (1975). *Secular Change and Isolate Divergence in the Aleutian Population System*. Ph.D. Dissertation, University of Connecticut, Storrs.

Harper, A.B. (1976). Aleut Life Expectancy and Adaptation. *Am. J. Phys. Anthropol.* 44: 183.

Hayden, B. (1975). The Carrying Capacity Dilemma. *In* A.C. Swedlund, Ed., *Population Studies in Archaeology and Biological Anthropology*. *Am. Antiq.* 40:11–21.

Hooper, C.L. (1897). *A Report on the Sea Otter Banks of Alaska*. Washington, D.C.: U.S. Govt. Printing Office.

Hrdlička, A.A. (1945). *The Aleutian and Commander Islands*. Philadelphia: Wistar Press.

Kelley, J.J., L.L. Longerich, and D.W. Wood. (1971). Effect of Upwelling, Mixing and High Primary Productivity on CO_2 Concentrations in Surface Waters of the Bering Sea. *J. Geophys. Res.* 76(36):8687–8693.

Kerley, E. (1965). The Microscopic Determination of Age in Human Bone. *Am. J. Anthropology.* 22:149–163.

Keyfitz, N., and W. Flieger. (1971). *Population Facts and Methods of Demography*. San Francisco: W.H. Freeman.

Kroeber, A.L. (1939). *Cultural and Natural Areas of Native North America*. Berkeley: University of California Press.

Laughlin, W.S. (1949). *The Physical Anthropology of Three Aleut Populations: Attu, Atka, and Nikolski*. Ph.D. Dissertation, Harvard University, Cambridge, MA.

Laughlin, W.S. (1963). Eskimos and Aleuts: Their Origins and Evolution. *Science* 142: 633–645.

Laughlin, W.S. (1968a). Adaptability and Human Genetics. *Proc. Natl. Acad. Sci. USA* 60:12–21.

Laughlin, W.S. (1968b). Hunting: An Integrating Biobehavior System and Its Evolutionary Importance. *In* R.B. Lee and I. Devore, Eds., *Man the Hunter*. Chicago: Aldine, pp. 304–320.

Laughlin, W.S. (1972). Ecology and Population Structure in the Arctic. *In* G.A. Harrison and A.J. Boyce, Eds., *The Structure of Human Populations*. Oxford: Clarendon Press, pp. 379–392.

Laughlin, W.S. (1975). Aleuts: Ecosystem, Holocene History and Siberian Origin. *Science* 189:507–515.

Laughlin, W.S. (1976). *Pribilof Aleut Claims Report*. Testimony presented May 10, 1976 to the Indian Claims Commission hearings on the claims of the Pribilof Aleuts. Washington, D.C.

Lotka, A.J. (1956). *Elements of Mathematical Biology*. New York: Dover.

Mazess, R.B. (1971). Estimation of Bone and Skeletal Weight by Direct Photon Absorptiometry. *Invest. Radiol.* 6:52–60.

Mazess, R.B. (1975). Biological Adaptation: Aptitudes and Acclimatization. *In* E.S. Watts, F.E. Johnston, and G.W. Lasker, Eds., *Biosocial Interrelations in Population Adaptation*. The Hague: Mouton.

Mazess, R.B., and W.F. Mather. (1975). Bone Mineral Content in Canadian Eskimos. *Hum. Biol.* 47:45–63.

Merbs, C.F. (1969). *Patterns of Activity Induced Pathology in a Canadian Eskimo Isolate*. Ph.D. Dissertation, University of Wisconsin, Madison.

Moore, J.A., A.C. Swedlund, and G.J. Armelagos. (1975). The Use of Life Tables in Paleodemography. *In* A.C. Swedlund, Ed., *Population Studies in Archaeology and Biological Anthropology. Am. Antiq.* 40:57–70.

Neel, J.V., and F.M. Salzano. (1964). A Prospectus for Genetic Studies of the American Indian. *Cold Spring Harbor Symp. Quant. Biol.* 29:85–98.

Pearson, K. (1901–1902). On the Change in Expectation of Life in Man During a Period of Circa 2000 Years. *Biometrika* 1:261–264.

Robert-Lamblin, J. (1976). The Ethnographic Contributions of Alphonse L. Pinart to the Eastern Aleutians. *In* W.S. Laughlin and A.B. Harper, Eds., *Aleut Population and Ecosystem Analysis*. Strausburg, Pa.: Dowden, Hutchinson and Ross. (In press.)

Rychkov, V.G., and V.A. Sheremetyeva. (1972). Population Genetics of the Aleuts of the Commander Islands. [In Russian.] *Vopr. Anthropol.* 40:45–70.

Schull, W.J., and F. Rothhammer. (1976). Analytic Methods for Genetic and Adaptational Studies. Paper prepared for *Origins and Affinities of the First Americans*. Burg Wartenstein Symposium, No. 72, August, 1976.

Smith, E.L., and W. Reddan. (1976). Physical Activity—A Modality for Bone Accretion in the Aged. Paper presented to the International Conference on Bone Mineral Measurement, New Orleans, La., January 1976.

Steffansson, V. (1958). Eskimo Longevity in Northern Alaska. *Science* 127: 16–19.

Stewart, T.D. (1931). Incidence of Separate Neural arch in the Lumbar Vertebrae of Eskimos. *Am. J. Phys. Anthropol.* 16: 51–62.

Stolnitz, G.J. (1956). *Life Tables from Limited Data: A Demographic Approach.* Princeton, N.J.: Princeton Univ. Press.

Thoday, J.M. (1953). Components of Fitness. *Symp. Soc. Exptl. Biol.* 7: 96–113.

Tingle, G.R. (1889). Testimony to the *Investigation of the Fur Seal and other Fisheries of Alaska.* Washington, D.C.: U.S. Govt. Printing Office, pp. 355–361.

Turner, C.G. (1967). *The Dentition of Arctic Peoples.* Ph.D. Dissertation, University of Wisconsin, Madison.

Ubelaker, D.H. (1974). *Reconstruction of Demographic Profiles from Ossuary Skeleton Samples.* Smithsonian Contributions to Anthropology No. 18. Washington, D.C.: U.S. Govt. Printing Office.

Ward, R.H., H. Gershowitz, M. Layrisse, and J.V. Neel. (1975). The Genetic Structure of a Tribal Population, the Yanomama Indians. XI. Gene Frequencies for 10 Blood Groups and ABH-Le Secretor Traits in the Yanomama and Their Neighbors. The Uniqueness of the Tribe. *Am. J. Hum. Genet.* 27: 1–30.

Weiss, K.M. (1973). *Demographic Models for Anthropology.* Am. Antiq. Memoirs No. 27.

Weiss, K.M. (1975). Demographic Disturbance and the Use of Life Tables in Anthropology. *In* A.C. Swedlund, Ed., *Population Studies in Archaeology and Biological Anthropology.* Am. Antiq. 40: 46–56.

Wright, S. (1932). The Roles of Mutation, Inbreeding, Crossbreeding and Selection in Evolution. *Proc. 6th Intl. Congr. Genet.* 1: 356–366.

Wynne-Edwards, V.C. (1962). *Animal Dispersion in Relation to Social Behavior.* New York: Hafner.

Index

Selection. *See* Natural selection
Selkups, 58–60, 67, 79, 81
Seminoles, 280
Sergeev, V. P., 76
Sergin, S. Ya., 23
Serologic. *See* particular system
Serum
   cholinesterase, 214
   lipoproteins, 120, 214
   protein, 119–20, 212, 214
Seward Peninsula, 23, 29, 46
Sex ratio, 306
   structure, 305
Shcheglova, M. S., 23
Shepard, F. P., 19
Sher, A. V., 27
Sheremetyeva, V. A., 138, 187, 317
Sherzer, J., 159, 167, 170–71
Shluger, S. A., 58, 76
Shors, 79, 83
Shpanberg Strait, 29, 32
Siberia, 1, 4, 7, 15, 17, 22, 26–9, 32–3, 47–9,
   57–86, 91, 98, 133, 186, 206
Siberian
   Mongoloids, 60–1, 65, 67, 70, 82, 85–6
   morphologic complex, 73–5, 79, 83, 86
   peoples, 2, 5, 57–86, 144, 161, 163, 185–88,
    190–91, 196–97, 200–2, 204–6
   Tatars, 58
Sierra, 243–44
Simons, M. J., 219
Simmons, R. T., 190, 211–12, 214
Simpson, N. E., 188, 190, 204–5
Sinnett, P., 227
Siouan language phylum, 141, 144, 149, 152–
   53, 155–59
Sioux, 109, 141, 144, 156, 168–69
Skinfold thickness, 278, 283, 285, 288
Slave, 109, 138, 153, 164–65, 190, 204
Smallpox, 258, 270, 322–23
Smirnova, N. S., 76
Smith, C. A. B., 125
Smith, E. L., 327
Smith, S. H., 276
Sneath, P. H. A., 126, 145–46, 201
Snow, C. E., 260
Sociocultural, 295
Socioeconomic, 294
Sokal, R. R., 126, 145–46, 201
Solomon Islands, 224–25, 232
Sources of variation, 249–50
South America, 1, 5, 47, 241
South American Indians, 108, 115–17, 187,
   225, 229, 265
Southwest culture area, 164, 167–69, 171–77
Soyots, 58–9, 61, 68, 78, 81, 83
Spees, E. K., 116
Spencer, R. F., 115, 167–68, 171
Spuhler, J. N., 3, 6, 135–36, 139, 142, 145–46,
   149, 153, 188, 200–1
Statistical analysis, 137, 143, 145, 191, 196,
   201, 314
Stefansson, V., 317
Steinberg, A. G., 119, 215
Steinberg, A. S., 119
Stewart, T. D., 9, 101, 241, 257, 259–60, 271
Stocker, S. W., 20, 35

Stolnitz, G. J., 314
Stone dishes, 93
Streten, N. A., 23
Stunkard, A., 285
Surplus of females, 298
Subarctic culture area, 163–67, 173–75, 185–
   87, 201, 204, 206
Sungir, 27
Superoxide dismatase, 222
Survivorship, 311–14, 321–23
Swanton, J. R., 136–37, 159, 170
Swedes, 130
Swinomish, 143, 158, 165–66
Syphilis, 258–59
Szathmary, E. J. E., 4, 7, 109, 118–20, 136,
   148, 185–86, 190

T
Taiwan, 226, 229
Tanis, R., 109, 115
Tanner, J. M., 276–77, 284
Tanoan, 142, 157
Tarahumara, 143, 171, 173–74
Tashian, R. E., 225
Tatsuoka, M. M., 146–47
Taylor, W. E., 185
Taylor, W. H., 227
Tectonic, 19
Telengets, 79
Tewa, 142, 173
Thalbitzer, W., 297
Thoday, J. M., 311
Thomas, J. W., 140, 190, 204
Thomas, R. B., 276
Thompson, E. A., 136
Thorsby, E., 223
Thorson, R. M., 31–2
Three-rooted first lower molars, 98, 100
Thule, 7, 159, 186
Tierra del Fuego, 50
Tingle, G., 325
Tittor, W., 116
Tiwa, 142, 173
Tlapacoya, 46, 49
Tlingit, 109, 140, 154, 159, 164–66, 188, 190–
   91, 201, 204
Tobias, P. V., 277
Todzhin, 190
Tolalars, 50, 60, 60, 190
Tokareva, T. Ya., 78
Toldt, C., 265
Tomirdiaro, S. V., 23
Tomtosova, R. F., 79
Towa, 142, 173
Trail Creek cave, 46
Transferrin variants, 7, 119, 214, 226, 232
Trio, 115, 280
Trofimova, T. A., 58, 78
Troup, B. M., 116
Tuberculosis, 259, 323
Tuchone, 139, 149, 164, 190, 204
Tungus-Manchu, 61, 65, 67–70, 72–4, 83
Turko-Mongolic, 61, 68, 72, 74
Turner, C. G., 94, 100, 316
Tuvin, 190
Tympanic plates, 98, 268